Linear
Ordinary
Differential
Equations

Linear Ordinary Differential Equations

Earl A. Coddington
Robert Carlson

siam.
Society for Industrial and Applied Mathematics

Philadelphia

Library of Congress Cataloging-in-Publication Data

Coddington, Earl A., 1920-
    Linear ordinary differential equations / Earl A. Coddington,
  Robert Carlson.
      p.   cm.
    Includes bibliographical references and index.
    ISBN 0-89871-388-9 (pbk.)
    1. Differential equations. I. Carlson, Robert, 1951-
  II. Title.
  QA372.C594 1997
  515'.352--dc21   96-53262

Dedicated to the memory of Earl A. Coddington, 1920–1991

Contents

Bibliography   **333**

Index   **335**

Preface

Differential equations are perhaps the most successful method discovered for modeling natural phenomena. Within this vast field, linear ordinary differential equations occupy a central role: in numerous examples the model is a linear system of equations, whereas in other cases a linear approximation is the essential tool for analyzing local behavior. The applications of linear ordinary differential equations extend to the study of partial differential equations, where they play a classical role in the construction of explicit solutions and a metaphorical role in guiding more advanced developments.

There are, of course, numerous elementary texts on differential equations and several works aimed at researchers. *Linear Ordinary Differential Equations* is designed as a textbook for mathematics, engineering, and science students at the end of their undergraduate studies or at the beginning of their graduate studies. Since the intended audience has an intermediate level of sophistication, the ideas of calculus, linear algebra, and elementary analysis are encountered frequently and explicitly, whereas more advanced topics such as Lebesgue integration and functional analysis are largely avoided. The good news is that a modest set of mathematical tools is adequate for a fairly deep study of differential equations.

The systematic study of linear differential equations illustrates the interrelationships among various mathematical subdisciplines. The existence of these relationships is a point easily missed in a typical, highly compartmentalized mathematics education. Our subject offers an excellent opportunity to interweave the threads of calculus, linear algebra, and analysis while exploring a field with wide applications in science and engineering. Since the main subject is systems of linear equations, the use of vectors and matrices is pervasive. Linear algebra is directly applied to the analysis of systems with constant or periodic coefficients and metaphorically applied as a guide in the study of eigenvalues and eigenfunction expansions. Another important theme is the use of power series. In addition to their usual role in the analysis of equations and special functions arising in applications, matrix-valued series are used for the development of the exponential function and related functions with matrix arguments. Our third main thread is the application of the methods of real analysis. In this area, most of the required material is reviewed, as it is needed to understand convergence of series solutions to equations, existence theorems for initial value problems, the asymptotic behavior of solutions, and the convergence of eigenfunction expansions.

The selection of material and the level of presentation should make the text a useful reference for students and professionals in science and engineering. The first chapter introduces linear systems of differential equations by presenting applications in life sciences, physics, and electrical engineering. These examples, and their generalizations, are revisited throughout the text.

Chapters 2 through 4 then develop the main applications of linear algebra to the study of systems of differential equations. After a review of linear algebra and a discussion of linear systems in general, there is an extensive treatment of systems of equations with constant coefficients. The analysis of constant coefficient systems includes a thorough discussion of the applications of the Jordan canonical form for matrices. Systems of equations with periodic coefficients are also treated in this part of the text.

In the first part of the book, power series are treated formally when they are encountered. Attention to analytical questions characterizes Chapters 5, 6, and 7. The essential material on convergence is reviewed and applied to the study of equations with analytic coefficients. Series techniques are then applied to systems with singular points. The general discussion of series methods is supplemented with a treatment of two classical examples, the Legendre equation and the Bessel equation. A study of convergence for the method of successive approximations concludes this portion of the book. This study provides a proof of the basic existence and uniqueness theorem, along with a proof of the continuous dependence of solutions on initial conditions and the coefficients of the equation.

Chapters 8 through 10 develop additional topics: the study of boundary value problems and an introduction to control theory. These topics have numerous applications and are important themes in the more advanced theory of differential equations. Boundary value problems are introduced with a study of the heat equation for a homogeneous rod. After an examination of self-adjoint eigenvalue problems, the convergence of eigenfunction expansions is treated. The final chapter offers an unusually elementary introduction to control theory, a central concern in modern technology.

We have tried to adhere to certain notational conventions, although on occasion these conventions are violated in the interest of notational simplicity. The real and complex numbers are denoted $\mathcal{R}$ and $\mathcal{C}$, respectively. Real and complex numbers, and functions with real or complex values, are normally denoted with lowercase letters, e.g., $x(t)$. Elements of $\mathcal{R}^n$, $\mathcal{C}^n$ are normally denoted with uppercase letters, e.g., X, Y; matrices are written in boldface, e.g., $\mathbf{A}$.

When new vocabulary is introduced, the term is presented in *italics*.

Robert Carlson
University of Colorado at Colorado Springs

Chapter 1

Simple Applications

1.1 Introduction

There are two attractive aspects to linear ordinary differential equations. On the one hand, the subject has a rich theory, with a constant interplay of ideas from calculus, linear algebra, and analysis. Large parts of this theory are accessible to students equipped with only an undergraduate education in mathematics. This theory also serves as an intellectual base camp from which less accessible regions of mathematics may be explored.

On the other hand, linear systems of ordinary differential equations arise in numerous scientific and engineering models. Our ability to explicitly solve equations in some cases or to provide a penetrating analysis when exact solutions are not available means that very detailed scientific and engineering predictions are possible. This introductory chapter describes several of these models. The main goal here is to point out the variety of contexts in which linear systems of ordinary differential equations arise and to observe that similar mathematical structures can occur in quite distinct technological guises.

Three such applications will be presented: compartment systems from biology and physiology, spring and mass systems from physics, and electric circuits. In these examples the biological, physical, or engineering behavior is modeled with a system of linear ordinary differential equations with constant coefficients. As we will see in subsequent chapters, such systems are amenable to a thorough analysis.

1.2 Compartment systems

Consider the problem of predicting the impact of a toxic waste discharge into a river and lake system. Suppose that a toxic chemical is accidentally discharged from a site upriver from Lake 1 (Figure 1.1). The toxic material leaks into the river at a constant rate for one day before the discharge can be stopped. We assume that the toxic material mixes quickly and thoroughly with the lake water as it enters each lake.

The toxic discharge will also have an impact on the second lake. A thought experiment, in which the water from Lake 1 barely trickles into Lake 2, suggests that the impact of the toxic discharge on Lake 2 should depend on the relationship between the rates of flow from Lake 1 into Lake 2 and from Lake 1 into the river. If the rate at which water flows from the first lake to the second lake is much slower than the rate at which water enters and leaves Lake 1 via the river, the contamination of Lake 2 should be minimal. These ideas can be validated with an explicit description of the history of toxic contamination for both lakes.

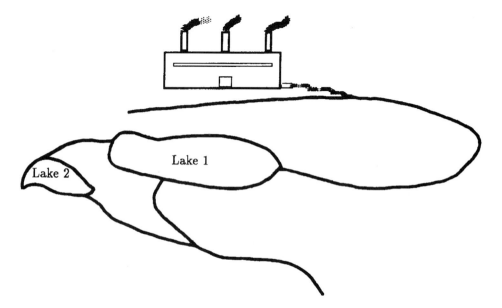

Figure 1.1: A system of lakes.

Some straightforward accounting of the flow of liquid into and out of each lake will yield a model of this contamination process. Denote by v_j the constant volume of liquid in lake j for $j = 1, 2$. Similarly, let τ_j denote the volume of toxic material in lake j. Let the constant rate of liquid flow into each lake be i_j, and let the constant rate of liquid flow from each lake to the river be o_j. Since the volumes v_j are conserved, $i_1 = i_2 + o_1$ and $i_2 = o_2$. The concentration of toxic material in each lake is $c_j = \tau_j / v_j$. Finally, assume that for the one-day period in which the toxic discharge is uncontrolled, the concentration c_0 of toxic material in the water flowing into Lake 1 and the corresponding rate $r(t) = c_0 i_1$ are constant. All of the volumes and rates are measured in some consistent set of units.

Assuming that liquid enters or leaves the river and lake system only by flowing downstream, we set up equations which account for the flow of water and toxin. Let 0 be the time when toxin starts to enter Lake 1. During the subsequent day the amount of toxin in Lake 1 is

$$\tau_1(t) = \int_0^t r(s)\,ds - \int_0^t c_1 i_2\,ds - \int_0^t c_1 o_1\,ds,$$

and the amount in Lake 2 is

$$\tau_2(t) = \int_0^t c_1 i_2\,ds - \int_0^t c_2 o_2\,ds.$$

Differentiation leads to the equations

$$\frac{d\tau_1}{dt} = r(t) - c_1 i_2 - c_1 o_1,$$

$$\frac{d\tau_2}{dt} = c_1 i_2 - c_2 o_2.$$

Using the conservation equation $i_1 = o_1 + i_2$ and the relationships $c_j = \tau_j/v_j$ for $j = 1, 2$, the equations become

$$\frac{d\tau_1}{dt} = r(t) - \frac{i_2}{v_1}\tau_1 - \frac{o_1}{v_1}\tau_1 = r(t) - \frac{i_1 \tau_1}{v_1}, \tag{1.1}$$

$$\frac{d\tau_2}{dt} = \frac{i_2}{v_1}\tau_1 - \frac{o_2}{v_2}\tau_2. \tag{1.2}$$

Of course at time 0 no toxin has accumulated in either lake, so

$$\tau_1(0) = 0 = \tau_2(0).$$

This system of equations may be handled by elementary techniques. Solving (1.1) first we find that during the day of toxic flow into Lake 1

$$\frac{d}{dt}\left[e^{i_1 t/v_1}\tau_1(t)\right] = e^{i_1 t/v_1}r(t).$$

Using the initial condition $\tau_1(0) = 0$ and the expression $r(t) = c_0 i_1$ for the input rate of toxin, this equation can be solved explicitly, giving

$$e^{i_1 t/v_1}\tau_1(t) = \int_0^t e^{i_1 s/v_1}r(s)\,ds = \int_0^t e^{i_1 s/v_1}c_0 i_1\,ds,$$

$$\tau_1(t) = \int_0^t e^{i_1(s-t)/v_1}c_0 i_1\,ds = c_0 v_1[1 - e^{-t i_1/v_1}].$$

Thus the concentration $c_1 = \tau_1/v_1$ in Lake 1 approaches the input concentration c_0.

Similarly, equation (1.2) may be solved for $\tau_2(t)$:

$$\frac{d}{dt}e^{o_2 t/v_2}\tau_2(t) = e^{o_2 t/v_2}\frac{i_2}{v_1}\tau_1(t),$$

$$e^{o_2 t/v_2} \tau_2(t) = \int_0^t e^{o_2 s/v_2} \frac{i_2}{v_1} \tau_1(s) \, ds,$$

$$\tau_2(t) = \int_0^t e^{o_2(s-t)/v_2} \frac{i_2}{v_1} \tau_1(s) \, ds.$$

Since $\tau_1(t)$ is known, it can be inserted into this last formula, giving an explicit description of $\tau_2(t)$.

At the end of the day, there is no additional discharge into Lake 1. For later times the system of differential equations is thus

$$\frac{d\tau_1}{dt} = -\frac{i_2}{v_1} \tau_1 - \frac{o_1}{v_1} \tau_1 = -\frac{i_1 \tau_1}{v_1}, \tag{1.3}$$

$$\frac{d\tau_2}{dt} = \frac{i_2}{v_1} \tau_1 - \frac{o_2}{v_2} \tau_2,$$

subject to the initial conditions that τ_j at time 1 for the second system is given by the solutions for the first system at time 1. The system (1.3) may be solved in the same manner as above.

The analysis of a complex system may involve the aggregation of a number of subsystems into a single compartment. Figure 1.2 shows a compartment system motivated by a serious medical problem. Drug therapies used to combat cancer or other diseases may involve the use of substances toxic to both the target cells and the other nontarget tissues. In this case there are three compartments: the circulatory system S_3, the target tissues S_1, and the nontarget tissues S_2, which are sensitive to the drug. A volume A of the drug is introduced into the circulatory system. Within each compartment it is assumed that the drug which has entered mixes quickly and uniformly. The body removes the drug from circulation by the filtering blood at a certain rate. Assume that all the drug entering the filter is removed from the blood. The problem is to determine whether the drug concentrations $c_j = \tau_j/v_j$ will become high enough to be beneficial in compartment one while not exceeding a dangerous level in compartment two.

To model this system, let v_j, denote the constant volume of fluid in each compartment, while $\tau_j(t)$ denotes the volume of drug in each compartment. Assume that fluid flows from the circulatory system into compartment $j = 1, 2$ at the same constant rate r_j as fluid flows back from the compartment to the circulation. The rate of fluid flow into the filter will be denoted r_3. The equation modeling the volume of drug in the circulatory compartment is then

$$\tau_3(t) = A - \int_0^t \left[\frac{r_1 + r_2 + r_3}{v_3} \right] \tau_3(s) \, ds + \int_0^t \frac{r_1}{v_1} \tau_1(s) + \frac{r_2}{v_2} \tau_2(s) \, ds.$$

Differentiation leads to

$$\frac{d\tau_3}{dt} = - \left[\frac{r_1 + r_2 + r_3}{v_3} \right] \tau_3 + \frac{r_1}{v_1} \tau_1 + \frac{r_2}{v_2} \tau_2, \quad \tau_3(0) = A.$$

Similar bookkeeping gives two more equations:

$$\frac{d\tau_1}{dt} = \frac{r_1}{v_3}\tau_3 - \frac{r_1}{v_1}\tau_1, \quad \tau_1(0) = 0,$$

$$\frac{d\tau_2}{dt} = \frac{r_2}{v_3}\tau_3 - \frac{r_2}{v_2}\tau_2, \quad \tau_2(0) = 0.$$

The analysis of this system will require a more sophisticated approach than the system (1.1) and (1.2). Not only are there more equations and unknowns, but also none of the equations can be solved alone.

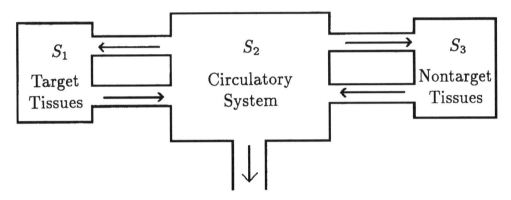

Figure 1.2: Organ systems.

1.3   Springs and masses

Linear systems of ordinary differential equations also arise in the study of mechanics. Consider a mass m attached to the end of a spring, as shown in Figure 1.3. If at time $t = 0$ the mass has position x_0 and velocity v_0, its subsequent displacement x from the equilibrium position is governed by Newton's law:

$$mx'' = F, \quad x'' = \frac{d^2 x}{dt^2},$$

where F represents the forces acting on the mass. In the usual spring and mass model, the net force acting on the mass is considered to be the sum of three terms. The first term is a restoring force $-kx$, which pulls the mass back toward the equilibrium position. The second term describes the damping force $-cx'$. The constants k and c are assumed to be nonnegative. The last term is a time-dependent force $f(t)$ which is independent of the position or velocity of the mass. Thus the linear spring model consists of a differential equation with initial conditions

$$mx'' + cx' + kx = f(t), \quad x(0) = x_0, \quad x'(0) = v_0. \tag{1.4}$$

The reader may recall that if $f(t)$ is any continuous function on the interval $[0, b)$, where $b > 0$, then there is a unique function $x(t)$ with two continuous derivatives on $[0, b)$ satisfying (1.4). Such existence and uniqueness theorems will be discussed in the next chapter and proven in Chapter 7. For the moment we will concentrate on cases with elementary solutions.

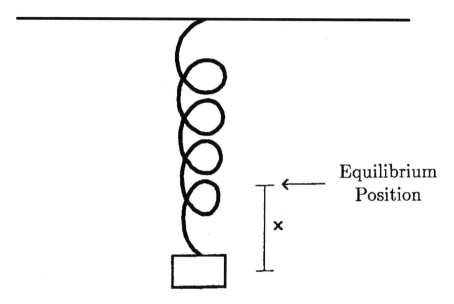

Figure 1.3: A mass and spring system.

It is often convenient to convert higher-order differential equations into systems of equations where no derivative beyond the first appears. In the case of the spring equation this can be achieved by setting $y_1 = x$ and $y_2 = x'$. The description of the motion of the spring then takes the form

$$y_1' = y_2,$$

$$y_2' = -\frac{k}{m}y_1 - \frac{c}{m}y_2 + \frac{f(t)}{m},$$

$$y_1(0) = x_0, \quad y_2(0) = v_0.$$

Introducing the matrix

$$\mathbf{A} = \begin{pmatrix} 0 & 1 \\ -k/m & -c/m \end{pmatrix}$$

and the vector-valued functions

$$Y = \begin{pmatrix} y_1 \\ y_2 \end{pmatrix}, \quad B(t) = \begin{pmatrix} 0 \\ f(t)/m \end{pmatrix},$$

this system of two equations may be written as

$$Y' - \mathbf{A}Y = B(t).$$

This is a first-order linear system for the vector-valued function Y as a function of the time t.

In the absence of the external forcing term $f(t)$, the equation for the spring becomes

$$mx'' + cx' + kx = 0, \quad x(0) = x_0, \quad x'(0) = v_0.$$

The corresponding system of first-order equations is

$$Y' - \mathbf{A}Y = 0. \tag{1.5}$$

For such a system of linear equations with constant coefficients, it is fruitful to look for solutions of the form

$$Y(t) = \begin{pmatrix} a \\ b \end{pmatrix} e^{\lambda t}.$$

When a function of this form is put into (1.5) the result is

$$\lambda \begin{pmatrix} a \\ b \end{pmatrix} e^{\lambda t} - \mathbf{A} \begin{pmatrix} a \\ b \end{pmatrix} e^{\lambda t} = 0,$$

which is equivalent to the eigenvalue problem

$$\mathbf{A} \begin{pmatrix} a \\ b \end{pmatrix} = \lambda \begin{pmatrix} a \\ b \end{pmatrix}.$$

As the reader will recall, the set of λ leading to nontrivial solutions of this equation is determined by the condition

$$\det(\lambda I - \mathbf{A}) = \det \begin{pmatrix} \lambda & -1 \\ k/m & \lambda + c/m \end{pmatrix} = 0$$

or

$$\lambda \left(\lambda + \frac{c}{m} \right) + \frac{k}{m} = 0.$$

In general there are two solutions to this equation:

$$\lambda_\pm = \frac{-c/m \pm \sqrt{(c/m)^2 - 4k/m}}{2}.$$

From the eigenvalue equation it follows that the nonzero eigenvectors will satisfy $b = \lambda a$. As long as $\lambda_+ \neq \lambda_-$, solutions of (1.5) can be found of the form

$$Y(t) = \begin{pmatrix} a_1 \\ b_1 \end{pmatrix} e^{\lambda_+ t} + \begin{pmatrix} a_2 \\ b_2 \end{pmatrix} e^{\lambda_- t}.$$

Checking the initial conditions, we see that

$$a_1 + a_2 = x_0, \quad b_1 + b_2 = \lambda_+ a_1 + \lambda_- a_2 = v_0.$$

Observe that if $c, m, k > 0$, then either $(c/m)^2 - 4k/m < 0$ or the magnitude of $\sqrt{(c/m)^2 - 4k/m}$ is less than c/m (or $\lambda_+ = \lambda_-$, which has been excluded for simplicity). In either case the eigenvalues $\lambda_\pm$ have negative real parts, so that solutions have limit 0 as $t \to +\infty$.

To simplify the subsequent discussion, suppose that $c = 0$, so that the spring has no damping. In this case the two eigenvalues are purely imaginary complex numbers:

$$\lambda_\pm = \pm i \sqrt{k/m}.$$

Thanks to the Euler formula

$$e^{i\theta} = \cos(\theta) + i \sin(\theta),$$

there are solutions of the equation

$$mx'' + kx = 0 \tag{1.6}$$

which are linear combinations of $\cos(\sqrt{k/m}\ t)$ and $\sin(\sqrt{k/m}\ t)$. In fact it is simple to verify that

$$x(t) = x_0 \cos(\sqrt{k/m}\ t) + \frac{v_0}{\sqrt{k/m}} \sin(\sqrt{k/m}\ t)$$

satisfies the equation and has the appropriate initial conditions.

It is also instructive to consider the undamped mass and spring with a nontrivial forcing term:

$$x'' + \frac{k}{m}x = \frac{a}{m}\sin(\omega t), \quad \omega > 0, \quad \omega^2 \neq \frac{k}{m}, \quad a \neq 0. \tag{1.7}$$

It will be convenient to let $\sqrt{k/m} = \omega_0 > 0$ so that two independent solutions of (1.6) have the form

$$x_1(t) = \cos(\omega_0 t), \quad x_2(t) = \sin(\omega_0 t).$$

Looking for a particular solution of (1.7) of the form $x_p = b \sin(\omega t)$, we find that

$$-\omega^2 b + \frac{k}{m}b = \frac{a}{m}$$

or

$$b = \frac{a}{m[k/m - \omega^2]}.$$

The general solution of (1.7) will have the form

$$x(t) = c \cos(\omega_0 t) + d \sin(\omega_0 t) + \frac{a}{m[\omega_0^2 - \omega^2]} \sin(\omega t).$$

Observe that although (1.7) is an equation whose coefficients are all periodic with period $2\pi/\omega$, the same is not necessarily true of the solutions. In fact, in most cases the only solution which has this period is the one with $c = 0 = d$ (see exercise 4).

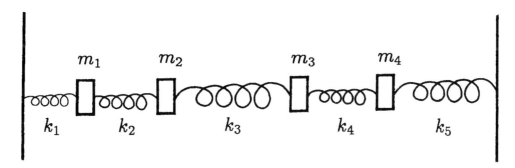

Figure 1.4: Several masses and springs.

Mechanical systems can easily lead to more complex systems of differential equations. Consider the system of springs and masses, without damping or external forcing, shown in Figure 1.4. If the masses are denoted $m_1, \ldots, m_4$, the connecting springs have spring constants $k_1, \ldots, k_5$, and the displacement of the jth mass from its equilibrium position is x_j, then the system of equations describing the motion of the masses is

$$m_1 x_1'' + k_1 x_1 - k_2 (x_2 - x_1) = 0,$$

$$m_j x_j'' + k_j (x_j - x_{j-1}) - k_{j+1}(x_{j+1} - x_j) = 0, \quad j = 2, 3,$$

$$m_4 x_4'' + k_4 (x_4 - x_3) + k_5 x_4 = 0.$$

1.4   Electric circuits

Linear systems of ordinary differential equations are also used to model electric circuits. A glance at any text on electric circuit analysis [2, 23] will show that this is an extensively developed subject, which we will certainly not treat in depth. Still, the important role that these circuits play in modern technology makes it worthwhile to develop a basic understanding of the models. Some simple examples will illustrate the relationship between circuit analysis and differential equations.

Electric circuit analysis has elements which have appeared both in compartment systems and in spring and mass systems. In compartment systems an essential role was played by the physical principle of conservation of mass. In circuit theory an analogous role is played by the conservation of electric charge; that is, the number of electrons in the circuit is not changing.

To introduce the ideas and terminology, consider the simple circuit shown in Figure 1.5. We will be interested in an equation for the *current* $i(t) = dq/dt$, which is the rate at which electric charge q passes a point in the circuit. Since charge is carried by electrons, this is the same as measuring the net rate at which electrons flow past a point in the circuit. The sign of the current will indicate the direction of net current flow.

Energy must be supplied to a circuit in order to have a current flow. The *voltage* $v(t)$ measures the energy per unit charge. In addition to a voltage source V, our circuit has two components, a resistor R and an inductor L, which determine how the current flow in the circuit depends on the voltage. The current flow through a resistor is given by the equation

$$v_R(t) = i(t)R, \quad R \geq 0,$$

where R is the *resistance* and v_R is the voltage drop across the resistor. Current flow through the inductor is given by the equation

$$v_L(t) = L\frac{di}{dt}, \quad L \geq 0,$$

where L is the inductance and again v_L is the voltage drop across the inductor. R and L are assumed constant.

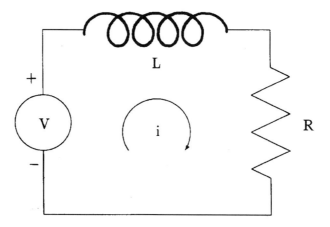

Figure 1.5: Circuit 1.

Conservation of energy requires that the energy supplied to the circuit must be equal to the energy used moving the electrons. This is the content of *Kirchhoff's voltage law*: at any time the sum of the voltages around any closed circuit is zero. Thus the equation for current in the circuit of Figure 1.5 is

$$v(t) - Li'(t) - Ri(t) = 0, \quad i' = \frac{di}{dt}.$$

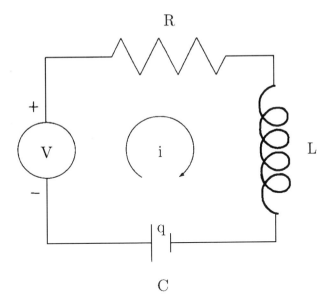

R

+

V   i   L

−

q

C

Figure 1.6: Circuit 2.

The circuit shown in Figure 1.6 has an additional component, a capacitor with constant *capacitance* C which can store charge $q(t)$. The stored charge q is given by

$$q(t) = Cv_C(t), \quad C \geq 0,$$

where $v_C(t)$ is the voltage drop across the capacitor. Applying Kirchhoff's law to circuit 2 leads to the equation

$$v(t) = Li'(t) + Ri(t) + \frac{q}{C}.$$

Differentiation of this equation gives

$$v'(t) = Li''(t) + Ri'(t) + \frac{1}{C}i(t).$$

Notice that this is precisely the form of the equation for the damped spring and mass system, with the external forcing corresponding to the derivative of the voltage.

The description of the third circuit, shown in Figure 1.7, is somewhat different since we are given a current source at I with current $i_0(t)$, and we wish to determine the voltage $v_2(t)$ that will be observed at V. The equations describing this circuit can be found with the use of *Kirchhoff's current law*, which says that because charge is conserved, the sum of the currents through any node must always be zero. Thus $i_0(t) = i_1(t) + i_2(t)$. We also know that the voltage drop across the capacitor $v_1(t)$ must satisfy $i_1(t) = Cv_1'(t)$. Similarly, in the part of the circuit through which the current $i_2(t)$ is flowing, Kirchhoff's voltage law implies that $v_1(t) = Li_2'(t) + Ri_2(t)$. (Compare this part of circuit

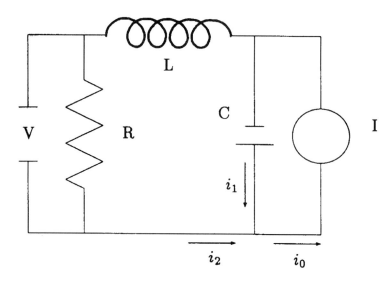

Figure 1.7: Circuit 3.

3 with circuit 1.) Differentiation of this last equation and substitution into the equation for i_1 leads to

$$i_0(t) = CLi_2''(t) + CRi_2'(t) + i_2, \qquad (1.8)$$

and the desired voltage is $v_2(t) = Ri_2(t)$.

Let's try to find a particular solution of equation (1.8) when $i_0(t) = \exp(i\omega t)$, where $\omega \in \mathcal{R}$. We optimistically look for a solution with a similar form and substitute $i_2(t) = z\exp(i\omega t)$. The result is

$$e^{i\omega t} = z[-CL\omega^2 + iCR\omega + 1]e^{i\omega t}.$$

Because the equation has constant coefficients, the existence of a particular solution $i_2(t) = z\exp(i\omega t)$ is equivalent to the algebraic condition

$$z[-k_1\omega^2 + ik_2\omega + 1] = 1, \quad k_1 = CL, \quad k_2 = CR.$$

Write $-k_1\omega^2 + ik_2\omega + 1$ in the polar form

$$-k_1\omega^2 + ik_2\omega + 1 = re^{i\phi},$$

with

$$r(\omega)^2 = (1 - k_1\omega^2)^2 + (k_2\omega)^2, \quad \phi(\omega) = \tan^{-1}\left(\frac{k_2\omega}{1 - k_1\omega^2}\right).$$

Then

$$z = \frac{1}{r}e^{-i\phi}$$

and the desired particular solution is

$$i_2(t) = a(\omega)e^{-i\phi(\omega)}e^{i\omega t},$$

with

$$a(\omega) = \left[\frac{1}{1 + (k_2^2 - 2k_1)\omega^2 + k_1^2\omega^4} \right]^{1/2}. \tag{1.9}$$

Since the equation (1.8) is linear, Euler's formula can be used to find a particular solution $i_2(t)$ for the real-valued input

$$i_0(t) = \cos(\omega t) = \frac{1}{2}[e^{i\omega t} + e^{-i\omega t}].$$

In this case

$$i_2(t) = [a(\omega)e^{-i\phi(\omega)}e^{i\omega t} + a(-\omega)e^{-i\phi(-\omega)}e^{-i\omega t}]/2 = a(\omega)\cos(\omega t - \phi).$$

1.5   Notes

In this chapter we have introduced some simple scientific problems which were modeled using linear differential equations. In each case the material presented was merely a brief introduction to an extensive subject. Discussions of the use of compartment systems in biology and medicine can be found in [4, 11, 20]. Springs and masses are important, if elementary, examples for physics (see [5, Vol. 1, Chaps. 21–25, and p. 49-4]). A physical approach to the study of electric circuits is also in [5, Vol. 2, Chap. 22]. A systematic development of the theory of linear electric circuits may be found in [2] or [23].

Additional scientific models will be introduced throughout the text, but many important applications are completely omitted. Some additional applications can be found in [18, pp. 119–152, 223–233] and [25, pp. 73–98, 191–216].

1.6   Exercises

1. Verify that the solution of the initial value problem

$$mx'' + kx = 0, \quad x(t_0) = x_0, \quad x'(t_0) = v_0$$

is given by

$$x(t) = x_0 \cos(\sqrt{k/m}\ [t - t_0]) + \frac{v_0}{\sqrt{k/m}} \sin(\sqrt{k/m}\ [t - t_0]).$$

2. Find the general solution of

$$x'' + \frac{k}{m}x = \frac{a}{m}\cos(\omega t).$$

(Hint: Use the approach following (1.7).) What is the general solution of

$$x'' + \frac{k}{m}x = \frac{a_1}{m}\cos(\omega t) + \frac{a_2}{m}\sin(\omega t)?$$

3. For any real numbers r_1 and r_2 there is a unique solution of the initial value problem

$$x'' + x = 0, \quad x(0) = r_1, \quad x'(0) = r_2.$$

Use this fact to show that

$$\sin(a)\cos(t) + \cos(a)\sin(t) = \sin(a + t),$$

$$\cos(a)\cos(t) - \sin(a)\sin(t) = \cos(a + t).$$

4. Show that if the general solution

$$x(t) = c\cos(\omega_0 t) + d\sin(\omega_0 t) + \frac{a}{m[\omega_0^2 - \omega^2]}\sin(\omega t), \quad a \neq 0, \quad \omega \neq \omega_0, \quad \omega \neq 0,$$

of (1.7) has period $2\pi/\omega$, then $c = 0 = d$ unless ω_0/ω is an integer.

5. Consider the solutions to the system of equations (1.1) and (1.2). Over the course of the day when toxins flow into the lakes, when will the concentration of toxin in Lake 2 be greatest? Before you start lengthy calculations, see if you can guess what the answer should be and then verify the result.

6. Solve the equations (1.3). For convenient but nontrivial values of the various constants, plot the solutions $\tau_1(t)$ and $\tau_2(t)$ for $t > 0$.

7. (a) Verify that the solution of the initial value problem

$$mx'' + kx = a\cos(\omega t), \quad x(0) = 0, \quad x'(0) = 0$$

is given by

$$x(t) = \frac{a}{m[\omega_0^2 - \omega^2]}[\cos(\omega t) - \cos(\omega_0 t)], \quad \omega_0^2 = \frac{k}{m}.$$

Show that this solution can also be expressed as

$$x(t) = \frac{2a}{m[\omega_0^2 - \omega^2]}\sin\left(\frac{\omega_0 - \omega}{2}t\right)\sin\left(\frac{\omega_0 + \omega}{2}t\right).$$

Sketch the graph of this function.

(b) Find a solution of

$$mx'' + kx = a\cos(\omega_0 t), \quad \omega_0^2 = \frac{k}{m},$$

of the form

$$x(t) = [c_1 + c_2 t]\cos(\omega_0 t) + [c_3 + c_4 t]\sin(\omega_0 t).$$

8. (a) Any solution of the equation

$$mx'' + kx = 0$$

also satisfies

$$mx'x'' + kxx' = 0.$$

By integration or otherwise, show that the energy of a solution

$$E(t) = \frac{m}{2}(x')^2 + \frac{k}{2}x^2$$

is a constant function of t.

(b) If you plot $(x(t), x'(t))$, what planar figure is obtained?

(c) Interpret

$$mx'' + cx' + kx = 0$$

as an equation for the rate of change of energy.

9. The flow of current in circuit 1 (Figure 1.5) is governed by the equation

$$Li' + Ri = v(t).$$

Find the solution if $v(t) = V_0$, a constant, and $i(0) = 0$. What physical situation might be modeled by this initial value problem?

10. Current flow in circuit 2 (Figure 1.6) with constant voltage is modeled by the equation

$$Li'' + Ri' + \frac{1}{C}i = 0.$$

Show that if $R = 0$, then the expression

$$g(t) = \frac{L}{2}(i')^2 + \frac{1}{2C}i^2$$

is a constant. (See exercise 8.)

11. The relationship between the input current $i_0(t)$ and the current $i_2(t)$ in circuit 3 (Figure 1.7) is determined by equation (1.8).

(a) Show that if

$$i_0(t) = CLi_2''(t) + CRi_2'(t) + i_2(t)$$

and

$$j_0(t) = CLj_2''(t) + CRj_2'(t) + j_2(t),$$

then for any complex numbers b and c

$$bi_0(t) + cj_0(t) = CL[bi_2''(t) + cj_2''(t)] + CR[bi_2'(t) + cj_2'(t)] + bi_2(t) + cj_2(t).$$

(b) Describe the behavior as $t \to \infty$ of solutions to the homogeneous equation

$$k_1 i_2'' + k_2 i_2' + i_2 = 0, \quad k_1, \ k_2 > 0.$$

(c) Sketch the graph of $a(\omega)$ from (1.9) if, say, $k_2^2 - 2k_1 > 0$.

(d) Describe the behavior of $i_2(t)$ as $t \to \infty$ if

$$i_0(t) = \sum_{k=1}^{K} c_k \cos(\omega_k t), \quad t > 0.$$

12. We consider the currents $i_1(t)$ and $i_2(t)$ for the circuit shown in Figure 1.8. If

$$\mathbf{A}_1 = \begin{pmatrix} L & 0 \\ -R_1 & R_1 + R_2 \end{pmatrix}, \quad \mathbf{A}_0 = \begin{pmatrix} R_1 & -R_1 \\ 0 & 1/C \end{pmatrix},$$

$$B(t) = \begin{pmatrix} V(t) \\ 0 \end{pmatrix},$$

and

$$i = \begin{pmatrix} i_1 \\ i_2 \end{pmatrix}, \quad i' = \begin{pmatrix} i_1' \\ i_2' \end{pmatrix},$$

show that the currents will satisfy the system of equations

$$\mathbf{A}_1 i' + \mathbf{A}_0 i = B(t),$$

which is a first-order linear system for the vector-valued function i as a function of time t.

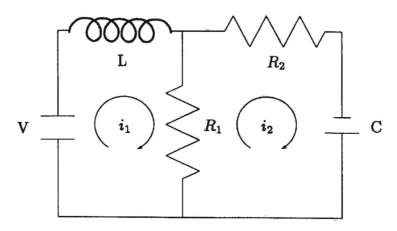

Figure 1.8: A circuit modeled with a system of differential equations.

13. The current i in a simple electrical circuit (see Figure 1.6) containing an inductance L, resistance R, capacitance C, and voltage v satisfies the equation

$$Li' + Ri + \frac{q}{C} = v, \quad i' = \frac{di}{dt},$$

where q is the charge on the capacitor. Since $i = q'$, we have

$$Lq'' + Rq' + \frac{q}{C} = v.$$

All the constants L, R, C are positive. Let $\Delta = R^2/L^2 - 4/LC$.

(a) Compute all solutions of

$$Lq'' + Rq' + \frac{q}{C} = 0 \tag{1.10}$$

for the three cases

$$\text{(i)} \quad \Delta > 0, \quad \text{(ii)} \quad \Delta = 0, \quad \text{(iii)} \quad \Delta < 0.$$

(b) Show that all solutions of (1.10) tend to zero as $t \to \infty$ in each of the cases (i), (ii), (iii). (The solutions are called transients.)

(c) In case (iii) show that every real solution q may be written as

$$q(t) = Ae^{\alpha t} \cos(\beta t - \delta),$$

where $A, \alpha, \beta, \delta \in \mathcal{R}$. (Hint: $\alpha = \frac{-R}{2L}$, $\beta = \sqrt{-\Delta}/2$.)

14. Consider

$$Lx = ax'' + bx' + cx, \quad a, b, c \in \mathcal{R},$$

with the characteristic polynomial p given by $p(\lambda) = a\lambda^2 + b\lambda + c$.

(a) Let $A, \omega \in \mathcal{R}$ and $p(i\omega) \neq 0$. Show that

$$Lx = Ae^{i\omega t} \tag{1.11}$$

has a particular solution y of the form

$$y(t) = \frac{A}{|p(i\omega)|} e^{i(\omega t - \alpha)}, \quad \alpha \in \mathcal{R}.$$

(Hint: Let $p(i\omega) = |p(i\omega)|e^{i\alpha}$.)

(b) If x satisfies (1.11) show that $u = \text{Re}(x)$, $v = \text{Im}(x)$, satisfy

$$Lu = A\cos(\omega t), \quad Lv = A\sin(\omega t).$$

(c) Using (a), (b) show that there is a particular solution q of

$$Lq'' + Rq' + \frac{q}{C} = E_0 \cos(\omega t),$$

where L, R, C, E_0, ω are positive constants, that has the form

$$q(t) = B\cos(\omega t - \alpha), \quad B, \alpha \in \mathcal{R}.$$

Determine B and α.

Chapter 2

Properties of Linear Systems

2.1 Introduction

The examples in Chapter 1 demonstrate that simple models may lead to a system of first-order differential equations. The study of toxic material flow through a system of lakes led directly to equations (1.1) and (1.2), which have the form

$$\frac{d\tau_1}{dt} = a_{11}\tau_1 + r(t),$$

$$\frac{d\tau_2}{dt} = a_{21}\tau_1 + a_{22}\tau_2.$$

In these equations the coefficients a_{11}, a_{21}, and a_{22} are known constants, $r(t)$ is a known function, and $\tau_1(t)$ and $\tau_2(t)$ are unknown functions whose behavior is to be determined.

The equation

$$mx'' + cx' + kx = f(t),$$

which arose in the model for the motion of a mass attached to a spring also led, although less directly, to a system of first-order equations. In that case the substitution

$$y_1 = x, \quad y_2 = x'$$

converted the single second-order equation to the system

$$y_1' = y_2,$$

$$y_2' = -\frac{k}{m}y_1 - \frac{c}{m}y_2 + \frac{f(t)}{m}.$$

Both of these cases are particular instances of a more general system of first-order differential equations which may be written as

$$\begin{pmatrix} x_1' \\ x_2' \end{pmatrix} = \begin{pmatrix} a_{11} & a_{12} \\ a_{21} & a_{22} \end{pmatrix} \begin{pmatrix} x_1 \\ x_2 \end{pmatrix} + \begin{pmatrix} b_1(t) \\ b_2(t) \end{pmatrix}.$$

The analysis of the spring and mass equations in Chapter 1 suggests that the study of the matrix

$$\mathbf{A} = \begin{pmatrix} a_{11} & a_{12} \\ a_{21} & a_{22} \end{pmatrix}$$

has a direct bearing on the understanding of such a system of equations. As we will begin to appreciate in this chapter, this connection between matrices and differential equations is only the first part of a network of deep and extensive links between linear algebra and the study of linear systems of differential equations.

The language, ideas, and theorems of linear algebra have a major role in the development of the theory of linear systems of ordinary differential equations. This chapter opens with a review of the basic ideas of linear algebra which should be familiar to the reader. Following this review, linear systems of differential equations are introduced. The existence and uniqueness results for such systems are presented. The remainder of the chapter explores the basic properties of solutions of systems of linear differential equations from the vector space viewpoint, with the existence and uniqueness theorem playing a central role.

2.2   Basic linear algebra

2.2.1   Vector spaces

Many of the results of differential equations are valid whether the real numbers or the complex numbers are the field of scalars for the vector spaces. For this reason $\mathcal{F}$ is used to denote a field which is either $\mathcal{R}$ or $\mathcal{C}$.

The set of ordered n-tuples $X = (x_1, \ldots, x_n)$, $x_j \in \mathcal{F}$, is denoted by $\mathcal{F}^n$. Elements $X \in \mathcal{F}^n$ are called *vectors* and x_j is called the *jth component* of X. The *zero vector* 0 is the one having all components zero, $0 = (0, \ldots, 0)$, and the negative of $X \in \mathcal{F}^n$ is $-X = (-x_1, \ldots, -x_n)$. Two vectors are *equal*, $X = Y$, if and only if $x_j = y_j$, $j = 1, \ldots, n$. The *sum* $X + Y$ is given by $X + Y = (x_1 + y_1, \ldots, x_n + y_n)$, and *scalar multiplication* of $X \in \mathcal{F}^n$ by a *scalar* $\alpha \in \mathcal{F}$ is defined by $\alpha X = (\alpha x_1, \ldots, \alpha x_n)$. With these definitions the following properties hold for all $X, Y, Z \in \mathcal{F}^n$:

(i) $(X + Y) + Z = X + (Y + Z),$

(ii) $X + Y = Y + X,$

(iii) $0 + X = X,$

(iv) $X + (-X) = 0,$

(v) $$\alpha(X + Y) = \alpha X + \alpha Y, \quad \alpha \in \mathcal{F},$$

(vi) $$(\alpha + \beta)X = \alpha X + \beta X, \quad \alpha, \beta \in \mathcal{F},$$

(vii) $$(\alpha\beta)X = \alpha(\beta X), \quad \alpha, \beta \in \mathcal{F},$$

(viii) $$1X = X.$$

Any set $\mathcal{V}$ with operations of addition and scalar multiplication (by elements of a field $\mathcal{F}$) satisfying (i)–(viii) is called a *vector space* over $\mathcal{F}$. (Of course the general definition applies when $\mathcal{F}$ is any field, not necessarily $\mathcal{R}$ or $\mathcal{C}$.) Thus $\mathcal{C}^n$ and $\mathcal{R}^n$ are vector spaces over $\mathcal{C}$ and $\mathcal{R}$, respectively.

A subset $\mathcal{B}$ of a vector space $\mathcal{V}$ over $\mathcal{F}$ is a *basis* for $\mathcal{V}$ if every $X \in \mathcal{V}$ can be written as a finite *linear combination* of elements of $\mathcal{B}$; that is,

$$X = \alpha_1 B_1 + \cdots + \alpha_k B_k, \quad \alpha_j \in \mathcal{F}, \quad B_j \in \mathcal{B},$$

and the set of elements $\mathcal{B}$ is *linearly independent*, which means that for any finite subset $B_1, \ldots, B_k \in \mathcal{B}$ the equation

$$\alpha_1 B_1 + \cdots + \alpha_k B_k = 0, \quad \alpha_j \in \mathcal{F},$$

implies $\alpha_1 = \cdots = \alpha_k = 0$. A vector space $\mathcal{V}$ is *finite dimensional* if it has a finite basis. If $\mathcal{V}$ is finite dimensional, any two bases for $\mathcal{V}$ have the same number of elements. This number is called the *dimension* of $\mathcal{V}$ and is denoted by $\dim(\mathcal{V})$.

If $\mathcal{V}$ is not finite dimensional we write $\dim(\mathcal{V}) = \infty$. The space $\mathcal{F}^n$ has the basis $\mathcal{B} = \{E_1, \ldots, E_n\}$, where E_j has all components zero except the jth, which is 1:

$$E_1 = (1, 0, \ldots, 0), \ E_2 = (0, 1, \ldots, 0), \ldots, \ E_n = (0, \ldots, 0, 1).$$

This is called the *standard basis* for $\mathcal{F}^n$. If $X = (x_1, \ldots, x_n) \in \mathcal{F}^n$ we have the representation $X = x_1 E_1 + \cdots + x_n E_n$ and $\dim(\mathcal{F}^n) = n$.

If $\mathcal{U}$, $\mathcal{V}$ are vector spaces over $\mathcal{F}$ such that $\mathcal{U} \subset \mathcal{V}$, we say $\mathcal{U}$ is a *subspace* of $\mathcal{V}$. If $V \in \mathcal{V}$, then the *translate* of $\mathcal{U}$ by V is the set

$$V + \mathcal{U} = \{V + U \mid U \in \mathcal{U}\}.$$

The translates of vector spaces are also called *affine spaces*.

For example, in $\mathcal{R}^2$ the set

$$\mathcal{U} = \{(x, y) \in \mathcal{R}^2 \mid y = 2x\} = \{(x, 2x) \in \mathcal{R}^2 \mid x \in \mathcal{R}\},$$

which is the straight line through the origin with slope 2, is a subspace of $\mathcal{V} = \mathcal{R}^2$. If $V = (3, -1)$ then the translate of $\mathcal{U}$ by V is the set

$$V + \mathcal{U} = \{(3 + x, -1 + 2x) \mid x \in \mathcal{R}\},$$

which is just the line through $(3, -1)$ with slope 2, so that $V + \mathcal{U}$ is the line through V parallel to $\mathcal{U}$.

2.2.2　Matrices

Suppose that R is any set, which for our purposes will usually be real or complex numbers. An $m \times n$ *matrix* $\mathbf{A}$ with *elements*, or *components*, $a_{ij} \in R$ is a function

$$\mathbf{A} : (i,j) \to \mathbf{A}(i,j) = a_{ij} \in R, \quad i = 1, \ldots, m, \quad j = 1, \ldots, n.$$

The set of all $m \times n$ matrices with elements in R is denoted by $M_{mn}(R)$. In case $m = n$, $\mathbf{A}$ is called a *square matrix* and $M_{nn}(R)$ is denoted by $M_n(R)$. In order to display the tabular nature of this function, we write

$$\mathbf{A} = \begin{pmatrix} a_{11} & \cdots & a_{1n} \\ \vdots & \cdots & \vdots \\ a_{m1} & \cdots & a_{mn} \end{pmatrix},$$

or more briefly $\mathbf{A} = (a_{ij})$, and say $\mathbf{A}$ has m rows and n columns, with a_{ij} being the element in the ith row and the jth column. To be useful the set R should be closed under addition, multiplication, and multiplication by elements of a field $\mathcal{F}$. A particular case where this occurs is when $R = \mathcal{F}$ itself.

If $\mathbf{A} = (a_{ij}), \mathbf{B} = (b_{ij}) \in M_{mn}(\mathcal{F})$, where $\mathcal{F}$ is a field, which could be either $\mathcal{R}$ or $\mathcal{C}$, then $\mathbf{A} = \mathbf{B}$ if and only if $a_{ij} = b_{ij}$; $i = 1, \ldots, m$; $j = 1, \ldots, n$. The *sum* $\mathbf{A} + \mathbf{B}$ is defined as the function $\mathbf{A} + \mathbf{B} : (i,j) \to a_{ij} + b_{ij} \in \mathcal{F}$, and we write

$$\mathbf{A} + \mathbf{B} = (a_{ij} + b_{ij}) = \begin{pmatrix} a_{11} + b_{11} & \cdots & a_{1n} + b_{1n} \\ \vdots & \cdots & \vdots \\ a_{m1} + b_{m1} & \cdots & a_{mn} + b_{mn} \end{pmatrix},$$

whereas *scalar multiplication* of $\mathbf{A}$ by $\alpha \in \mathcal{F}$ is defined as $\alpha \mathbf{A} : (i,j) \to \alpha a_{ij} \in \mathcal{F}$, and we write

$$\alpha \mathbf{A} = (\alpha a_{ij}) = \begin{pmatrix} \alpha a_{11} & \cdots & \alpha a_{1n} \\ \vdots & \cdots & \vdots \\ \alpha a_{m1} & \cdots & \alpha a_{mn} \end{pmatrix}.$$

The zero $m \times n$ matrix, with all elements zero, is denoted by 0_{mn}; if there is no chance of confusion it is simply written as 0. The negative $-\mathbf{A}$ of $\mathbf{A} = (a_{ij}) \in M_{mn}(\mathcal{F})$ is given by $-\mathbf{A} = (-a_{ij})$. With these definitions $M_{mn}(\mathcal{F})$ is a vector space over $\mathcal{F}$ with dimension mn.

Certain matrices may be multiplied. If $\mathbf{A} = (a_{ij}) \in M_{mn}(\mathcal{F}), \mathbf{B} = (b_{ij}) \in M_{np}(\mathcal{F})$, then the *product* $\mathbf{AB} = \mathbf{C} = (c_{ik}) \in M_{mp}(\mathcal{F})$ is defined by

$$c_{ik} = a_{i1}b_{1k} + \cdots + a_{in}b_{nk}, \quad i = 1, \ldots, m, \quad k = 1, \ldots, p.$$

If $\mathbf{A}$, $\mathbf{B}$, and $\mathbf{C}$ are any matrices with elements in $\mathcal{F}$ for which the products below make sense, then

(i)　　　　　　　　　　　$(\mathbf{AB})\mathbf{C} = \mathbf{A}(\mathbf{BC}),$

(ii) $$\mathbf{A}(\mathbf{B} + \mathbf{C}) = \mathbf{A}\mathbf{B} + \mathbf{A}\mathbf{C},$$

(iii) $$(\mathbf{A} + \mathbf{B})\mathbf{C} = \mathbf{A}\mathbf{C} + \mathbf{B}\mathbf{C},$$

(iv) $$\alpha(\mathbf{A}\mathbf{B}) = (\alpha\mathbf{A})\mathbf{B} = \mathbf{A}(\alpha\mathbf{B}), \quad \alpha \in \mathcal{F}.$$

The *identity matrix* $\mathbf{I}_n \in M_n(\mathcal{F})$ is the one with ones down the principal diagonal and zeros elsewhere:

$$\mathbf{I}_n = (\delta_{ij}) = \begin{pmatrix} 1 & 0 & \dots & 0 \\ 0 & 1 & \dots & 0 \\ 0 & 0 & \dots & 0 \\ 0 & 0 & \dots & 1 \end{pmatrix},$$

where the *Kronecker* δ has $\delta_{ij} = 1$ when $i = j$ and $\delta_{ij} = 0$ when $i \neq j$ for $i, j = 1, \dots, n$. The identity matrix has the property that $\mathbf{I}_n\mathbf{A} = \mathbf{A}\mathbf{I}_n = \mathbf{A}$ for all $\mathbf{A} \in M_n(\mathcal{F})$. A matrix $\mathbf{A} \in M_n(\mathcal{F})$ is said to be *invertible* if there is a matrix $\mathbf{A}^{-1}$, the *inverse* of $\mathbf{A}$, such that

$$\mathbf{A}\mathbf{A}^{-1} = \mathbf{A}^{-1}\mathbf{A} = \mathbf{I}_n.$$

If $\mathbf{B} \in M_n(\mathcal{F})$ is such that $\mathbf{A}\mathbf{B} = \mathbf{I}_n$, then $\mathbf{B}\mathbf{A} = \mathbf{I}_n$ and $\mathbf{B} = \mathbf{A}^{-1}$.

The vector space $\mathcal{F}^n$ will now be identified with the set of all $n \times 1$ matrices $M_{n1}(\mathcal{F})$ with elements in $\mathcal{F}$. Thus for $X = (x_1, \dots, x_n) \in \mathcal{F}^n$ we put

$$X = \begin{pmatrix} x_1 \\ \vdots \\ x_n \end{pmatrix}.$$

The matrix $\mathbf{A} \in M_{mn}(\mathcal{F})$ determines a function $f_{\mathbf{A}} : \mathcal{F}^n \to \mathcal{F}^m$ defined by $f_{\mathbf{A}}(X) = \mathbf{A}X$. This function is *linear*,

$$f_{\mathbf{A}}(\alpha X + \beta Y) = \alpha f_{\mathbf{A}}(X) + \beta f_{\mathbf{A}}(Y),$$

and its range and nullspace are respectively defined as

$$Ran(\mathbf{A}) = \{Y \in \mathcal{F}^m \mid Y = \mathbf{A}X, \ X \in \mathcal{F}^n\}$$

and

$$N(\mathbf{A}) = \{X \in \mathcal{F}^n \mid \mathbf{A}X = 0\}.$$

These are vector spaces over $\mathcal{F}$. Defining the *rank* of $\mathbf{A}$ as $\dim(Ran(\mathbf{A}))$ and the *nullity* of $\mathbf{A}$ as $\dim(N(\mathbf{A}))$, there is the important result

$$\text{rank}(\mathbf{A}) + \text{nullity}(\mathbf{A}) = n.$$

2.2.3   Vector spaces of functions

If $I \subset \mathcal{R}$ is an interval and $\mathcal{F}$ is a field which is either $\mathcal{C}$ or $\mathcal{R}$, the set $F(I, \mathcal{F})$ of all functions $f : I \to \mathcal{F}$ is a vector space over $\mathcal{F}$ with addition and scalar multiplication defined by

$$(f + g)(t) = f(t) + g(t),$$

$$(\alpha f)(t) = \alpha f(t).$$

The zero function, denoted by 0, is the one whose value at each $t \in I$ is 0, and the negative $-f$ of f is given by $-f : t \in I \to -f(t) \in \mathcal{F}$. The subset $C(I, \mathcal{F})$ of all continuous functions is also a vector space over $\mathcal{F}$, since $f + g$ and αf are in $C(I, \mathcal{F})$ if $f, g \in C(I, \mathcal{F})$, $\alpha \in \mathcal{F}$.

Now consider functions defined on an interval $I \subset \mathcal{R}$ whose values are matrices. Addition and scalar multiplication of such functions are defined by

$$(\mathbf{A} + \mathbf{B})(t) = \mathbf{A}(t) + \mathbf{B}(t),$$

$$(\alpha \mathbf{A})(t) = \alpha \mathbf{A}(t).$$

The zero function, denoted by 0, is given by $t \in I \to 0_{mn} \in M_{mn}(\mathcal{F})$, while the negative $-\mathbf{A}$ of $\mathbf{A}$ is defined as $t \in I \to -\mathbf{A}(t) \in M_{mn}(\mathcal{F})$. With these definitions $F(I, M_{mn}(\mathcal{F}))$ is a vector space over $\mathcal{F}$. If $\mathbf{C} \in F(I, M_{mp}(\mathcal{F}))$, then the product $\mathbf{AC} \in F(I, M_{mp}(\mathcal{F}))$ is given by

$$(\mathbf{AC})(t) = \mathbf{A}(t)\mathbf{C}(t).$$

Each $\mathbf{A} \in F(I, M_{mn}(\mathcal{F}))$ determines mn functions $a_{ij} \in F(I, \mathcal{F})$, the elements of $\mathbf{A}$, where

$$\mathbf{A}(t) = (a_{ij}(t)).$$

A function $\mathbf{A} : I \subset \mathcal{R} \to M_{mn}(\mathcal{F})$ is said to be *continuous* at $c \in I$ if each of the functions $a_{ij}(t)$ is continuous at c. A function $\mathbf{A} : I \to M_{mn}(\mathcal{F})$ is said to be *differentiable* at $c \in I$ and has the *derivative* $\mathbf{A}'(c)$ at c, if each of the functions $a_{ij}(t)$ is differentiable at c, and then $\mathbf{A}'(c) = (a'_{ij}(c))$.

The usual rules are valid. If $\mathbf{A}, \mathbf{B} : I \subset \mathcal{R} \to M_{mn}(\mathcal{F})$ are differentiable at $c \in I$, then so are $\mathbf{A} + \mathbf{B}$ and $\alpha \mathbf{A}$ for $\alpha \in \mathcal{F}$ and $(\mathbf{A} + \mathbf{B})'(c) = \mathbf{A}'(c) + \mathbf{B}'(c)$ whereas $(\alpha \mathbf{A})'(c) = \alpha \mathbf{A}'(c)$. If, in addition, $\mathbf{C} : I \subset \mathcal{R} \to M_{np}(\mathcal{F})$ is differentiable at c, then $\mathbf{AC}$ is also and $(\mathbf{AC})'(c) = \mathbf{A}'(c)\mathbf{C}(c) + \mathbf{A}(c)\mathbf{C}'(c)$.

The set of matrix-valued functions which are continuous at each $t \in I$ is denoted $C(I, M_{mn}(\mathcal{F}))$. If $\mathbf{A}$ is differentiable at each $t \in I$, then its *derivative* $\mathbf{A}' = (a'_{ij})$ is the function

$$\mathbf{A}' : t \in I \to \mathbf{A}'(t) = (a'_{ij}(t)) \in M_{mn}(\mathcal{F}).$$

Higher derivatives $\mathbf{A}^{(k)}$, $k = 2, 3, \ldots$, are defined as usual. $C^k(I, M_{mn}(\mathcal{F}))$ will denote the set of all $\mathbf{A} : I \to M_{mn}(\mathcal{F})$ such that $\mathbf{A}^{(k)} \in C(I, M_{mn}(\mathcal{F}))$, and $C^\infty(I, M_{mn}(\mathcal{F}))$ will denote those $\mathbf{A}$ which are in $C^k(I, M_{mn}(\mathcal{F}))$ for all $k = 1, 2, 3 \dots$.

Let $I = [a, b]$ be a compact interval. A function $\mathbf{A} = (a_{ij}) : I \to M_{mn}(\mathcal{F})$ is said to be *integrable* on I if each a_{ij} is integrable on I, and in this case we define its *integral* to be the $m \times n$ matrix

$$\int_a^b \mathbf{A}(t) \, dt = \left(\int_a^b a_{ij}(t) \, dt \right).$$

Thus the ijth element of the integral of $\mathbf{A}$ is the integral of the ijth element of $\mathbf{A}$. If $\mathbf{A} \in C(I, M_{mn}(\mathcal{F}))$ then $\mathbf{A}$ is integrable. The usual integration rules are valid. If $\mathbf{A}, \mathbf{B} : I \to M_{mn}(\mathcal{F})$ are integrable on I and $\alpha, \beta \in \mathcal{F}$, then $\alpha \mathbf{A} + \beta \mathbf{B}$ is integrable on I and

$$\int_a^b (\alpha \mathbf{A} + \beta \mathbf{B})(t) \, dt = \alpha \int_a^b \mathbf{A}(t) \, dt + \beta \int_a^b \mathbf{B}(t) \, dt.$$

2.3   First-order systems

2.3.1   Introduction

An equation of the form

$$\mathbf{A}_1(t)X' + \mathbf{A}_0(t)X = B_0(t), \quad t \in I, \tag{2.1}$$

where $\mathbf{A}_1(t)$, $\mathbf{A}_0(t)$ are continuous $n \times n$ matrix-valued functions and $B_0(t)$ is a continuous n vector-valued function, is called a *first-order linear system* of differential equations. Here I is an arbitrary real interval, and $\mathbf{A}_0$, $\mathbf{A}_1$, and B_0 may be either real or complex valued. The equations presented in the introduction to this chapter are examples of such systems.

A *solution* of (2.1) on I is a differentiable function $X : I \to \mathcal{F}^n$ such that

$$\mathbf{A}_1(t)X'(t) + \mathbf{A}_0(t)X(t) = B_0(t), \quad t \in I.$$

If B_0 is not the zero function then (2.1) is called a *nonhomogeneous system*, whereas the equation $\mathbf{A}_1(t)X' + \mathbf{A}_0(t)X = 0$ is called a *homogeneous system*. The study of (2.1) is easier if we can solve for X' and (2.1) takes the form

$$X' = \mathbf{A}(t)X + B(t), \quad t \in I. \tag{2.2}$$

Here $\mathbf{A} \in C(I, M_n(\mathcal{F}))$ and $B \in C(I, \mathcal{F}^n)$; that is, $\mathbf{A}$ and B are continuous matrix- and vector-valued functions, respectively.

The following fundamental result is established in Chapter 7 (Theorem 7.4).

Existence and uniqueness theorem for linear systems: *Given any $\tau \in I$ and $\xi \in \mathcal{F}^n$, there exists a unique solution X of (2.2) on I such that $X(\tau) = \xi$.*

The condition $X(\tau) = \xi$ is called an *initial condition* and the above result gives the existence and uniqueness of solutions to the *initial value problem*

$$X' = \mathbf{A}(t)X + B(t), \quad X(\tau) = \xi, \quad t \in I. \tag{2.3}$$

Although the proof of the existence and uniqueness theorem is deferred, we will use it frequently in studying the solutions of (2.2). This study begins with the homogeneous system. Given the solutions of the homogeneous system, the corresponding nonhomogeneous system can be solved by integration. Linear equations, or systems of equations, of order $n > 1$ can be reduced to a system of the form (2.2).

2.3.2 First-order homogeneous systems

Consider the homogeneous system

$$X' = \mathbf{A}(t)X, \quad t \in I, \tag{H}$$

where $\mathbf{A} \in C(I, M_n(\mathcal{F}))$ and I is any real interval. Given any $\tau \in I$ and $\xi \in \mathcal{F}^n$ there exists a unique solution $X(t)$ of (H) on I satisfying $X(\tau) = \xi$. The zero function

$$X(t) = \begin{pmatrix} 0 \\ \vdots \\ 0 \end{pmatrix}, \quad t \in I,$$

is a solution of (H); it is called the *trivial solution*. If $X(t)$ is any solution of (H) such that $X(\tau) = 0$ for some $\tau \in I$, then uniqueness requires that $X(t) = 0$ for all $t \in I$, so that X is the trivial solution. This observation leads to the following result describing the structure of the set S of all solutions of (H) on I.

Theorem 2.1: *The set S of solutions of (H) on I is an n-dimensional vector space over $\mathcal{F}$.*

Proof: If $X(t), Y(t) \in S$, and $\alpha, \beta \in \mathcal{F}$ then

$$X'(t) = \mathbf{A}(t)X(t), \quad Y'(t) = \mathbf{A}(t)Y(t), \quad t \in I.$$

Since differentiation is linear,

$$(\alpha X + \beta Y)'(t) = \alpha X'(t) + \beta Y'(t)$$

$$= \alpha \mathbf{A}(t)X(t) + \beta \mathbf{A}(t)Y(t) = \mathbf{A}(t)[\alpha X + \beta Y](t).$$

Thus $\alpha X(t) + \beta Y(t) \in S$, so that $S \subset C(I, \mathcal{F}^n)$ is a vector space over $\mathcal{F}$.

A basis consisting of n elements is constructed as follows. Fix $t \in 1$, let $F_1, \ldots, F_n \in \mathcal{F}^n$ be a basis for $\mathcal{F}^n$, and let $X_j(t) \in S$ satisfy the initial condition

$X_j(\tau) = F_j$. To see that $X_1, \ldots, X_n$ is a basis for S, suppose first that there are constants $c_1, \ldots, c_n \in \mathcal{F}$ such that

$$c_1 X_1(t) + \cdots + c_n X_n(t) = \begin{pmatrix} 0 \\ \vdots \\ 0 \end{pmatrix},$$

the zero function on I. Evaluating this sum at τ gives

$$c_1 X_1(\tau) + \cdots + c_n X_n(\tau) = c_1 F_1 + \cdots + c_n F_n = 0.$$

Since $F_1, \ldots, F_n$ is a linearly independent set, $c_1 = \cdots = c_n = 0$, showing that $X_1, \ldots, X_n$ is a linearly independent set in S.

Let $X \in S$ and suppose $X(\tau) = \xi$. Since $F_1, \ldots, F_n$ is a basis for $\mathcal{F}^n$, ξ can be written as

$$\xi = c_1 F_1 + \cdots + c_n F_n$$

for unique constants $c_n \in \mathcal{F}$. If $Y \in S$ is the solution

$$Y = c_1 X_1 + \cdots + c_n X_n,$$

then $Y(\tau) = \xi = X(\tau)$. Solutions to initial value problems (2.3) are unique, so $X = Y = c_1 X_1 + \cdots + c_n X_n$, which shows that $X_1, \ldots, X_n$ spans S and is thus a basis. $\square$

Recall that in any finite-dimensional vector space of dimension n, any set of n linearly independent elements forms a basis.

Theorem 2.2: *Let $\tau \in I$, and let $X_1, \ldots, X_k$ be any k solutions of (H) on I. Then $X_1, \ldots, X_k$ is a linearly independent set in S if and only if $X_1(\tau) = F_1, \ldots, X_k(\tau) = F_k$ is a linearly independent set in $\mathcal{F}^n$. In particular, if $k = n$, then $X_1, \ldots, X_n$ is a basis for S if and only if $X_1(\tau) = F_1, \ldots, X_n(\tau) = F_n$ is a basis for $\mathcal{F}^n$.*

Proof: The argument used in the proof of Theorem 2.1 shows that if $F_1, \ldots, F_k$ is linearly independent in $\mathcal{F}^n$, then $X_1, \ldots, X_k$ is linearly independent in S.

Conversely, let $X_1, \ldots, X_k$ be linearly independent in S, and suppose

$$c_1 F_1 + \cdots + c_k F_k = 0$$

for some $c_j \in \mathcal{F}$. The solution $X = c_1 X_1 + \cdots + c_k X_k$ satisfies $X(\tau) = 0$, which implies $X = 0 = c_1 X_1 + \cdots + c_k X_k$, and hence $c_1 = \cdots = c_k = 0$. Consequently, $F_1, \ldots, F_k$ is a linearly independent set in $\mathcal{F}^n$. $\square$

A particular basis for $\mathcal{F}^n$ is the *standard basis* $E_1, \ldots, E_n$, where E_j has all components zero except the jth, which is one:

$$E_1 = \begin{pmatrix} 1 \\ 0 \\ \vdots \\ 0 \end{pmatrix}, \ldots, \quad E_n = \begin{pmatrix} 0 \\ 0 \\ \vdots \\ 1 \end{pmatrix}.$$

A basis for S can be found by choosing $X_j \in S$ such that $X_j(\tau) = E_j$ for some fixed $\tau \in I$. In this case, if

$$\xi = \begin{pmatrix} \xi_1 \\ \vdots \\ \xi_n \end{pmatrix}$$

and $X(\tau) = \xi$, then $X = \xi_1 X_1 + \cdots + \xi_n X_n$.

As an example, consider the system of two equations for $t \in \mathcal{R}$:

$$x_1' = 2x_1,$$

$$x_2' = 3x_2,$$

which can be written as

$$X' = \mathbf{A}X, \quad \mathbf{A} = \begin{pmatrix} 2 & 0 \\ 0 & 3 \end{pmatrix}, \quad X = \begin{pmatrix} x_1 \\ x_2 \end{pmatrix}. \tag{2.4}$$

Clearly, $x_1(t) = c_1 e^{2t}$ and $x_2(t) = c_2 e^{3t}$, where c_1, c_2 are arbitrary constants. In particular,

$$U(t) = \begin{pmatrix} \exp(2t) \\ 0 \end{pmatrix}, \quad V(t) = \begin{pmatrix} 0 \\ \exp(3t) \end{pmatrix},$$

give two solutions U, V of the system, and since $U(0) = E_1$, $V(0) = E_2$, we know U, V is a basis for the solutions. Every solution X of (2.4) has the form $X = c_1 U + c_2 V$; that is,

$$X(t) = c_1 U(t) + c_2 V(t) = \begin{pmatrix} c_1 \exp(2t) \\ c_2 \exp(3t) \end{pmatrix},$$

where c_1 and c_2 are constants.

For a second example, consider the system

$$x_1' = x_2,$$

$$x_2' = -x_1$$

or

$$X' = \mathbf{A}X, \quad \mathbf{A} = \begin{pmatrix} 0 & 1 \\ -1 & 0 \end{pmatrix}, \quad X = \begin{pmatrix} x_1 \\ x_2 \end{pmatrix}. \tag{2.5}$$

It is easy to check that

$$U(t) = \begin{pmatrix} \cos(t) \\ -\sin(t) \end{pmatrix}, \quad V(t) = \begin{pmatrix} \sin(t) \\ \cos(t) \end{pmatrix}$$

are solutions of this system satisfying $U(0) = E_1$, $V(0) = E_2$, and hence U, V is a basis for the solutions.

There is a slightly more sophisticated way of expressing the results of Theorems 2.1 and 2.2. Let $\tau \in I$ be fixed, and for any $\xi \in \mathcal{F}^n$ denote by $X_\xi(t)$ that solution of (H) such that $X_\xi(\tau) = \xi$.

Theorem 2.3: *The map $S : \xi \in \mathcal{F}^n \to X_\xi \in S$ is an isomorphism; that is, it is linear, one to one, and onto S. Therefore, since $\mathcal{F}^n$ is n dimensional over $\mathcal{F}$, so is S.*

The proof of this result is left as an exercise.

A basis $X_1, \ldots, X_n$ for S can be identified with the $n \times n$ matrix $\mathbf{X}(t) = (X_1, \ldots, X_n)$ with columns $X_1, \ldots, X_n$. The basis $\mathbf{X}$ is also called a *fundamental matrix* for (H). If

$$X_j = \begin{pmatrix} x_{1j} \\ \vdots \\ x_{nj} \end{pmatrix}, \quad j = 1, \ldots, n,$$

then the ijth element of $\mathbf{X}(t)$ is $x_{ij}(t)$, $\mathbf{X} = (x_{ij})$. A linear combination $c_1 X_1 + \cdots + c_n X_n$ of the basis elements can then be written succinctly as $\mathbf{X}C$, where

$$C = \begin{pmatrix} c_1 \\ \vdots \\ c_n \end{pmatrix},$$

and each $X \in S$ can be written uniquely as $X = \mathbf{X}C$ for some $C \in \mathcal{F}^n$. The particular basis $\mathbf{X} = (X_1, \ldots, X_n)$ such that $X_j(\tau) = E_j$ is the one satisfying $\mathbf{X}(\tau) = \mathbf{I}_n$, the $n \times n$ identity matrix.

In the example (2.4), $\mathbf{X} = (U, V)$ is the basis given by

$$\mathbf{X}(t) = \begin{pmatrix} \exp(2t) & 0 \\ 0 & \exp(3t) \end{pmatrix},$$

whereas for (2.5), $\mathbf{X} = (U, V)$ is defined via

$$\mathbf{X}(t) = \begin{pmatrix} \cos(t) & \sin(t) \\ -\sin(t) & \cos(t) \end{pmatrix}.$$

In both cases we have $\mathbf{X}(0) = \mathbf{I}_2$. The solution $X(t)$ of (2.5) satisfying

$$X(0) = \begin{pmatrix} 1 \\ -1 \end{pmatrix}$$

is given by

$$X(t) = \mathbf{X}(t) \begin{pmatrix} 1 \\ -1 \end{pmatrix} = U(t) - V(t) = \begin{pmatrix} \cos(t) - \sin(t) \\ -\sin(t) - \cos(t) \end{pmatrix}.$$

More generally, an $n \times n$ matrix $\mathbf{X} = (X_1, \ldots, X_n)$ with columns $X_1, \ldots, X_n$ is a *solution matrix* for (H) if each $X_j \in S$. Extending the previous notation, we will write $\mathbf{X} \in S$. Such an $\mathbf{X}$ is a solution of the matrix equation

$$\mathbf{X}' = \mathbf{A}(t)\mathbf{X}$$

in the sense that

$$\mathbf{X}'(t) = \mathbf{A}(t)\mathbf{X}(t), \quad t \in I,$$

since the jth column of the two sides of this equation are the same,

$$X'_j(t) = \mathbf{A}(t)X_j(t).$$

Theorem 2.4: *A solution matrix* $\mathbf{X}$ *for* (H) *is a basis for* S *if and only if* $\mathbf{X}(t)$ *is invertible for all* $t \in I$.

Proof: Suppose $\mathbf{X}(t)$ is invertible for all $t \in I$, and let $\mathbf{X}C = 0$ for some $C \in \mathcal{F}^n$. Then $C = \mathbf{X}^{-1}(t)\mathbf{X}(t)C = 0$, showing that $\mathbf{X}$ is linearly independent and hence a basis for S.

Conversely, suppose $\mathbf{X}$ is a basis for S. Let τ be arbitrary in I, and suppose that $C \in \mathcal{F}^n$ with $\mathbf{X}(\tau)C = 0$. The solution $X(t) = \mathbf{X}(t)C$ vanishes at τ, so $X(t) = \mathbf{X}(t)C = 0$. Since $\mathbf{X}$ is a basis for S, we must have $C = 0$. Thus $\mathbf{X}(\tau)$ is an $n \times n$ matrix whose only null vector is 0, so that $\mathbf{X}(\tau)$ is invertible. $\square$

If $\mathbf{X}$ is a solution matrix for (H) and $\mathbf{C} \in M_n(\mathcal{F})$, then $\mathbf{Y} = \mathbf{X}\mathbf{C}$ is also a solution matrix for

$$\mathbf{Y}'(t) = \mathbf{X}'(t)\mathbf{C} = \mathbf{A}(t)\mathbf{X}(t)\mathbf{C} = \mathbf{A}(t)\mathbf{Y}(t).$$

The next result shows that if $\mathbf{X}$ is a basis for S, then every $\mathbf{Y} \in S$ has this form.

Theorem 2.5: *Suppose* $\mathbf{X}$ *is a basis for* S. *If* $\mathbf{C} \in M_n(\mathcal{F})$ *then* $\mathbf{Y} = \mathbf{X}\mathbf{C} \in S$. *Conversely, given any solution matrix* $\mathbf{Y}$ *for* (H), *there exists a* $\mathbf{C} \in M_n(\mathcal{F})$ *such that* $\mathbf{Y} = \mathbf{X}\mathbf{C}$. *The solution matrix* $\mathbf{Y} = \mathbf{X}\mathbf{C}$ *is a basis for* S *if and only if* $\mathbf{C}$ *is invertible.*

Proof: We have shown that $\mathbf{Y} = \mathbf{X}\mathbf{C} \in S$ for any $\mathbf{C} \in M_n(\mathcal{F})$. If $\mathbf{Y} \in S$, $\tau \in I$, and $\mathbf{C} = \mathbf{X}^{-1}(\tau)\mathbf{Y}(\tau)$, then $\mathbf{Y}$ and $\mathbf{X}\mathbf{C}$ are solution matrices having the same initial value $\mathbf{Y}(\tau)$, and uniqueness then implies $\mathbf{Y} = \mathbf{X}\mathbf{C}$. Now $\mathbf{Y} = \mathbf{X}\mathbf{C}$ is a basis if and only if $\mathbf{X}\mathbf{C}$ is invertible, and since $\mathbf{X}$ is invertible this is true if and only if $\mathbf{C}$ is invertible. $\square$

2.3.3   The Wronskian

A matrix $\mathbf{C}$ is invertible if and only if its determinant $\det(\mathbf{C})$ is not zero. If $\mathbf{X} = (X_1, \ldots, X_n)$ is any solution matrix for the homogeneous system (H), the *Wronskian* $W_\mathbf{X}$ of $\mathbf{X}$ is the function

$$W_\mathbf{X}(t) = \det(\mathbf{X}(t)), \quad t \in I.$$

In the example (2.4)

$$\mathbf{X}(t) = \begin{pmatrix} \exp(2t) & 0 \\ 0 & \exp(3t) \end{pmatrix}, \quad W_\mathbf{X}(t) = \exp(5t),$$

and in example (2.5)

$$\mathbf{X}(t) = \begin{pmatrix} \cos(t) & \sin(t) \\ -\sin(t) & \cos(t) \end{pmatrix}, \quad W_\mathbf{X}(t) = 1.$$

Recall that the trace of a square matrix $\mathbf{A} = (a_{ij})$ is defined to be the sum of the diagonal elements, $\operatorname{tr}(\mathbf{A}) = a_{11} + a_{22} + \cdots + a_{nn}$.

Theorem 2.6: *Let $\mathbf{X}$ be a solution matrix for (H), and let $W_{\mathbf{X}} = \det(\mathbf{X})$. Then*

$$W_{\mathbf{X}}(t) = W_{\mathbf{X}}(\tau) \exp\left(\int_{\tau}^{t} \operatorname{tr}(\mathbf{A})(s)\, ds\right), \quad t, \tau \in I. \tag{2.6}$$

Hence $W_{\mathbf{X}}(t) \neq 0$ for all $t \in I$ if and only if $W_{\mathbf{X}}(\tau) \neq 0$ for one $\tau \in I$.

Proof: Using the definition of the determinant

$$W_{\mathbf{X}} = \sum_{p} sgn(p) x_{1p(1)} x_{2p(2)} \cdots x_{np(n)},$$

where the sum is over all permutations $p = (p(1), \ldots, p(n))$ of $(1, \ldots, n)$. Here $sgn(p)$ is the sign of the permutation, $sgn(p) = \pm 1$ if p is even or odd, respectively [1]. Using the product rule for differentiation, $W'_{\mathbf{X}}$ is the sum of n determinants:

$$W'_{\mathbf{X}} = w_1 + \cdots + w_n,$$

where w_i is obtained from $W_{\mathbf{X}}$ by replacing the ith row $x_{i1}, \ldots, x_{in}$ by the derivatives $x'_{i1}, \ldots, x'_{in}$. For example,

$$w_1 = \det \begin{pmatrix} x'_{11} & x'_{12} & \cdots & x'_{1n} \\ x_{21} & x_{22} & \cdots & x_{2n} \\ \vdots & & & \\ x_{n1} & x_{n2} & \cdots & x_{nn} \end{pmatrix}.$$

The equation $\mathbf{X}' = \mathbf{A}\mathbf{X}$ implies that

$$x'_{ik} = \sum_{j=1}^{n} a_{ij} x_{jk}$$

and hence

$$x'_{11} = \sum_{j=1}^{n} a_{1j} x_{j1}, \ldots, \quad x'_{1n} = \sum_{j=1}^{n} a_{1j} x_{jn}. \tag{2.7}$$

The value of w_1 remains unchanged if we add to the first row a constant multiple of another row. Adding to the first row $-a_{12}$ times the second row, $-a_{13}$ times the third row, etc., we see that

$$w_1 = \det \begin{pmatrix} a_{11}x_{11} & a_{11}x_{12} & \cdots & a_{11}x_{1n} \\ x_{21} & x_{22} & \cdots & x_{2n} \\ \vdots & \vdots & \cdots & \vdots \\ x_{n1} & x_{n2} & \cdots & x_{nn} \end{pmatrix}$$

in view of (2.7). Similarly, $w_j = a_{jj} W_X$, so that

$$W'_{\mathbf{X}} = (a_{11} + \cdots + a_{nn}) W_{\mathbf{X}} = \operatorname{tr}(\mathbf{A}) W_{\mathbf{X}}.$$

From this it follows that (2.6) is valid since both sides of the equation satisfy $W'_{\mathbf{X}} = (\operatorname{tr}\mathbf{A}) W_{\mathbf{X}}$ and have the value $W_{\mathbf{X}}(\tau)$ at $t = \tau$. $\square$

2.3.4   First-order nonhomogeneous systems

Now consider the nonhomogeneous system

$$X' = \mathbf{A}(t)X + B(t), \qquad\qquad (NH)$$

where $\mathbf{A} \in C(I, M_n(\mathcal{F}))$, $B \in C(I, \mathcal{F}^n)$, and I is an arbitrary real interval. Suppose $X_p(t)$ is a particular solution of (NH) and $X(t)$ is any other solution. Then $U(t) = X(t) - X_p(t)$ satisfies the corresponding homogeneous equation

$$U'(t) = X'(t) - X_p'(t) = \mathbf{A}(t)[X(t) - X_p(t)] = \mathbf{A}(t)U(t).$$

Conversely, given any solution $X_p(t)$ of (NH) and a solution $U(t)$ of (H), the function $X(t) = X_p(t) + U(t)$ is a solution of (NH), since

$$X'(t) = X_p'(t) + U'(t) = \mathbf{A}X_p(t) + B(t) + \mathbf{A}U(t) = \mathbf{A}X(t) + B(t).$$

These simple facts are summarized in the next theorem.

Theorem 2.7: *The set S' of all solutions of (NH) on I is an affine space*

$$S' = X_p + S = \{X \mid X = X_p + U, \ U \in S\},$$

where X_p is a particular solution of (NH).

As an example, consider the system, defined for $t \in \mathcal{R}$,

$$x_1' = x_2 + 1, \qquad\qquad (2.8)$$

$$x_2' = -x_1 + 2,$$

which can be written as

$$X' = \mathbf{A}X + B, \quad X = \begin{pmatrix} x_1 \\ x_2 \end{pmatrix}, \quad \mathbf{A} = \begin{pmatrix} 0 & 1 \\ -1 & 0 \end{pmatrix}, \quad B = \begin{pmatrix} 1 \\ 2 \end{pmatrix}.$$

The corresponding homogeneous system is just (2.5), which has a basis

$$\mathbf{X}(t) = \begin{pmatrix} \cos(t) & \sin(t) \\ -\sin(t) & \cos(t) \end{pmatrix}.$$

Since

$$X_p(t) = \begin{pmatrix} 2 \\ -1 \end{pmatrix}$$

is a particular solution, solutions X of (2.8) are given by

$$X(t) = X_p(t) + \mathbf{X}(t)C = \begin{pmatrix} 2 + c_1 \cos(t) + c_2 \sin(t) \\ -1 - c_1 \sin(t) - c_2 \cos(t) \end{pmatrix},$$

where c_1, c_2 are arbitrary constants. The solution X satisfying $X(0) = 0$ is given by

$$X(t) = \begin{pmatrix} 2 - 2\cos(t) + \sin(t) \\ -1 + 2\sin(t) + \cos(t) \end{pmatrix}$$

since $X(0) = X_p(0) + X(0)C = X_p(0) + C = 0$ implies

$$C = \begin{pmatrix} c_1 \\ c_2 \end{pmatrix} = -X_p(0) = \begin{pmatrix} -2 \\ 1 \end{pmatrix}.$$

Provided a basis $\mathbf{X}$ for (H) is known, the solutions of (NH) can be found by integration. This is done by the *variation of parameters* method (also called *variation of constants*). If $\mathbf{X}$ is a basis for (H) every solution X of (H) can be written as $X = \mathbf{X}C$ for $C \in \mathcal{F}^n$. By analogy, consider solutions of (NH) of the form $X(t) = \mathbf{X}(t)C(t)$, where C is a function $C : I \to \mathcal{F}^n$. Then

$$X'(t) = \mathbf{X}'(t)C(t) + \mathbf{X}(t)C'(t) = \mathbf{A}(t)\mathbf{X}(t)C(t) + \mathbf{X}(t)C'(t) = \mathbf{A}(t)X(t) + B(t)$$

if and only if

$$\mathbf{X}(t)C'(t) = B(t), \quad \text{or} \quad C'(t) = \mathbf{X}^{-1}(t)B(t), \quad t \in I.$$

This equation for C has solutions

$$C(t) = \xi + \int_\tau^t \mathbf{X}^{-1}(s)B(s) \, ds,$$

where $\tau \in I$ and ξ is a constant vector in $\mathcal{F}^n$. Thus $X = \mathbf{X}C$ given by

$$X(t) = \mathbf{X}(t)\xi + \mathbf{X}(t) \int_\tau^t \mathbf{X}^{-1}(s)B(s) \, ds, \quad t \in I,$$

is a solution of (NH). Choosing $\mathbf{X}$ so that $\mathbf{X}(\tau) = \mathbf{I}_n$, we see that $X(\tau) = \xi$. This establishes the following result.

Theorem 2.8: *Let $\mathbf{X}$ be a basis for the solutions of (H). There exists a solution X of (NH) of the form $X = \mathbf{X}C$, where C is any differentiable function satisfying $\mathbf{X}C' = B$. If $\mathbf{X}$ is the basis such that $\mathbf{X}(\tau) = \mathbf{I}_n$, the solution X of (NH) satisfying $X(\tau) = \xi$ is given by*

$$X(t) = \mathbf{X}(t)\xi + \mathbf{X}(t) \int_\tau^t \mathbf{X}^{-1}(s)B(s) \, ds, \quad t \in I. \tag{2.9}$$

Formula (2.9) is called the *variation of parameters formula*. Notice in particular that

$$X_p(t) = \mathbf{X}(t) \int_\tau^t \mathbf{X}^{-1}(s)B(s) \, ds, \quad t \in I$$

satisfies (NH) with $X_p(\tau) = 0$.

The variation of parameters formula requires inversion of $\mathbf{X}$. Recall that for any invertible square matrix $\mathbf{X} = (x_{ij})$,

$$\mathbf{X}^{-1} = \frac{adj(\mathbf{X})}{\det(\mathbf{X})}$$

where $adj(\mathbf{X}) = (\xi_{ij})$ and ξ_{ij} is the cofactor of x_{ji} in $\mathbf{X}$. Thus ξ_{ij} is $(-1)^{i+j}$ times the determinant obtained from $\mathbf{X}$ by crossing out the jth row and ith column in $\mathbf{X}$. In the 2×2 case we have

$$\mathbf{X} = \begin{pmatrix} x_{11} & x_{12} \\ x_{21} & x_{22} \end{pmatrix}, \quad \mathbf{X}^{-1} = \frac{1}{x_{11}x_{22} - x_{12}x_{21}} \begin{pmatrix} x_{22} & -x_{12} \\ -x_{21} & x_{11} \end{pmatrix}.$$

Consider the example (2.8) with the basis

$$\mathbf{X}(t) = \begin{pmatrix} \cos(t) & \sin(t) \\ -\sin(t) & \cos(t) \end{pmatrix}$$

for the homogeneous system $X' = \mathbf{A}X$. In this case

$$\mathbf{X}^{-1}(t) = \begin{pmatrix} \cos(t) & -\sin(t) \\ \sin(t) & \cos(t) \end{pmatrix},$$

and therefore the solution X_p of $X' = \mathbf{A}X + B$ satisfying $X_p(0) = 0$ is given by

$$X_p(t) = \mathbf{X}(t) \int_\tau^t \mathbf{X}^{-1}(s)B(s) \, ds$$

$$= \begin{pmatrix} \cos(t) & \sin(t) \\ -\sin(t) & \cos(t) \end{pmatrix} \int_0^t \begin{pmatrix} \cos(s) & -\sin(s) \\ \sin(s) & \cos(s) \end{pmatrix} \begin{pmatrix} 1 \\ 2 \end{pmatrix} ds$$

$$= \begin{pmatrix} \cos(t) & \sin(t) \\ -\sin(t) & \cos(t) \end{pmatrix} \begin{pmatrix} \sin(t) + 2\cos(t) - 2 \\ -\cos(t) + 2\sin(t) + 1 \end{pmatrix} = \begin{pmatrix} 2 - 2\cos(t) + \sin(t) \\ -1 + 2\sin(t) + \cos(t) \end{pmatrix}.$$

2.4   Higher-order equations

2.4.1   Linear equations of order n

Differential equations often involve derivatives $x^{(n)}$ with $n > 1$. The largest positive integer n such that $x^{(n)}$ appears in an equation is called the *order* of the differential equation. The study of a linear equation of order n on an interval I,

$$x^{(n)} + a_{n-1}(t)x^{(n-1)} + \cdots + a_0(t)x = b(t), \quad t \in I, \tag{2.10}$$

where $a_j, b \in C(I, \mathcal{F})$, can be reduced to the study of an associated first-order system. A *solution* of (2.10) on I is a function $x : I \to \mathcal{F}$ having n (continuous) derivatives on I such that

$$x^{(n)}(t) + a_{n-1}(t)x^{(n-1)}(t) + \cdots + a_0(t)x(t) = b(t), \quad t \in I.$$

There is a standard first-order system associated with (2.10). If $x(t)$ is a solution of (2.10), the n functions

$$y_1(t) = x(t), \quad y_2(t) = x'(t), \ldots, \quad y_n(t) = x^{(n-1)}(t)$$

will satisfy the system of equations

$$y_1' = y_2,$$

$$y_2' = y_3,$$

$$\vdots$$

$$y_{n-1}' = y_n,$$

$$y_n' = -a_0 y_1 - a_1 y_2 - \cdots - a_{n-1} y_n + b.$$

This system can be written in the form

$$Y' = \mathbf{A}(t)Y + B(t), \quad Y(t) = \begin{pmatrix} y_1 \\ \vdots \\ y_n \end{pmatrix}, \quad t \in I, \tag{2.11}$$

if

$$\mathbf{A} = \begin{pmatrix} 0 & 1 & 0 & \dots & 0 \\ 0 & 0 & 1 & \dots & 0 \\ \vdots & \vdots & \vdots & \dots & \vdots \\ 0 & 0 & 0 & \dots & 1 \\ -a_0 & -a_1 & -a_2 & \dots & -a_{n-1} \end{pmatrix}, \quad B = \begin{pmatrix} 0 \\ \vdots \\ 0 \\ b \end{pmatrix} = bE_n. \tag{2.12}$$

Conversely, if Y satisfies (2.11) on I, its first component $y_1 = x$ is such that $y_j = x^{(j-1)}$, $j = 1, \dots, n$, and x satisfies (2.10) on I. The system (2.11), with $\mathbf{A}$, B given by (2.12), is called the *first-order system associated* with (2.10).

For example, the first-order system associated with the equation

$$x'' + x = e^t$$

is

$$y_1' = y_2,$$

$$y_2' = -y_1 + e^t$$

or

$$Y' = \mathbf{A}Y + B(t), \quad Y = \begin{pmatrix} y_1 \\ y_2 \end{pmatrix}, \quad \mathbf{A} = \begin{pmatrix} 0 & 1 \\ -1 & 0 \end{pmatrix}, \quad B(t) = \begin{pmatrix} 0 \\ e^t \end{pmatrix}.$$

The coefficients in (2.11) satisfy $\mathbf{A} \in C(I, M_n(\mathcal{F}))$, $B \in C(I, \mathcal{F}^n)$, and therefore the existence and uniqueness theorem for linear systems (see Theorem 7.4) applied to (2.11) yields the following result.

Existence and uniqueness theorem for linear nth-order equations: *Given any $\tau \in I$ and*

$$\xi = \begin{pmatrix} \xi_1 \\ \vdots \\ \xi_n \end{pmatrix} \in \mathcal{F}^n,$$

there exists a unique solution x of (2.10) *on I satisfying*

$$x^{(j-1)}(\tau) = \xi_j, \quad j = 1, \ldots, n.$$

For any $x \in C^{n-1}(I, \mathcal{F})$ we define the function $\tilde{x} \in C(I, \mathcal{F}^n)$ by

$$\tilde{x} = \begin{pmatrix} x \\ x^{(1)} \\ \vdots \\ x^{(n-1)} \end{pmatrix}.$$

Thus x is a solution of (2.10) if and only if $y = \tilde{x}$ is a solution of (2.11).

If $b(t) \neq 0$ for some $t \in I$ we say (2.10) is a *nonhomogeneous equation* (NH_n), and the equation

$$x^{(n)} + a_{n-1}(t)x^{(n-1)} + \cdots + a_0(t)x = 0 \qquad (H_n)$$

is the corresponding *homogeneous equation*. The system associated with (H_n) is the homogeneous one

$$Y' = \mathbf{A}(t)Y. \qquad (2.13)$$

A solution $x(t)$ of (H_n) is *trivial* if $x(t) = 0$ for all $t \in I$. One consequence of the theorem above is that if x is a solution of (H_n) with $\tilde{x}(\tau) = 0$ for some $\tau \in I$, then $x(t) = 0$ for all $t \in I$. Theorems 2.1 and 2.2 have the following consequences.

Theorem 2.9: *The set of all solutions S_n of (H_n) on I is an n-dimensional vector space over $\mathcal{F}$. A set of k solutions $x_1(t), \ldots, x_k(t)$ of (H_n) is linearly independent if and only if for each $\tau \in I$, $\tilde{x}_1(\tau), \ldots, \tilde{x}_k(\tau)$ is linearly independent in $\mathcal{F}^n$.*

Proof: The latter statement requires checking, but it follows directly from the observation that $x_1, \ldots, x_k \in S$ is a linearly independent set if and only if $\tilde{x}_1, \ldots, \tilde{x}_k$ is a linearly independent set. In fact, the map $x \in S_n \to \tilde{x} \in S$ is an isomorphism. $\square$

We identify any set of n solutions $x_1(t), \ldots, x_n(t) \in S_n$ with the $1 \times n$ matrix

$$\mathbf{X}(t) = (x_1, \ldots, x_n),$$

writing $\mathbf{X} \in S_n$ if each $x_j \in S_n$. The matrix-valued function $\mathbf{X}$ is a basis for S_n if and only if

$$\tilde{\mathbf{X}} = (\tilde{x}_1, \ldots, \tilde{x}_n),$$

which is a solution matrix for (2.13), is a basis.

As an example, a basis for the solutions of the equation

$$x'' + x = 0$$

is given by

$$\mathbf{X}(t) = (\cos(t), \sin(t)),$$

and
$$\tilde{\mathbf{X}}(t) = \begin{pmatrix} \cos(t) & \sin(t) \\ -\sin(t) & \cos(t) \end{pmatrix}$$

is a basis for the associated system

$$Y' = \mathbf{A}Y, \quad Y = \begin{pmatrix} y_1 \\ y_2 \end{pmatrix}, \quad \mathbf{A} = \begin{pmatrix} 0 & 1 \\ -1 & 0 \end{pmatrix}.$$

The *Wronskian* $W_{\mathbf{X}}(t)$ of $\mathbf{X}(t) = (x_1, \ldots, x_n) \in S_n$ is defined to be the function

$$W_{\mathbf{X}} = \det(\tilde{\mathbf{X}}(t)) = \det \begin{pmatrix} x_1 & \cdots & x_n \\ x_1^{(1)} & \cdots & x_n^{(1)} \\ \vdots & \cdots & \vdots \\ x_1^{(n-1)} & \cdots & x_n^{(n-1)} \end{pmatrix}.$$

Theorems 2.4 and 2.6 now imply the following result.

Theorem 2.10: $\mathbf{X} = (x_1, \ldots, x_n) \in S_n$ *is a basis for S_n if and only if $\tilde{\mathbf{X}}(t)$ is invertible for all $t \in I$; that is, if and only if $W_{\mathbf{X}}(t) \neq 0$ for all $t \in I$. The Wronskian satisfies*

$$W_{\mathbf{X}}(t) = W_{\mathbf{X}}(\tau) \exp \left(-\int_\tau^t a_{n-1}(s)\, ds \right), \quad t, \tau \in I,$$

so that $W_{\mathbf{X}}(t) \neq 0$ for all $t \in I$ if and only if $W_{\mathbf{X}}(\tau) \neq 0$ for any $\tau \in I$.

Proof: The formula for $W_{\mathbf{X}}$ follows from (2.6), with $\mathbf{A}$ replaced by the $\mathbf{A}$ in (2.12), so that $\mathrm{tr}(\mathbf{A}) = -a_{n-1}$. $\square$

2.4.2 Nonhomogeneous linear equations of order n

If S_n' denotes the set of all solutions to (NH_n) on I, an application of Theorem 2.7 gives the following theorem.

Theorem 2.11: *The set of solutions S_n' of (NH_n) on I is the affine space*

$$S_n' = x_p + S_n = \{x \mid x = x_p + u, \ u \in S_n\},$$

where x_p is a particular solution of (NH_n).

For example, it is easy to check that

$$x_p(t) = e^t/2$$

is a solution of the equation

$$x'' + x = e^t. \tag{2.14}$$

Since a basis for the solutions of

$$x'' + x = 0$$

is given by $\mathbf{X}(t) = (\cos(t), \sin(t))$, the solutions x of (2.14) are given by

$$x(t) = e^t/2 + c_1 \cos(t) + c_2 \sin(t),$$

where c_1, c_2 are arbitrary constants.

A particular solution of (NH_n) can be obtained by using Theorem 2.8 on the system (2.11) associated with (NH_n). If $\mathbf{X} = (x_1, \ldots, x_n)$ is a basis for the solutions S_n of

$$x^{(n)} + a_{n-1}(t)x^{(n-1)} + \cdots + a_0(t)x = 0, \quad t \in I, \qquad (H_n)$$

then $\tilde{\mathbf{X}} = (\tilde{x}_1, \ldots, \tilde{x}_n)$ is a basis for the solutions S of the system

$$Y' = \mathbf{A}(t)Y. \qquad (H)$$

Theorem 2.8 says that the nonhomogeneous system

$$Y' = \mathbf{A}(t)Y + B(t), \quad B(t) = bE_n, \qquad (NH)$$

has a solution $Y(t)$ of the form $Y(t) = \tilde{\mathbf{X}}C$, where C is any differentiable function satisfying $\tilde{\mathbf{X}}C' = B$. The first component $x = \mathbf{X}C$ of Y will then be a solution of the nonhomogeneous equation

$$x^{(n)} + a_{n-1}(t)x^{(n-1)} + \cdots + a_0(t)x = b(t), \quad t \in I. \qquad (NH_n)$$

In terms of the components $c_1, \ldots, c_n$ of C, the equation $\tilde{\mathbf{X}}C' = B$ becomes the system of linear equations

$$x_1 c_1' + \cdots + x_n c_n' = 0, \qquad (2.15)$$

$$x_1^{(1)} c_1' + \cdots + x_n^{(1)} c_n' = 0,$$

$$\vdots$$

$$x_1^{(n-2)} c_1' + \cdots + x_n^{(n-2)} c_n' = 0,$$

$$x_1^{(n-1)} c_1' + \cdots + x_n^{(n-1)} c_n' = b.$$

One C such that C' satisfies (2.15), or $\tilde{\mathbf{X}}C' = bE_n$, is

$$C(t) = \int_\tau^t \tilde{\mathbf{X}}^{-1}(s)b(s)E_n \, ds, \quad \tau, t \in I.$$

Thus a particular solution x of (NH_n) on I is given by

$$x(t) = \mathbf{X}(t) \int_\tau^t \tilde{\mathbf{X}}^{-1}(s)b(s)E_n \, ds, \quad \tau, t \in I.$$

We have been assuming that a_n, the coefficient of $x^{(n)}$ in the equation (NH_n), is identically equal to 1. If $a_n(t) \neq 0$ for $t \in I$ and if $a_n \in C(I, \mathcal{F})$, the more general equation

$$a_n(t)x^{(n)} + a_{n-1}(t)x^{(n-1)} + \cdots + a_0(t)x = b(t), \quad t \in I,$$

can be reduced to the case $a_n(t) = 1$ by dividing by $a_n(t)$. This replaces $b(t)$ by $b(t)/a_n(t)$, so the equations (2.15) become $\tilde{\mathbf{X}}C' = bE_n/a_n$, and the formula for a particular solution becomes

$$x(t) = \mathbf{X}(t) \int_\tau^t \tilde{\mathbf{X}}^{-1}(s) \frac{b(s)}{a_n(s)} E_n \, ds, \quad \tau, t \in I. \tag{2.16}$$

Theorem 2.12: *Let* $\mathbf{X} = (x_1, \ldots, x_n)$ *be a basis for the solutions* S_n *of*

$$a_n(t)x^{(n)} + a_{n-1}(t)x^{(n-1)} + \cdots + a_0(t)x = 0, \quad t \in I.$$

There exists a solution x *of*

$$a_n(t)x^{(n)} + a_{n-1}(t)x^{(n-1)} + \cdots + a_0(t)x = b(t), \quad t \in I,$$

of the form $x = \mathbf{X}C$, *where* C *is a differentiable function satisfying* $\tilde{\mathbf{X}}C' = (b/a_n)E_n$. *Such a solution* x *is given by* (2.16).

Consider the example

$$x^{(3)} - x^{(1)} = t, \tag{NH_3}$$

whose corresponding homogeneous equation is

$$x^{(3)} - x^{(1)} = 0. \tag{H_3}$$

(H_3) has solutions of the form $x(t) = e^{\lambda t}$ if $\lambda^3 - \lambda = 0$ or $\lambda = 0, 1, -1$. Thus

$$\mathbf{X}(t) = (1, e^t, e^{-t})$$

gives a set of three solutions of (H_3). Since

$$\tilde{\mathbf{X}}(t) = \begin{pmatrix} 1 & e^t & e^{-t} \\ 0 & e^t & -e^{-t} \\ 0 & e^t & e^{-t} \end{pmatrix} \quad \text{and} \quad W_{\mathbf{X}}(t) = 2,$$

$\mathbf{X}$ is a basis for the solutions of (H_3). Thus there is a solution x of (NH_3) of the form

$$x(t) = c_1(t) + e^t c_2(t) + e^{-t}c_3(t), \tag{2.17}$$

where c_1, c_2, c_3 satisfy

$$c_1' + e^t c_2' + e^{-t}c_3' = 0,$$

$$e^t c_2' - e^{-t}c_3' = 0,$$

$$e^t c_2' + e^{-t}c_3' = t.$$

Solving this system of equations for c_1', c_2', and c_3' and then integrating, we can choose

$$c_1(t) = -t^2/2, \quad c_2(t) = -(t+1)e^{-t}/2, \quad c_3(t) = (t-1)e^t/2.$$

The solution x given by (2.17) is then $x(t) = -(t^2/2) - 1$. But since -1 is a solution of (H_3), another particular solution of (NH_3) is $x(t) = -t^2/2$. Every solution of (NH_3) has the form

$$x(t) = -t^2/2 + a_1 + a_2e^t + a_3e^{-t}, \quad a_j \in \mathcal{F}.$$

To compute the solution x_p satisfying $\tilde{x}_p(0) = 0$, we must satisfy the three equations

$$x_p(0) = a_1 + a_2 + a_3 = 0,$$
$$x_p'(0) = a_2 - a_3 = 0,$$
$$x_p''(0) = -1 + a_2 + a_3 = 0,$$

which have as a solution $a_1 = -1, a_2 = 1/2, a_3 = 1/2$. Then

$$x_p(t) = -t^2/2 - 1 + e^t/2 + e^{-t}/2 = -t^2/2 - 1 + \cosh(t).$$

2.5 Notes

There are many books covering the basic material in linear algebra; one sample is [1]. More comprehensive treatments, which include topics that will be used in Chapter 3, can be found in [7, 9, 10].

So far we have only needed some elementary calculus of vector-valued functions. More sophisticated ideas for studying vector-valued functions will appear in subsequent chapters. The text [17] covers the necessary material and is more user friendly than the classic [22].

2.6 Exercises

1. Let $\mathbf{A}, \mathbf{B} \in M_2(\mathcal{C})$, where

$$\mathbf{A} = \begin{pmatrix} 1 & -2i \\ 2i & 4 \end{pmatrix}, \quad \mathbf{B} = \begin{pmatrix} -i & 1+i \\ 1 & -1 \end{pmatrix}.$$

Compute

$$\text{(a) } \mathbf{A} + \mathbf{B}, \quad \text{(b) } \mathbf{AB}.$$

Find a basis for $Ran(\mathbf{A})$, and compute $\dim(Ran(\mathbf{A}))$. Similarly, find a basis for $N(\mathbf{A})$, and compute $\dim(N(\mathbf{A}))$. Verify that $\dim(Ran(\mathbf{A})) + \dim(N(\mathbf{A})) = 2$.

2. If $\mathbf{A} \in M_n(\mathcal{F})$, show that the nullspace $N(\mathbf{A})$ and range $Ran(\mathbf{A})$ are vector spaces over $\mathcal{F}$.

3. Suppose that $\mathbf{A} \in M_n(\mathcal{F})$. Show that $\mathbf{A}$ is invertible if and only if the function $x \in \mathcal{F}^n \to \mathbf{A}x \in \mathcal{F}^n$ is one to one and onto.

4. If $\mathbf{A} \in M_{mn}(\mathcal{F})$, show that the function $x \in \mathcal{F}^n \to \mathbf{A}x \in \mathcal{F}^m$ is linear. Show conversely that every linear function $L : \mathcal{F}^n \to \mathcal{F}^m$ can be written as multiplication on the left by a matrix $\mathbf{A}$.

5. Show that $f_1, \ldots, f_n$ is a basis for $\mathcal{F}^n$ if and only if there is an invertible matrix $\mathbf{A} \in M_n(\mathcal{F})$ such that $f_j = \mathbf{A}e_j$, where $e_1, \ldots, e_n$ is the standard basis for $\mathcal{F}^n$.

6. Suppose that for each $t \in I$ the matrix-valued function $\mathbf{A}_1(t)$ in (2.1) is invertible and that $\mathbf{A}_1^{-1}(t) \in C(I, M_n(\mathcal{F}))$. Show that (2.1) can be rewritten in the same form as (2.2).

7. Let $I = [0, 1]$. Are the following sets of functions $x_1, x_2 \in C(I, \mathcal{C}^2)$ linearly independent? Give a proof in each case. Note that here

$$E_1 = \begin{pmatrix} 1 \\ 0 \end{pmatrix}, \quad E_2 = \begin{pmatrix} 0 \\ 1 \end{pmatrix}.$$

(a) $$x_1(t) = e^t E_1, \quad x_2(t) = e^{-t} E_2,$$

(b) $$x_1(t) = \begin{pmatrix} t \\ 0 \end{pmatrix}, \quad x_2(t) = \begin{pmatrix} t^2 \\ t^2 \end{pmatrix},$$

(c) $$x_1(t) = \begin{pmatrix} \sin(t) \\ 2t \end{pmatrix}, \quad x_2(t) = \begin{pmatrix} e^{it} - e^{-it} \\ 4it \end{pmatrix},$$

(d) $$x_1(t) = tE_1, \quad x_2(t) = t^2 E_1.$$

8. Find the basis $\mathbf{X} = (U, V)$ for the system

$$x_1' = -2x_1,$$

$$x_2' = x_2$$

on $\mathcal{R}$ such that $\mathbf{X}(0) = \mathbf{I}_2$.

9. Find the basis $\mathbf{X} = (U, V)$ for the system

$$x_1' = -2x_1,$$

$$x_2' = x_1 - 2x_2$$

on $\mathcal{R}$ such that $\mathbf{X}(0) = \mathbf{I}_2$. (Hint: Solve for $x_1(t)$ and then solve for $x_2(t)$.)

10. (a) Verify that $\mathbf{X} = (U, V)$, where

$$U(t) = e^{-t} \begin{pmatrix} \cos(t) \\ -\sin(t) \end{pmatrix}, \quad V(t) = e^{-t} \begin{pmatrix} \sin(t) \\ \cos(t) \end{pmatrix}$$

is a basis for the solutions of

$$x_1' = -x_1 + x_2, \tag{2.18}$$

$$x_2' = -x_1 - x_2$$

on $\mathcal{R}$.

(b) Compute that solution X of (2.18), satisfying

$$X(0) = \begin{pmatrix} 1 \\ 1 \end{pmatrix}.$$

11. Show that the coefficient matrix $\mathbf{A}$ in

$$X' = \mathbf{A}(t)X, \quad \mathbf{A} \in C(I, M_n(\mathcal{F})) \tag{H}$$

is uniquely determined by a basis $\mathbf{X}$ for the solutions of (H). (Hint: $\mathbf{X}' = \mathbf{A}\mathbf{X}$.)

12. Prove Theorem 2.3.

13. Let $\mathbf{X}$ be a differentiable $n \times n$ matrix-valued function on a real interval I, and assume that $\mathbf{X}(t) \in M_n(\mathcal{F})$ is invertible for all $t \in I$. Show that $\mathbf{X}^{-1}$ is differentiable on I and

$$(\mathbf{X}^{-1})' = -\mathbf{X}^{-1}\mathbf{X}'\mathbf{X}^{-1}.$$

(Hint: $\mathbf{X}^{-1} = adj(\mathbf{X})/\det(\mathbf{X})$, where $adj(\mathbf{X}) = (\mathbf{X}_{ij})$ and $(\mathbf{X}_{ij})$ is the cofactor of x_{ji} in X. Both $adj(\mathbf{X})$ and $\det(\mathbf{X})$ are differentiable, and $\det(\mathbf{X}) \neq 0$. Use the identity $\mathbf{X}\mathbf{X}^{-1} = \mathbf{I}_n$.)

14. Recall that if $\mathbf{X} = (x_{ij})$, the *adjoint* matrix is $\mathbf{X}^* = (\overline{x_{ji}})$.

(a) If $\mathbf{X}$ is a basis for

$$X' = \mathbf{A}(t)X, \quad t \in I, \tag{H}$$

where $\mathbf{A} \in C(I, M_n(\mathcal{F}))$, show that $\mathbf{Y} = (\mathbf{X}^*)^{-1}$ is a basis for

$$Y' = -\mathbf{A}^*(t)Y, \quad t \in I. \tag{H^*}$$

(Hint: Use exercise 13.) The equation (H^*) is called the *adjoint equation* to (H).

(b) If $\mathbf{X}$ is a basis for (H), show that $\mathbf{Y}$ is a basis for (H^*) if and only if $\mathbf{Y}^*\mathbf{X} = \mathbf{C}$, where $\mathbf{C}$ is a constant invertible matrix.

(c) If $\mathbf{A}(t)$ is *skew hermitian*, that is, $\mathbf{A}^*(t) = -\mathbf{A}(t)$ for $t \in I$, we say (H) is *selfadjoint*. In this case show that $\mathbf{X}^*\mathbf{X} = \mathbf{C}$, a constant invertible matrix, for any basis $\mathbf{X}$ of (H).

15. Let $\mathbf{A}, \mathbf{B} \in C(I, M_n(\mathcal{F}))$, where I is a real interval. Consider the matrix differential equation

$$\mathbf{Z}' = \mathbf{A}(t)\mathbf{Z} + \mathbf{Z}\mathbf{B}(t), \quad t \in I. \tag{2.19}$$

(a) Let $\mathbf{X}$ be a basis for $X' = \mathbf{A}(t)X$ and $\mathbf{Y}$ a basis for $Y' = B^*(t)Y$. Show that for any $\mathbf{C} \in M_n(\mathcal{F})$ the matrix-valued function $\mathbf{Z} = \mathbf{X}\mathbf{C}\mathbf{Y}^*$ satisfies (2.19).

(b) Conversely, show that if $\mathbf{Z}$ satisfies (2.19), then for some constant $n \times n$ matrix $\mathbf{C}$ we have $\mathbf{Z} = \mathbf{X}\mathbf{C}\mathbf{Y}^*$. (Hint: Let $\mathbf{U} = \mathbf{X}^{-1}\mathbf{Z}(\mathbf{Y}^*)^{-1}$ and differentiate.)

16. (a) Verify that

$$\mathbf{X}(t) = \begin{pmatrix} t & t^3 \\ 1 & 3t^2 \end{pmatrix}, \quad 0 < t < \infty,$$

gives a basis $\mathbf{X}$ for the system

$$X' = \mathbf{A}(t)X, \quad \mathbf{A}(t) = \frac{1}{t^2}\begin{pmatrix} 0 & t^2 \\ -3 & 3t \end{pmatrix}, \quad 0 < t < \infty.$$

(b) Compute $W_{\mathbf{X}}(t) = \det(\mathbf{X}(t))$, and verify that

$$W_{\mathbf{X}}(t) = W_{\mathbf{X}}(1)\exp\left[\int_1^t \operatorname{tr}(\mathbf{A}(s)) \, ds\right].$$

17. Let $\mathbf{X} = (U, V)$ where

$$U(t) = \begin{pmatrix} t^2 \\ t \end{pmatrix}, \quad V(t) = \begin{pmatrix} t^3 \\ t^2 \end{pmatrix}, \quad t \in \mathcal{R}.$$

(a) Show that $\mathbf{X}$ is linearly independent on $\mathcal{R}$.
(b) Note that $\det(\mathbf{X}) = 0$ for $t \in \mathcal{R}$. Does this contradict (a)? Why?

18. Suppose $\mathbf{X}$ is a basis for the system

$$X' = \mathbf{A}(t)X, \quad t \in I,$$

where $\mathbf{A} \in C(I, M_n(\mathcal{C}))$.

(a) If $\operatorname{tr}(\mathbf{A}(t)) = 0$ for $t \in I$, show that $W_{\mathbf{X}}(t) = c$, a constant.
(b) If $\mathbf{A}^*(t) = -\mathbf{A}(t)$ for $t \in I$, show that $|W_{\mathbf{X}}(t)| = c$, a constant.
(c) If $\mathbf{A} \in C(I, M_n(\mathcal{R}))$ and $\mathbf{A}^*(t) = -\mathbf{A}(t)$ for $t \in I$, show that $W_{\mathbf{X}}(t) = c$, a constant.

19. Consider the complex system

$$X' = \mathbf{A}(t)X, \quad \mathbf{A} \in C(I, M_n(\mathcal{C})), \tag{2.20}$$

on an interval I. If $\mathbf{A} = (a_{jk})$, then

$$a_{jk}(t) = b_{jk}(t) + ic_{jk}(t),$$

where $b_{jk}(t), c_{jk}(t) \in \mathcal{R}$, so that $\mathbf{A} = \mathbf{B} + i\mathbf{C}$, where $\mathbf{B} = (b_{jk})$, $\mathbf{C} = (c_{jk})$. Similarly, if X is a solution of (2.20) on I, $X(t) = U(t) + iV(t)$, where $U(t), V(t) \in \mathcal{R}^n$.

(a) Show that $X = U + iV$ is a solution of the complex system (2.20) if and only if

$$W = \begin{pmatrix} U \\ V \end{pmatrix}$$

is a solution of the real system

$$W' = \mathbf{A}(t)W, \quad \mathbf{A} = \begin{pmatrix} \mathbf{B} & -\mathbf{C} \\ \mathbf{C} & \mathbf{B} \end{pmatrix}. \tag{2.21}$$

(b) Show that $\mathbf{X} = \mathbf{U} + i\mathbf{V}$, where $\mathbf{U}, \mathbf{V} \in C(I, M_n(\mathcal{R}))$ is a solution matrix of (2.20) if and only if

$$\mathbf{W} = \begin{pmatrix} \mathbf{U} & -\mathbf{V} \\ \mathbf{V} & \mathbf{U} \end{pmatrix}$$

is a solution matrix of (2.21).

(c) Show that $\det(\mathbf{W}) = |\det(\mathbf{X})|^2$, so that $\mathbf{X}$ is a basis of (2.20) if and only if $\mathbf{W}$ is a basis of (2.21).

20. Consider the nonhomogeneous system

$$x_1' = -2x_1 + t, \tag{NH}$$

$$x_2' = x_1 - 2x_2 + t^2$$

on $\mathcal{R}$, or $X' = \mathbf{A}X + B(t)$, where

$$X = \begin{pmatrix} x_1 \\ x_2 \end{pmatrix}, \quad \mathbf{A} = \begin{pmatrix} -2 & 0 \\ 1 & -2 \end{pmatrix}, \quad B(t) = \begin{pmatrix} t \\ t^2 \end{pmatrix}.$$

The basis $\mathbf{X}$ for the corresponding homogeneous system $X' = \mathbf{A}X$ satisfying $\mathbf{X}(0) = \mathbf{I}_2$ was computed in exercise 9.

(a) Find all solutions of (NH). (Hint: Solve first for $x_1(t)$.)

(b) Find that solution Z of (NH) satisfying

$$Z(0) = \begin{pmatrix} 0 \\ 0 \end{pmatrix}.$$

(c) Compute X_p, where

$$X_p(t) = \mathbf{X}(t) \int_0^t \mathbf{X}^{-1}(s) B(s) \, ds.$$

(d) What is the relation between Z and X_p?

21. Consider the nonhomogeneous system

$$X' = \mathbf{A}(t)X + B(t), \quad \mathbf{A} \in C(I, M_n(\mathcal{F})), \quad B \in C(I, \mathcal{F}^n) \tag{NH}$$

on some interval I, together with the corresponding homogeneous system

$$X' = \mathbf{A}(t)X. \tag{H}$$

For $t, \tau \in I$, let $\mathbf{S}(t, \tau) \in M_n(\mathcal{F})$ be the unique matrix satisfying

$$\frac{\partial \mathbf{S}}{\partial t}(t, \tau) = \mathbf{A}(t)\mathbf{S}(t, \tau), \quad \mathbf{S}(\tau, \tau) = \mathbf{I}_n.$$

Show that
    (a) if $\mathbf{X}$ is any basis for (H) then

$$\mathbf{S}(t,\tau) = \mathbf{X}(t)\mathbf{X}^{-1}(\tau), \quad t,\tau \in I;$$

    (b) $\mathbf{S}(t,\tau)\mathbf{S}(\tau,\sigma) = \mathbf{S}(t,\sigma), \quad t,\tau,\sigma \in I;$
    (c) $\mathbf{S}^{-1}(t,\tau) = \mathbf{S}(\tau,t), \quad t,\tau \in I;$
    (d) the solution X of (NH) satisfying $X(\tau) = \xi$ is given by

$$X(t) = \mathbf{S}(t,\tau)\xi + \int_{\tau}^{t} \mathbf{S}(t,s)B(s)\,ds, \quad t \in I.$$

22. (a) Let $\mathbf{X}$ be a basis for the solutions of (H_n) and suppose $a_{n-1}(t) = 0$ for $t \in I$. Show that $W_{\mathbf{X}}(t) = c$, a constant.
    (b) Find the basis $\mathbf{X}$ for

$$x'' - x = 0, \quad t \in \mathcal{R},$$

satisfying $\tilde{\mathbf{X}}(0) = \mathbf{I}_2$. (Hint: Try solutions of the form $x(t) = e^{\lambda t}$.)
    (c) Compute $W_{\mathbf{X}}$ for the $\mathbf{X}$ in (b).
23. Consider the homogeneous equation

$$x''' - 5x'' + 6x' = 0 \tag{H_3}$$

on $\mathcal{R}$.
    (a) Show that $\mathbf{X} = (x_1, x_2, x_3)$ given by

$$\mathbf{X}(t) = (1, e^{2t}, e^{3t}), \quad t \in \mathcal{R},$$

is a basis for (H_3).
    (b) Find that solution u of (H_3) satisfying $\tilde{u}(0) = E_2$; that is, $u(0) = 0$, $u'(0) = 1$, $u''(0) = 0$.
    (c) What is the first-order system

$$Y' = \mathbf{A}Y \tag{H}$$

associated with (H_3)?
    (d) Compute $\tilde{\mathbf{X}}$, a basis for (H).
    (e) Compute $W_{\mathbf{X}} = \det(\tilde{\mathbf{X}})$.
    (f) Verify that

$$W_{\mathbf{X}}(t) = W_{\mathbf{X}}(0)\exp\left(\int_0^t 5\,ds\right) = e^{5t}W_{\mathbf{X}}(0), \quad t \in \mathcal{R}.$$

24. Consider the nonhomogeneous equation

$$x''' - 5x'' + 6x' = e^t, \quad t \in \mathcal{R}. \tag{NH_3}$$

(a) Write down the first-order system

$$Y' = \mathbf{A}Y + B \qquad\qquad (NH)$$

associated with (NH_3).

(b) Using the basis $\mathbf{X}$ given in exercise 23(a), for (H_3) find a solution x of (NH_3) of the form $x = c_1 x_1 + c_2 x_2 + c_3 x_3$, where the c_j are differentiable functions.

(c) Find that solution x_p of (NH_3) satisfying $\tilde{x}_p(0) = 0$.

(d) Find that solution x of (NH_3) satisfying $\tilde{x}(0) = E_2$. (Note: $x = u + x_p$, where u is given in exercise 23(b) and x_p is given in (c) above.)

25. (a) Let $\mathbf{X} = (x_1, x_2)$ be a basis for the solutions of the second-order equation

$$x'' + a_1(t)x' + a_0(t)x = 0, \quad t \in I, \qquad\qquad (H_2)$$

where $a_j \in C(I, \mathcal{F})$. Show that the variation of parameters formula (2.16) for solutions x of the nonhomogeneous equation

$$x'' + a_1(t)x' + a_0(t)x = b(t), \quad t \in I, \qquad\qquad (NH_2)$$

where $b \in C(I, \mathcal{F})$ takes the form

$$x(t) = \xi_1 x_1(t) + \xi_2 x_2(t) + \int_\tau^t \frac{x_1(s)x_2(t) - x_1(t)x_2(s)}{W_{\mathbf{X}}(s)} b(s) \, ds, \qquad (2.22)$$

where $\xi_j \in \mathcal{F}$, $W_{\mathbf{X}}(s) = x_1(s)x_2'(s) - x_1'(s)x_2(s)$.

(b) Show that the function x_p given by

$$x_p(t) = \int_\tau^t \frac{x_1(s)x_2(t) - x_1(t)x_2(s)}{W_{\mathbf{X}}(s)} b(s) \, ds$$

satisfies $\tilde{x}_p(\tau) = 0$; that is, $x_p(\tau) = 0$, $x_p'(\tau) = 0$.

(c) If $\mathbf{X}$ is the basis satisfying $\tilde{\mathbf{X}}(\tau) = \mathbf{I}_2$, show that the x given in (2.22) satisfies $\tilde{x}(\tau) = \xi$; that is, $x(\tau) = \xi_1$, $x'(\tau) = \xi_2$.

26. Consider the nonhomogeneous equation

$$a_n(t)x^{(n)} + a_{n-1}(t)x^{(n-1)} + \cdots + a_0(t)x = b(t), \qquad\qquad (NH_n)$$

where the $a_j, b \in C(I, \mathcal{F})$ for some interval I, along with the corresponding homogeneous equation

$$a_n(t)x^{(n)} + a_{n-1}(t)x^{(n-1)} + \cdots + a_0(t)x = 0. \qquad\qquad (H_n)$$

Suppose $\mathbf{X} = (x_1, \ldots, x_n)$ is a basis for (H_n).

(a) Show that the particular solution x given by (2.16) satisfies $\tilde{x}(\tau) = 0$.

(b) For fixed $s \in I$ let

$$x_s(t) = \frac{\mathbf{X}(t)\tilde{\mathbf{X}}^{-1}(s)}{a_n(s)} E_n, \quad t \in I.$$

Show that x_s satisfies (H_n) and

$$\tilde{x}_s(s) = \frac{1}{a_n(s)} E_n.$$

Thus if $k(t, s) = x_s(t)$ for $t, s \in I$, then

$$x(t) = \int_\tau^t k(t, s) b(s) \ ds,$$

and k is independent of the basis $\mathbf{X}$ chosen.

(c) If $\mathbf{X}$ is the basis for (H_n) satisfying $\tilde{\mathbf{X}}(\tau) = \mathbf{I}_n$, show that the x given by

$$x(t) = \mathbf{X}(t)\xi + \int_\tau^t k(t, s) b(s) \ ds, \quad t \in I,$$

satisfies (NH_n), $\tilde{x}(\tau) = \xi$.

27. All the results in sections 2.4.1 and 2.4.2 extend to systems of m linear equations of order n,

$$X^{(n)} + \mathbf{A}_{n-1}(t) X^{(n-1)} + \cdots + \mathbf{A}_0(t) X = B(t), \quad t \in I, \tag{2.23}$$

where

$$\mathbf{A}_j \in C(I, M_m(\mathcal{F})), \quad j = 0, 1, \ldots, n-1, \quad B \in C(I, \mathcal{F}^m).$$

and a *solution* of this equation is a function $X : I \to \mathcal{F}^m$ having n derivatives on I and satisfying (2.23).

(a) Show that (2.23) may be rewritten as the first-order system

$$Y' = \mathcal{A}(t)Y + \mathcal{B}(t), \quad t \in I,$$

where

$$\mathcal{A} = \begin{pmatrix} 0 & \mathbf{I}_m & 0 & \cdots & 0 \\ 0 & 0 & \mathbf{I}_m & \cdots & 0 \\ \vdots & \vdots & \vdots & \cdots & \vdots \\ 0 & 0 & 0 & \cdots & \mathbf{I}_m \\ -\mathbf{A}_0 & -\mathbf{A}_1 & -\mathbf{A}_2 & \cdots & -\mathbf{A}_{n-1} \end{pmatrix}, \quad \mathcal{B} = \begin{pmatrix} 0 \\ \vdots \\ 0 \\ B \end{pmatrix},$$

and the zeros represent $m \times m$ zero matrices.

(b) Show that the set of solutions of $Y' = \mathcal{A}(t)Y$ is an nm-dimensional vector space over $\mathcal{F}$.

(c) Formulate the existence and uniqueness theorem for (2.23).

Chapter 3

Constant Coefficients

3.1   Introduction

The previous chapter introduced the basic algebraic structure of the set of solutions to a linear system of ordinary differential equations. In particular the variation of parameters technique reduces the problem of solving a general nonhomogeneous system to the more restricted problem of solving the associated homogeneous system. In this chapter we will focus on the first-order linear homogeneous system

$$X' = \mathbf{A}X \tag{3.1}$$

with a constant coefficient matrix $\mathbf{A} \in M_n(\mathcal{F})$. There are several reasons for directing our attention to systems with constant coefficients. First, as Chapter 1 demonstrated, such equations arise in the study of many basic applications. Second, this class of problems is susceptible to a complete analysis, which depends only on the linear algebraic properties of the coefficient matrix $\mathbf{A}$. Finally, the analysis of such systems is a model for the study of more sophisticated problems.

The coefficient matrix $\mathbf{A}$ in (3.1) can be considered as a constant function in $C(R, M_n(\mathcal{F}))$, so the existence and uniqueness theorem for linear systems (see Theorem 7.4) implies that given any $\tau \in \mathcal{R}$, $\xi \in \mathcal{F}^n$, there is a unique $X(t) \in C(\mathcal{R}, \mathcal{F}^n)$ satisfying (3.1) on $\mathcal{R}$, with $X(\tau) = \xi$. Thus the set S of solutions of (3.1) are functions defined on the whole real axis with values in $\mathcal{F}^n$.

All such systems may be solved by an algebraic analysis of the matrix $\mathbf{A}$. If $n = 1$ and $\mathbf{A} = a \in \mathcal{F}$, the equation has the simple general solution

$$x(t) = \alpha e^{at}, \quad \alpha \in \mathcal{F}.$$

Note that $x(0) = \alpha$, and hence the solution x satisfying $x(0) = 1$ is given by $x(t) = e^{at}$. A similarly straightforward system arises if the $n \times n$ matrix $\mathbf{A}$ is diagonal:

$$\mathbf{A} = \begin{pmatrix} a_{11} & 0 & 0 & \ldots & 0 \\ 0 & a_{22} & 0 & \ldots & 0 \\ \vdots & \vdots & \vdots & \ldots & \vdots \\ 0 & 0 & 0 & \ldots & a_{nn} \end{pmatrix}.$$

In this case it is easy to verify that the vector-valued function

$$X(t) = \begin{pmatrix} \xi_1 e^{a_{11}t} \\ \vdots \\ \xi_n e^{a_{nn}t} \end{pmatrix} = \begin{pmatrix} e^{a_{11}t} & 0 & 0 & \ldots & 0 \\ 0 & e^{a_{22}t} & 0 & \ldots & 0 \\ \vdots & \vdots & \vdots & \ldots & \vdots \\ 0 & 0 & 0 & \ldots & e^{a_{nn}t} \end{pmatrix} \begin{pmatrix} \xi_1 \\ \vdots \\ \xi_n \end{pmatrix}$$

satisfies

$$X'(t) = \mathbf{A}X(t), \quad X(0) = \begin{pmatrix} \xi_1 \\ \vdots \\ \xi_n \end{pmatrix}.$$

This explicit solution of systems with a diagonal matrix $\mathbf{A}$ extends to the much richer class of systems whose coefficient matrix is *similar* to a diagonal matrix. That is, suppose there is an invertible matrix $\mathbf{Q}$ with column vectors $\alpha_j \in \mathcal{F}^n$,

$$\mathbf{Q} = (\alpha_1, \ldots, \alpha_n),$$

and a diagonal matrix

$$\mathbf{J} = \begin{pmatrix} \lambda_1 & 0 & 0 & \ldots & 0 \\ 0 & \lambda_2 & 0 & \ldots & 0 \\ \vdots & \vdots & \vdots & \ldots & \vdots \\ 0 & 0 & 0 & \ldots & \lambda_n \end{pmatrix}$$

such that

$$\mathbf{A} = \mathbf{Q}\mathbf{J}\mathbf{Q}^{-1}.$$

In this case the system (3.1) may be rewritten as

$$[\mathbf{Q}^{-1}X]' = \mathbf{J}[\mathbf{Q}^{-1}X].$$

The functions $Y = [\mathbf{Q}^{-1}X]$ satisfy a system with a diagonal coefficient matrix, so

$$Y(t) = \begin{pmatrix} e^{\lambda_1 t} & 0 & 0 & \ldots & 0 \\ 0 & e^{\lambda_2 t} & 0 & \ldots & 0 \\ \vdots & \vdots & \vdots & \ldots & \vdots \\ 0 & 0 & 0 & \ldots & e^{\lambda_n t} \end{pmatrix} \begin{pmatrix} \xi_1 \\ \vdots \\ \xi_n \end{pmatrix}$$

satisfies

$$Y'(t) = \mathbf{J}Y(t), \quad Y(0) = \begin{pmatrix} \xi_1 \\ \vdots \\ \xi_n \end{pmatrix}.$$

But this implies that $X(t) = \mathbf{Q}Y$ satisfies

$$X'(t) = \mathbf{Q}\mathbf{J}\mathbf{Q}^{-1}X(t) = \mathbf{A}X(t), \quad X(0) = \mathbf{Q}\begin{pmatrix}\xi_1 \\ \vdots \\ \xi_n\end{pmatrix}.$$

Since there are $n \times n$ matrices which are not similar to a diagonal matrix, this approach will not work in every case. The appropriate generalization of these ideas, which is applicable to all constant coefficient systems (3.1), will lead us to consider the *Jordan canonical form* of a matrix $\mathbf{A}$.

As an alternative to this linear algebraic analysis of (3.1), there is an analytic approach which always provides an explicit solution. The price we pay for this greater generality is that the algebraic structure of $\mathbf{A}$, which governs the behavior of solutions, is often concealed. Recall that e^{at} has a power series expansion

$$e^{at} = 1 + at + \frac{a^2t^2}{2!} + \cdots = \sum_{k=0}^{\infty} \frac{a^kt^k}{k!} \tag{3.2}$$

which is convergent for all $t \in \mathcal{R}$. It is remarkable that a similar series provides solutions for (3.1).

By analogy with the solution (3.2) in the case $n = 1$, consider solutions X of (3.1) satisfying $X(0) = \xi$ which have a power series representation

$$X(t) = C_0 + C_1 t + C_2 t^2 + \cdots = \sum_{k=0}^{\infty} C_k t^k, \quad C_k \in \mathcal{F}^n. \tag{3.3}$$

Insertion of this series into (3.1) yields

$$X'(t) = C_1 + 2C_2 t + 3C_3 t^2 + \cdots = \sum_{k=0}^{\infty}(k+1)C_{k+1}t^k,$$

$$\mathbf{A}X(t) = \mathbf{A}C_0 + \mathbf{A}C_1 t + \mathbf{A}C_2 t^2 + \cdots = \sum_{k=0}^{\infty} \mathbf{A}C_k t^k.$$

Hence $X'(t) = \mathbf{A}X(t)$ if and only if these two power series are the same; that is,

$$(k+1)C_{k+1} = \mathbf{A}C_k, \quad k = 0, 1, 2, \ldots.$$

Since $X(0) = \xi = C_0$, we see that $C_1 = \mathbf{A}\xi$, $2C_2 = \mathbf{A}^2\xi$, and generally

$$C_k = \frac{\mathbf{A}^k}{k!}\xi, \quad k = 0, 1, 2, \ldots.$$

(Note that $\mathbf{A}^0 = \mathbf{I}_n$.) This formal power series argument suggests that the solution $X(t)$ satisfying $X(0) = \xi$ has the power series representation

$$X(t) = \xi + \mathbf{A}t\xi + \mathbf{A}^2t^2\xi/2! + \cdots = \sum_{k=0}^{\infty} \frac{\mathbf{A}^k t^k}{k!}\xi, \quad t \in \mathcal{R}. \tag{3.4}$$

A basis for (3.1) may be obtained by taking solutions $X_j(t)$ with $X_j(0) = E_j$, the standard basis for $\mathcal{F}^n$. Then the basis $\mathbf{X_A} = (X_1, \ldots, X_n)$ satisfies $\mathbf{X_A}(0) = (E_1, \ldots, E_n) = \mathbf{I}_n$. Letting $\xi = E_j$ in (3.4) we get

$$X_j(t) = \sum_{k=0}^{\infty} \frac{\mathbf{A}^k t^k}{k!} E_j.$$

Since $X_\mathbf{A}(t)E_j = X_j(t)$ is just the jth column of $X_\mathbf{A}(t)$, it follows that

$$\mathbf{X_A}(t) = \sum_{k=0}^{\infty} \frac{\mathbf{A}^k t^k}{k!}. \tag{3.5}$$

Since this chapter emphasizes the use of tools from linear algebra, the additional analysis required to justify this power series approach is deferred until Chapter 5, where the general study of power series solutions for differential equations is considered. Here we simply note that the series (3.5) converges for all $t \in \mathcal{R}$ and does in fact define a basis for (3.1). It is natural to define the exponential function $e^{\mathbf{A}t}$ to be this series,

$$\exp(\mathbf{A}t) = e^{\mathbf{A}t} = \sum_{k=0}^{\infty} \frac{\mathbf{A}^k t^k}{k!}, \quad \mathbf{A} \in M_n(\mathcal{F}), \quad t \in \mathcal{R}.$$

The next result summarizes this discussion.

Theorem 3.1: *The basis $\mathbf{X_A}$ for the solutions S of (3.1) with $\mathbf{X_A}(0) = \mathbf{I}_n$ is given by*

$$\mathbf{X_A}(t) = e^{\mathbf{A}t}, \quad t \in \mathcal{R}.$$

The solution X of (3.1) such that $X(0) = \xi \in \mathcal{F}^n$ is given by

$$X(t) = e^{\mathbf{A}t}\xi.$$

For any fixed $\tau \in \mathcal{R}$ the function $\mathbf{Y}$ given by

$$\mathbf{Y}(t) = \mathbf{X_A}(t - \tau) = e^{\mathbf{A}(t-\tau)}$$

satisfies

$$\mathbf{Y}'(t) = \mathbf{X_A}'(t - \tau) = \mathbf{A}\mathbf{X_A}(t - \tau) = \mathbf{A}\mathbf{Y}(t),$$

and $\mathbf{Y}(\tau) = \mathbf{I}_n$. Hence $\mathbf{Y}$ is also a basis for (3.1), and the solution X of (3.1) satisfying $X(\tau) = \xi$ can be written as

$$X(t) = \mathbf{Y}(t)\xi = e^{\mathbf{A}(t-\tau)}\xi.$$

As an example, let

$$\mathbf{A} = \begin{pmatrix} 2 & 0 \\ 0 & 3 \end{pmatrix}, \quad X = \begin{pmatrix} x_1 \\ x_2 \end{pmatrix},$$

so that the system (3.1) is

$$x_1' = 2x_1,$$

$$x_2' = 3x_2.$$

It is easy to check that

$$\mathbf{A}^k = \begin{pmatrix} 2^k & 0 \\ 0 & 3^k \end{pmatrix}, \quad k = 0, 1, 2, \ldots,$$

and hence

$$e^{\mathbf{A}t} = \sum_{k=0}^{\infty} \frac{\mathbf{A}^k t^k}{k!} = \sum_{k=0}^{\infty} \begin{pmatrix} 2^k t^k / k! & 0 \\ 0 & 3^k t^k / k! \end{pmatrix} = \begin{pmatrix} e^{2t} & 0 \\ 0 & e^{3t} \end{pmatrix}.$$

The solution X satisfying $X(0) = \xi$ is given by

$$X(t) = e^{\mathbf{A}t}\xi = \begin{pmatrix} e^{2t} & 0 \\ 0 & e^{3t} \end{pmatrix} \begin{pmatrix} \xi_1 \\ \xi_2 \end{pmatrix} = \begin{pmatrix} e^{2t}\xi_1 \\ e^{3t}\xi_2 \end{pmatrix}, \quad \xi = \begin{pmatrix} \xi_1 \\ \xi_2 \end{pmatrix},$$

as we already know.

As a second example, consider

$$\mathbf{A} = \begin{pmatrix} 0 & 1 \\ -1 & 0 \end{pmatrix}.$$

Now $\mathbf{A}^2 = -\mathbf{I}_2$, $\mathbf{A}^3 = -\mathbf{A}$, $\mathbf{A}^4 = \mathbf{I}_2$, $\mathbf{A}^5 = \mathbf{A}$, $\ldots$, and, in general,

$$\mathbf{A}^{2k} = (-1)^k \mathbf{I}_2, \quad \mathbf{A}^{2k+1} = (-1)^k \mathbf{A}, \quad k = 0, 1, 2, \ldots.$$

Thus

$$e^{\mathbf{A}t} = [1 - t^2/2! + t^4/4! - \cdots]\mathbf{I}_2 + [t - t^3/3! + t^5/5! - \cdots]\mathbf{A}$$

$$= \cos(t)\,\mathbf{I}_2 + \sin(t)\,\mathbf{A} = \begin{pmatrix} \cos(t) & \sin(t) \\ -\sin(t) & \cos(t) \end{pmatrix}.$$

The analysis of the example (2.5) has already verified that the columns of this matrix form the basis $X_{\mathbf{A}}$ for the solutions of

$$x_1' = x_2,$$

$$x_2' = -x_1,$$

satisfying $\mathbf{X}_{\mathbf{A}}(0) = \mathbf{I}_2$.

3.2 Properties of the exponential of a matrix

The exponential of a matrix behaves like an ordinary exponential function in some ways. The following properties are consequences of the fact that $\mathbf{X_A}(t) = e^{\mathbf{A}t}$ is the unique $\mathbf{X} \in C^1(\mathcal{R}, M_n(\mathcal{F}))$ satisfying

$$\mathbf{X}' = \mathbf{A}\mathbf{X}, \quad \mathbf{X}(0) = \mathbf{I}_n.$$

Theorem 3.2: *If $s, t \in \mathcal{R}$, $\mathbf{A} \in M_n(\mathcal{F})$, then*

(i)
$$e^{\mathbf{A}(s+t)} = e^{\mathbf{A}s}e^{\mathbf{A}t} = e^{\mathbf{A}t}e^{\mathbf{A}s},$$

(ii)
$$(e^{\mathbf{A}t})^{-1} = e^{-\mathbf{A}t},$$

(iii)
$$\det(e^{\mathbf{A}t}) = e^{tr(\mathbf{A})t}.$$

Proof: If $\mathbf{X_A}(t) = e^{\mathbf{A}t}$, then

$$\mathbf{X_A}(t+s) = e^{\mathbf{A}(s+t)}, \quad \mathbf{X_A}(t)\mathbf{X_A}(s) = e^{\mathbf{A}t}e^{\mathbf{A}s}.$$

Letting $s \in \mathcal{R}$ be fixed, put $\mathbf{Y}(t) = \mathbf{X_A}(t+s)$ and $\mathbf{Z}(t) = \mathbf{X_A}(t)\mathbf{X_A}(s)$. The functions $\mathbf{Y}$ and $\mathbf{Z}$ are solutions of $X' = \mathbf{A}X$, since

$$\mathbf{Y}'(t) = \mathbf{X}'_{\mathbf{A}}(t+s) = \mathbf{A}\mathbf{X_A}(t+s) = \mathbf{A}\mathbf{Y}(t),$$

$$\mathbf{Z}'(t) = \mathbf{X}'_{\mathbf{A}}(t)\mathbf{X_A}(s) = \mathbf{A}\mathbf{X_A}(t)\mathbf{X_A}(s) = \mathbf{A}\mathbf{Z}(t).$$

Moreover, $\mathbf{Y}(0) = \mathbf{Z}(0) = \mathbf{X_A}(s)$, and uniqueness implies that $\mathbf{Y}(t) = \mathbf{Z}(t)$ for all $t \in \mathcal{R}$, which is just

$$e^{\mathbf{A}(s+t)} = e^{\mathbf{A}t}e^{\mathbf{A}s}.$$

Interchanging the roles of s and t we get the two equalities in (i). Now (i) with $s = -t$ implies that

$$\mathbf{I}_n = e^{-\mathbf{A}t}e^{\mathbf{A}t} = e^{\mathbf{A}t}e^{-\mathbf{A}t},$$

which is (ii). As for (iii), note that the Wronskian $\det(\mathbf{X_A})$ of $\mathbf{X_A}$ satisfies (see Theorem 2.6)

$$\det(\mathbf{X_A}(t)) = \det(\mathbf{X_A}(0)) \exp\left[\int_0^t tr(\mathbf{A}) \, ds\right]$$

or

$$\det(e^{\mathbf{A}t}) = e^{tr(\mathbf{A})t},$$

which is (iii). $\square$

The analogy with the ordinary exponential function is not complete, however. For $\mathbf{A}, \mathbf{B} \in M_n(\mathcal{F})$ it may happen that

$$e^{\mathbf{A}+\mathbf{B}} \neq e^{\mathbf{A}}e^{\mathbf{B}}.$$

An example is provided by taking

$$\mathbf{A} = \begin{pmatrix} 1 & 0 \\ 0 & 0 \end{pmatrix}, \quad \mathbf{B} = \begin{pmatrix} 0 & 0 \\ 1 & 0 \end{pmatrix}.$$

Since $\mathbf{A}^2 = \mathbf{A}$ we have $e^{\mathbf{A}} = \mathbf{I}_2 + (e-1)\mathbf{A}$, or

$$e^{\mathbf{A}} = \begin{pmatrix} e & 0 \\ 0 & 1 \end{pmatrix},$$

and since $\mathbf{B}^2 = 0$, $e^{\mathbf{B}} = \mathbf{I}_2 + \mathbf{B}$, or

$$e^{\mathbf{B}} = \begin{pmatrix} 1 & 0 \\ 1 & 1 \end{pmatrix}.$$

Now

$$\mathbf{C} = \mathbf{A} + \mathbf{B} = \begin{pmatrix} 1 & 0 \\ 1 & 0 \end{pmatrix},$$

and $\mathbf{C}^2 = \mathbf{C}$ implies that $e^{\mathbf{C}} = \mathbf{I}_2 + (e-1)\mathbf{C}$, or

$$e^{\mathbf{A}+\mathbf{B}} = \begin{pmatrix} e & 0 \\ e-1 & 1 \end{pmatrix}.$$

On the other hand,

$$e^{\mathbf{A}} e^{\mathbf{B}} = \begin{pmatrix} e & 0 \\ 1 & 1 \end{pmatrix} \neq e^{\mathbf{A}+\mathbf{B}}, \quad e^{\mathbf{B}} e^{\mathbf{A}} = \begin{pmatrix} e & 0 \\ e & 1 \end{pmatrix} \neq e^{\mathbf{A}+\mathbf{B}}.$$

Notice that in this example $\mathbf{AB} = 0 \neq \mathbf{BA} = \mathbf{B}$. What is true is that if $\mathbf{AB} = \mathbf{BA}$, that is, if $\mathbf{A}$ and $\mathbf{B}$ commute, then

$$e^{\mathbf{A}+\mathbf{B}} = e^{\mathbf{A}} e^{\mathbf{B}} = e^{\mathbf{B}} e^{\mathbf{A}}.$$

Theorem 3.3: *The following are equivalent:*

(i)
$$\mathbf{AB} = \mathbf{BA}, \quad \mathbf{A}, \mathbf{B} \in M_n(\mathcal{F}),$$

(ii)
$$e^{\mathbf{A}t} \mathbf{B} = \mathbf{B} e^{\mathbf{A}t}, \quad t \in \mathcal{R},$$

(iii)
$$e^{\mathbf{A}t} e^{\mathbf{B}t} = e^{\mathbf{B}t} e^{\mathbf{A}t} = e^{(\mathbf{A}+\mathbf{B})t}, \quad t \in \mathcal{R}.$$

Proof: We will show that (i) $\Rightarrow$ (ii) $\Rightarrow$ (iii) $\Rightarrow$ (i). If $\mathbf{X}_{\mathbf{A}}(t) = e^{\mathbf{A}t}$ and (i) holds, then

$$(\mathbf{B}\mathbf{X}_{\mathbf{A}})' = \mathbf{B}\mathbf{X}_{\mathbf{A}}' = \mathbf{B}\mathbf{A}\mathbf{X}_{\mathbf{A}} = \mathbf{A}(\mathbf{B}\mathbf{X}_{\mathbf{A}}),$$

and therefore $\mathbf{B}\mathbf{X}_{\mathbf{A}}(t) = \mathbf{X}_{\mathbf{A}}(t)\mathbf{C}$ for some matrix $\mathbf{C} \in M_n(\mathcal{F})$. Evaluating at $t = 0$ gives $\mathbf{B} = \mathbf{C}$, which is (ii).

If (ii) holds, then $\mathbf{Y} = \mathbf{X_A X_B}$ satisfies

$$\mathbf{Y}' = \mathbf{X'_A X_B} + \mathbf{X_A X'_B} = \mathbf{AX_A X_B} + \mathbf{X_A B X_B}$$

$$= \mathbf{A(X_A X_B)} + \mathbf{B(X_A X_B)} = (\mathbf{A} + \mathbf{B})\mathbf{Y}.$$

Since

$$\mathbf{X'_{A+B}} = (\mathbf{A} + \mathbf{B})\mathbf{X_{A+B}}$$

and

$$\mathbf{X_{A+B}}(0) = \mathbf{I}_n = \mathbf{Y}(0),$$

uniqueness implies $\mathbf{X_{A+B}} = \mathbf{Y} = \mathbf{X_A X_B}$, which is (iii).

Finally, suppose (iii) is valid. Differentiation gives $\mathbf{X'_A} = \mathbf{AX_A}$ and $\mathbf{X''_A} = \mathbf{AX'_A} = \mathbf{A^2 X_A}$, implying that $\mathbf{X'_A}(0) = \mathbf{AX_A}(0) = \mathbf{A}$ and $\mathbf{X''_A}(0) = \mathbf{A^2}$. Thus

$$\mathbf{X''_{A+B}}(0) = (\mathbf{A} + \mathbf{B})^2 = \mathbf{A^2} + \mathbf{AB} + \mathbf{BA} + \mathbf{B^2} = (\mathbf{X_A X_B})''(0)$$

$$= \mathbf{X''_A}(0)\mathbf{X_B}(0) + 2\mathbf{X'_A}(0)\mathbf{X'_B}(0) + \mathbf{X_A}(0)\mathbf{X''_B}(0) = \mathbf{A^2} + 2\mathbf{AB} + \mathbf{B^2},$$

or $\mathbf{AB} = \mathbf{BA}$, which is (i). □

As an example, if

$$\mathbf{A} = \begin{pmatrix} 3 & 1 \\ -1 & 3 \end{pmatrix} = 3\mathbf{I}_2 + \mathbf{J}, \quad \mathbf{J} = \begin{pmatrix} 0 & 1 \\ -1 & 0 \end{pmatrix},$$

then $(3\mathbf{I}_2)\mathbf{J} = \mathbf{J}(3\mathbf{I}_2) = 3\mathbf{J}$ and

$$e^{\mathbf{A}t} = e^{3\mathbf{I}_2 t} e^{\mathbf{J}t} = (e^{3t}\mathbf{I}_2)e^{\mathbf{J}t} = e^{3t}\begin{pmatrix} \cos(t) & \sin(t) \\ -\sin(t) & \cos(t) \end{pmatrix}.$$

The basis $\mathbf{X_A}$ for the solutions of

$$x'_1 = 3x_1 + x_2,$$

$$x'_2 = -x_1 + 3x_2,$$

satisfying $\mathbf{X_A}(0) = \mathbf{I}_2$, is thus given by

$$\mathbf{X_A}(t) = e^{\mathbf{A}t} = \begin{pmatrix} e^{3t}\cos(t) & e^{3t}\sin(t) \\ -e^{3t}\sin(t) & e^{3t}\cos(t) \end{pmatrix}.$$

3.3   Nonhomogeneous systems

The variation of parameters formula (2.9) can be applied to the system

$$X' = \mathbf{A}X + B(t), \quad \mathbf{A} \in M_n(\mathcal{F}), \quad B \in C(I, \mathcal{F}^n), \tag{3.6}$$

where I is a real interval. The basis for (3.1) satisfying $\mathbf{X}(\tau) = \mathbf{I}_n$ is $\mathbf{X}(t) = e^{\mathbf{A}(t-\tau)}$, so that

$$\mathbf{X}(t)\mathbf{X}^{-1}(s) = e^{\mathbf{A}(t-\tau)}e^{-\mathbf{A}(s-\tau)} = e^{\mathbf{A}(t-s)},$$

and Theorem 2.8 yields the following result.

Theorem 3.4: *The solution $X(t)$ of (3.6) satisfying $X(\tau) = \xi$ is given by*

$$X(t) = e^{\mathbf{A}(t-\tau)}\xi + \int_\tau^t e^{\mathbf{A}(t-s)}B(s)\,ds, \quad t, \tau \in I, \quad \xi \in \mathcal{F}^n. \tag{3.7}$$

3.4   Structure of the solution space

3.4.1   A special case

Although the formula $\mathbf{X_A}(t) = e^{\mathbf{A}t}$ is a very elegant representation for a basis of solutions for the system $X' = \mathbf{A}X$ and (3.7) is a very pretty formula for a solution of (3.6), we are still left with the problem of computing $e^{\mathbf{A}t}$ in order to determine the precise structure of the set S of solutions of $X' = \mathbf{A}X$. To gain some intuition, let us return to the case $n = 1$, where the equation is $x' = ax$, $a \in \mathcal{F}$. Here every solution has the form $x(t) = e^{at}\alpha$, where $\alpha \in \mathcal{F}$. If

$$L(x) = x' - ax, \quad x \in C^1(\mathcal{R}, \mathcal{F}),$$

then for any complex number λ,

$$L(e^{\lambda t}\alpha) = e^{\lambda t}(\lambda - a)\alpha,$$

and this is zero for $\alpha \neq 0$ if and only if $\lambda = a$.

The analogue of this idea provides a fruitful method for discovering solutions of (3.1). Defining

$$L(X) = X' - \mathbf{A}X, \quad X \in C^1(\mathcal{R}, \mathcal{F}^n),$$

let us try to find a solution of (3.1) of the form $e^{\lambda t}\alpha$, where $\lambda \in \mathcal{F}$, $\alpha \in \mathcal{F}^n$, $\alpha \neq 0$. It follows from

$$L(e^{\lambda t}\alpha) = \lambda e^{\lambda t}\alpha - \mathbf{A}e^{\lambda t}\alpha = e^{\lambda t}(\lambda\alpha - \mathbf{A}\alpha)$$

that $L(e^{\lambda t}\alpha) = 0$, or $e^{\lambda t}\alpha$ is a solution of (3.1), if and only if

$$\mathbf{A}\alpha = \lambda\alpha \quad \text{or} \quad (\lambda\mathbf{I}_n - \mathbf{A})\alpha = 0. \tag{3.8}$$

The constant vector $\alpha = 0$ is a solution of (3.8), which leads to the trivial solution of (3.1). For a nontrivial solution we must find $\alpha \neq 0$ satisfying (3.8). An element $\lambda \in \mathcal{F}$ such that (3.8) has a solution $\alpha \in \mathcal{F}^n$, $\alpha \neq 0$, is called an *eigenvalue* of $\mathbf{A}$ in $\mathcal{F}$, and any $\alpha \neq 0$ satisfying (3.8) is called an *eigenvector* of $\mathbf{A}$ for λ.

The equation (3.8) is a system of n homogeneous linear equations for the n components $\alpha_1, \ldots, \alpha_n$ of α, and there is a nontrivial solution if and only if the determinant $\det(\lambda \mathbf{I}_n - \mathbf{A})$ is 0. If $\mathbf{A} = (a_{ij})$, then

$$\det(\lambda \mathbf{I}_n - \mathbf{A}) = \det \begin{pmatrix} \lambda - a_{11} & -a_{12} & \cdots & -a_{1n} \\ -a_{21} & \lambda - a_{22} & \cdots & -a_{2n} \\ \vdots & \vdots & \cdots & \vdots \\ -a_{n1} & -a_{n2} & \cdots & \lambda - a_{nn} \end{pmatrix},$$

so that $P_{\mathbf{A}}(\lambda) = \det(\lambda \mathbf{I}_n - \mathbf{A})$ is a polynomial in λ of degree n, with the coefficient of λ^n being 1. This polynomial $P_{\mathbf{A}}$ is called the *characteristic polynomial* of $\mathbf{A}$, and its roots $\lambda_j \in \mathcal{F}$ are the eigenvalues of $\mathbf{A}$.

Matrices with all entries real may have no real eigenvalues. For example, if

$$\mathbf{A} = \begin{pmatrix} 0 & 1 \\ -1 & 0 \end{pmatrix} \in M_2(\mathcal{R}),$$

then

$$P_{\mathbf{A}}(\lambda) = \det \begin{pmatrix} \lambda & -1 \\ 1 & \lambda \end{pmatrix} = \lambda^2 + 1.$$

This characteristic polynomial has the roots $i, -i$, which are not in $\mathcal{R}$. For this reason it is convenient to assume that $\mathcal{F} = \mathcal{C}$, $\mathbf{A} \in M_n(\mathcal{C})$, since the fundamental theorem of algebra guarantees that every polynomial of degree $n \geq 1$ with complex coefficients has n complex roots, counting multiplicities. Note that any $\mathbf{A} \in M_n(\mathcal{R})$ can be considered as an $\mathbf{A} \in M_n(\mathcal{C})$.

For $\mathbf{A} \in M_n(\mathcal{C})$ let $\lambda_1, \ldots, \lambda_k \in \mathcal{C}$ be its distinct eigenvalues, the distinct roots of $p_{\mathbf{A}}$. The polynomial $p_{\mathbf{A}}$ has a product representation

$$p_{\mathbf{A}}(\lambda) = (\lambda - \lambda_1)^{m_1} \cdots (\lambda - \lambda_k)^{m_k},$$

where the *algebraic multiplicity* m_j for each eigenvalue λ_j is uniquely determined, and $m_1 + \cdots + m_k = n$. For each λ_j let $\alpha_j \in \mathcal{C}^n$ be an eigenvector of $\mathbf{A}$ for λ_j, that is, $\mathbf{A}\alpha_j = \lambda_j \alpha_j$. The computations above show that the functions

$$X_j(t) = e^{\lambda_j t} \alpha_j, \quad j = 1, \ldots, k, \quad t \in \mathcal{R}, \tag{3.9}$$

are solutions of $X' = \mathbf{A}X$.

If $\mathbf{A}$ has n distinct eigenvalues, then the eigenvectors $\alpha_1, \ldots, \alpha_n$ are linearly independent and hence form a basis for $\mathcal{C}^n$ (see exercise 20). In this special case the X_j defined by (3.9) give a basis for S since $X_j(0) = \alpha_j$ and the structure of S is known.

Theorem 3.5: *If $\mathbf{A} \in M_n(\mathcal{C})$ has n distinct eigenvalues $\lambda_1, \ldots, \lambda_n$ with corresponding eigenvectors $\alpha_1, \ldots, \alpha_n$, then the set of functions*

$$X_j(t) = e^{\lambda_j t}\alpha_j, \quad j = 1, \ldots, n,$$

is a basis for S.

Suppose that $\mathbf{Q}$ and $\mathbf{J}$ denote the $n \times n$ matrices

$$\mathbf{Q} = (\alpha_1, \ldots, \alpha_n), \quad \mathbf{J} = \mathrm{diag}(\lambda_1, \ldots, \lambda_n).$$

Here $\mathbf{J} = \mathrm{diag}(\lambda_1, \ldots, \lambda_n)$ denotes a diagonal matrix

$$\mathbf{J} = \begin{pmatrix} \lambda_1 & 0 & 0 & \cdots & 0 \\ 0 & \lambda_2 & 0 & \cdots & 0 \\ \vdots & \vdots & \vdots & \cdots & \vdots \\ 0 & 0 & 0 & \cdots & \lambda_n \end{pmatrix}$$

with $\lambda_1, \ldots, \lambda_n$ down the main diagonal and all other elements zero. The jth column of the basis $\mathbf{X}(t) = (X_1(t), \ldots, X_n(t))$ is just the jth column of $\mathbf{Q}$ multiplied by $\exp(\lambda_j t)$, which means that the basis in Theorem 3.5 is

$$\mathbf{X}(t) = \mathbf{Q}e^{\mathbf{J}t}, \quad e^{\mathbf{J}t} = \mathrm{diag}(e^{\lambda_1 t}, \ldots, e^{\lambda_n t}).$$

Several examples will illustrate these ideas. First, let

$$\mathbf{A} = \begin{pmatrix} 1 & 2 \\ 2 & -2 \end{pmatrix}.$$

The characteristic polynomial is

$$p_{\mathbf{A}}(\lambda) = \det \begin{pmatrix} \lambda - 1 & -2 \\ -2 & \lambda + 2 \end{pmatrix} = \lambda^2 + \lambda - 6 = (\lambda + 3)(\lambda - 2),$$

so that $\mathbf{A}$ has eigenvalues $\lambda_1 = -3$, $\lambda_2 = 2$. The components c_1, c_2 of an eigenvector α for an eigenvalue λ satisfy

$$(\lambda - 1)c_1 - 2c_2 = 0, \quad -2c_1 + (\lambda + 2)c_2 = 0.$$

Thus for $\lambda_1 = -3$ we obtain

$$-4c_1 - 2c_2 = 0, \quad -2c_1 - c_2 = 0,$$

or $c_2 = -2c_1$. Eigenvectors for $\lambda = -3$ and $\lambda = 2$ are, respectively,

$$\begin{pmatrix} 1 \\ -2 \end{pmatrix}, \quad \begin{pmatrix} 2 \\ 1 \end{pmatrix}.$$

The corresponding solutions of $X' = \mathbf{A}X$ are given by

$$U(t) = e^{-3t} \begin{pmatrix} 1 \\ -2 \end{pmatrix}, \quad V(t) = e^{2t} \begin{pmatrix} 2 \\ 1 \end{pmatrix}.$$

Therefore a basis for the solutions of the system

$$x_1' = x_1 + 2x_2,$$

$$x_2' = 2x_1 - 2x_2$$

is given by

$$\mathbf{X}(t) = (U(t), V(t)) = \begin{pmatrix} e^{-3t} & 2e^{2t} \\ -2e^{-3t} & e^{2t} \end{pmatrix}, \quad t \in \mathcal{R}.$$

If

$$\mathbf{A} = \begin{pmatrix} 0 & 1 \\ -1 & 0 \end{pmatrix},$$

then $p_{\mathbf{A}}(\lambda) = \lambda^2 + 1$, whose roots are $\lambda_1 = i$, $\lambda_2 = -i$. Eigenvectors for λ_1, λ_2 are

$$\begin{pmatrix} 1 \\ i \end{pmatrix}, \quad \begin{pmatrix} i \\ 1 \end{pmatrix},$$

respectively, with corresponding solutions U, V of $X' = \mathbf{A}X$ being

$$U(t) = e^{it} \begin{pmatrix} 1 \\ i \end{pmatrix}, \quad V(t) = e^{-it} \begin{pmatrix} i \\ 1 \end{pmatrix}.$$

Thus a basis $\mathbf{X}$ for $X' = \mathbf{A}X$ is given by

$$\mathbf{X}(t) = \begin{pmatrix} e^{it} & ie^{-it} \\ ie^{it} & e^{-it} \end{pmatrix}.$$

We observed at the end of section 3.1 that another basis is $\mathbf{X_A}$, where

$$\mathbf{X_A}(t) = \begin{pmatrix} \cos(t) & \sin(t) \\ -\sin(t) & \cos(t) \end{pmatrix}.$$

As a third example, let

$$\mathbf{A} = \begin{pmatrix} -2 & 1 \\ 0 & -2 \end{pmatrix}.$$

Then $p_{\mathbf{A}}(\lambda) = (\lambda+2)^2$ with one root $\lambda = -2$ having multiplicity $m = 2$. Every eigenvector for $\lambda = -2$ is a nonzero multiple of

$$E_1 = \begin{pmatrix} 1 \\ 0 \end{pmatrix},$$

and a solution U of $X' = \mathbf{A}X$ is given by

$$U(t) = e^{-2t} \begin{pmatrix} 1 \\ 0 \end{pmatrix} = \begin{pmatrix} e^{-2t} \\ 0 \end{pmatrix}.$$

In order to find a basis for $X' = \mathbf{A}X$ another solution is needed. This problem is discussed in general in the next section.

3.4.2   The general case

If $\mathbf{A}$ has fewer than n linearly independent eigenvectors, we are still faced with the problem of finding more solutions. Their discovery will hinge on a result from algebra, the primary decomposition theorem. (See [9] for an explicit presentation of this result, or [7, Theorem 10.2.2]. Reference [10] also has related material.) To see how this result comes into play, let us return to the equation

$$L(e^{\lambda t}\alpha) = e^{\lambda t}(\lambda\alpha - \mathbf{A}\alpha), \tag{3.10}$$

which is valid for all $\lambda \in \mathcal{C}$, $t \in \mathcal{R}$, $\alpha \in \mathcal{C}^n$. Pick a second vector $\beta \in \mathcal{C}^n$. Since differentiation of $e^{\lambda t}\beta$ with respect to λ commutes with differentiation with respect to t, equation (3.10) gives

$$\frac{\partial}{\partial\lambda}L(e^{\lambda t}\beta) = L\left(\frac{\partial}{\partial\lambda}e^{\lambda t}\beta\right) = L(te^{\lambda t}\beta).$$

Also

$$\frac{\partial}{\partial\lambda}L(e^{\lambda t}\beta) = \frac{\partial}{\partial\lambda}[e^{\lambda t}(\lambda\beta - \mathbf{A}\beta)] = te^{\lambda t}(\lambda\beta - \mathbf{A}\beta) + e^{\lambda t}\beta,$$

which gives

$$L(te^{\lambda t}\beta) = e^{\lambda t}[\beta + t(\lambda\beta - \mathbf{A}\beta)]. \tag{3.11}$$

Adding the equations (3.10) and (3.11), we obtain

$$L(e^{\lambda t}(\alpha + t\beta)) = e^{\lambda t}[(\lambda\alpha - \mathbf{A}\alpha + \beta) + t(\lambda\beta - \mathbf{A}\beta)].$$

Thus $e^{\lambda t}(\alpha + t\beta)$ is a solution if and only if

$$(\mathbf{A} - \lambda\mathbf{I}_n)\alpha = \beta \tag{3.12}$$

and

$$(\mathbf{A} - \lambda\mathbf{I}_n)\beta = 0. \tag{3.13}$$

Equations (3.12) and (3.13) imply that

$$(\mathbf{A} - \lambda\mathbf{I}_n)^2\alpha = 0. \tag{3.14}$$

Conversely, if α satisfies (3.14) and β is defined by (3.12), then (3.13) is valid. If λ is an eigenvalue of $\mathbf{A}$ with algebraic multiplicity two we would like to find α, β satisfying (3.14), (3.12). If α_1 is an eigenvector for λ, then (3.14) and (3.12) are satisfied with $\beta = 0$. This gives the solution $X_1(t) = e^{\lambda t}\alpha_1$ obtained earlier. The primary decomposition theorem guarantees that there is a second choice of α satisfying (3.14), say $\alpha = \alpha_2$, such that α_1, α_2 is a linearly independent set. Then β defined by (3.12) with $\alpha = \alpha_2$ gives rise to a second solution $X_2(t) = e^{\lambda t}(\alpha_2 + t\beta)$, and X_1, X_2 is a linearly independent set since α_1, α_2, which are their values at $t = 0$, are linearly independent. Both X_1 and X_2 may be written in a common form:

$$X_1(t) = e^{\lambda t}[\alpha_1 + t(A - \lambda\mathbf{I}_n)\alpha_1],$$

$$X_2(t) = e^{\lambda t}[\alpha_2 + t(A - \lambda \mathbf{I}_n)\alpha_2].$$

Before stating the primary decomposition theorem we introduce some terminology. If λ is an eigenvalue of $\mathbf{A} \in M_n(\mathcal{C})$, the *eigenspace* $\mathcal{E}(\mathbf{A}, \lambda)$ of $\mathbf{A}$ for λ is the vector space

$$\mathcal{E}(\mathbf{A}, \lambda) = N(\mathbf{A} - \lambda \mathbf{I}_n);$$

that is, $\mathcal{E}(\mathbf{A}, \lambda)$ is the set of all eigenvectors of $\mathbf{A}$ for λ together with the zero vector. The *generalized eigenspace* $\mathcal{F}(\mathbf{A}, \lambda)$ of $\mathbf{A}$ for λ is the vector space

$$\mathcal{F}(\mathbf{A}, \lambda) = N((\mathbf{A} - \lambda \mathbf{I}_n)^m), \quad \text{some } m > 0,$$

which is the set of all $\alpha \in \mathcal{C}^n$ satisfying

$$(\mathbf{A} - \lambda \mathbf{I}_n)^m \alpha = 0 \quad \text{for some} \quad m = 1, 2, 3, \ldots.$$

Note that $\mathcal{E}(\mathbf{A}, \lambda) \subset \mathcal{F}(\mathbf{A}, \lambda)$.

Primary decomposition theorem: *Let* $\mathbf{A} \in M_n(\mathcal{C})$ *have distinct eigenvalues* $\lambda_1, \ldots, \lambda_k$ *with corresponding algebraic multiplicities* $m_1, \ldots, m_k$, *where* $m_1 + \cdots + m_k = n$. *Then*

$$\dim(\mathcal{F}(\mathbf{A}, \lambda_j)) = m_j, \quad j = 1, \ldots, k,$$

and

$$\mathcal{C}^n = \mathcal{F}(\mathbf{A}, \lambda_1) \oplus \cdots \oplus \mathcal{F}(\mathbf{A}, \lambda_k),$$

a direct sum in the sense that each $\xi \in \mathcal{C}^n$ *can be written uniquely in the form*

$$\xi = \xi_1 + \cdots + \xi_k, \quad \xi_j \in \mathcal{F}(\mathbf{A}, \lambda_j).$$

A direct consequence of this theorem is the following characterization of the structure of S in the general case.

Theorem 3.6: *Suppose that* $\mathbf{A} \in M_n(\mathcal{C})$ *has distinct eigenvalues* $\lambda_1, \ldots, \lambda_k$ *with algebraic multiplicities* $m_1, \ldots, m_k$, *respectively. If* $\xi \in \mathcal{C}^n$ *and* $\xi = \xi_1 + \cdots + \xi_k$, *where* $\xi_j \in \mathcal{F}(\mathbf{A}, \lambda_j)$, *then the solution* X *of* $X' = \mathbf{A}X$ *satisfying* $X(0) = \xi$ *is given by*

$$X(t) = e^{\mathbf{A}t}\xi = e^{\lambda_1 t}P_1(t) + \cdots + e^{\lambda_k t}P_k(t), \tag{3.15}$$

where P_j *is a vector polynomial in* t *of degree at most* $m_j - 1$,

$$P_j(t) = \left[\mathbf{I}_n + t(\mathbf{A} - \lambda_j \mathbf{I}_n) + \frac{t^2}{2!}(\mathbf{A} - \lambda_j \mathbf{I}_n)^2 \right.$$

$$\left. + \cdots + \frac{t^{m_j - 1}}{(m_j - 1)!}(\mathbf{A} - \lambda_j \mathbf{I}_n)^{m_j - 1} \right] \xi_j. \tag{3.16}$$

Proof: The primary decomposition theorem guarantees that $\xi \in \mathcal{C}^n$ can be represented uniquely as

$$\xi = \xi_1 + \cdots + \xi_k, \quad \xi_j \in \mathcal{F}(\mathbf{A}, \lambda_j).$$

Thus

$$X(t) = e^{\mathbf{A}t}\xi = e^{\mathbf{A}t}\xi_1 + \cdots + e^{\mathbf{A}t}\xi_k,$$

and we have

$$e^{\mathbf{A}t}\xi_j = e^{\lambda_j \mathbf{I}_n t + (\mathbf{A} - \lambda_j \mathbf{I}_n)t}\xi_j = e^{\lambda_j \mathbf{I}_n t}e^{(\mathbf{A} - \lambda_j \mathbf{I}_n)t}\xi_j = e^{\lambda_j t}e^{(\mathbf{A} - \lambda_j \mathbf{I}_n)t}\xi_j.$$

But since

$$(\mathbf{A} - \lambda_j \mathbf{I}_n)^{m_j}\xi_j = 0,$$

it follows that

$$(\mathbf{A} - \lambda_j \mathbf{I}_n)^p \xi_j = 0, \quad p \geq m_j,$$

and

$$e^{(\mathbf{A} - \lambda_j \mathbf{I}_n)t}\xi_j = \sum_{p=0}^{m_j - 1} \frac{t^p}{p!}(\mathbf{A} - \lambda_j \mathbf{I}_n)^p \xi_j = P_j(t),$$

where $P_j(t)$ is given in (3.16). Thus

$$e^{\mathbf{A}t}\xi_j = e^{\lambda_j t}P_j(t), \quad j = 1, \ldots, k,$$

and

$$X(t) = e^{\lambda_1 t}P_1(t) + \cdots + e^{\lambda_k t}P_k(t),$$

which is (3.15). $\square$

Finding a basis for the solutions of $X' = \mathbf{A}X$ is now reduced to finding a basis for each of the spaces $\mathcal{F}(\mathbf{A}, \lambda_j)$.

Theorem 3.7: *If for each $j = 1, \ldots, k$ and $i = 1, \ldots, m_j$, the vectors α_{ij} form a basis for $\mathcal{F}(\mathbf{A}, \lambda_j)$, then the n vectors α_{ij} give a basis for $\mathcal{C}^n$, and the X_{ij} given by*

$$X_{ij}(t) = e^{\lambda_j t}P_{ij}(t), \quad j = 1, \ldots, k \quad \text{and} \quad i = 1, \ldots, m_j,$$

$$P_{ij}(t) = \sum_{p=0}^{m_j - 1} \frac{t^p}{p!}(\mathbf{A} - \lambda_j \mathbf{I}_n)^p \alpha_{ij}$$

constitute a basis for S.

Proof: Putting $\xi = \alpha_{ij}$ in Theorem 3.6, X_{ij} is the solution of $X' = \mathbf{A}X$ satisfying $X_{ij}(0) = \alpha_{ij}$. The n solutions X_{ij} give a basis for S since the α_{ij} are linearly independent and form a basis for $\mathcal{C}^n$. To see this, suppose there are constants $c_{ij} \in \mathcal{C}$ such that

$$\sum_{j=1}^{k}\sum_{i=1}^{m_j} c_{ij}\alpha_{ij} = 0.$$

With

$$\alpha_j = \sum_{i=1}^{m_j} c_{ij}\alpha_{ij},$$

we have

$$\alpha = \alpha_1 + \cdots + \alpha_k = 0, \quad \alpha_j \in \mathcal{F}(\mathbf{A}, \lambda_j),$$

and the primary decomposition theorem, implies that each $\alpha_j = 0$. Then the linear independence of the α_{ij}, $i = 1, \ldots, m_j$, implies that all $c_{ij} = 0$. □

Corollary 3.8: *If*

$$\dim \ \mathcal{F}(\mathbf{A}, \lambda_j) = \dim \ \mathcal{E}(\mathbf{A}, \lambda_j) = m_j, \quad j = 1, \ldots, k,$$

then a basis for S exists of the form

$$X_{ij}(t) = e^{\lambda_j t}\alpha_{ij}, \quad i = 1, \ldots, m_j, \quad j = 1, \ldots, k,$$

where $\alpha_{ij} \in \mathcal{C}^n$, $i = 1, \ldots, m_j$, is a basis for $\mathcal{E}(\mathbf{A}, \lambda_j)$.

This corollary applies in the simple case $\mathbf{A} = \mathbf{I}_n$. The matrix $\mathbf{A}$ has a single eigenvalue $\lambda = 1$ with multiplicity n. The equation $(\mathbf{A} - \mathbf{I}_n)\alpha = 0$ for eigenvectors α with $\lambda = 1$ has n linearly independent solutions $E_1, \ldots, E_n$. A basis is then given by

$$\mathbf{X}(t) = (e^t E_1, \ldots, e^t E_n) = e^t \mathbf{I}_n = e^{t\mathbf{I}_n}.$$

Corollary 3.9: *If $\mathbf{A} \in M_n(\mathcal{C})$ has n distinct eigenvalues $\lambda_1, \ldots, \lambda_n$ with corresponding eigenvectors $\alpha_1, \ldots, \alpha_n$, then $\mathbf{X} = (X_1, \ldots, X_n)$, where*

$$X_j(t) = e^{\lambda_j t}\alpha_j, \quad j = 1, \ldots, n,$$

is a basis for S. We have

$$\mathbf{A} = \mathbf{Q}\mathbf{J}\mathbf{Q}^{-1}, \quad \mathbf{Q} = (\alpha_1, \ldots, \alpha_n), \quad \mathbf{J} = \mathrm{diag}(\lambda_1, \ldots, \lambda_n)$$

and

$$e^{\mathbf{A}t} = \mathbf{Q}e^{\mathbf{J}t}\mathbf{Q}^{-1}, \quad e^{\mathbf{J}t} = \mathrm{diag}(e^{\lambda_1 t}, \ldots, e^{\lambda_n t}).$$

Proof: This is the special case of Corollary 3.8 with $k = n$, $m_j = 1$ for each $j = 1, \ldots, n$. The equations

$$\mathbf{A}\alpha_j = \lambda_j\alpha_j, \quad j = 1, \ldots, n,$$

can be written as

$$\mathbf{A}\mathbf{Q} = \mathbf{Q}\mathbf{J}, \quad \mathbf{Q} = (\alpha_1, \ldots, \alpha_n), \quad \mathbf{J} = (\lambda_1 E_1, \ldots, \lambda_n E_n),$$

and hence $\mathbf{A} = \mathbf{Q}\mathbf{J}\mathbf{Q}^{-1}$. The basis $\mathbf{X} = (X_1, \ldots, X_n)$, where

$$\mathbf{X}(t) = (e^{\lambda_1 t}\alpha_1, \ldots, e^{\lambda_n t}\alpha_n),$$

can then be written as

$$\mathbf{X}(t) = \mathbf{Q}e^{\mathbf{J}t}, \quad e^{\mathbf{J}t} = (e^{\lambda_1 t}E_1, \ldots, e^{\lambda_n t}E_n).$$

Since $\mathbf{X_A}(t) = e^{\mathbf{A}t}$ is also a basis for S, we must have $\mathbf{X_A}(t) = \mathbf{Q}e^{\mathbf{J}t}\mathbf{C}$ for some invertible $\mathbf{C} \in M_n(\mathcal{C})$. Putting $t = 0$ we get $\mathbf{I}_n = \mathbf{Q}\mathbf{C}$, or $\mathbf{C} = \mathbf{Q}^{-1}$, and hence $e^{\mathbf{A}t} = \mathbf{Q}e^{\mathbf{J}t}\mathbf{Q}^{-1}$. □

3.4.3   Some examples

Theorems 3.6 and 3.7 give a prescription for solving $X' = \mathbf{A}X$. We first determine the eigenvalues of $\mathbf{A}$, together with their multiplicities. If λ is an eigenvalue of multiplicity m, there are m linearly independent solutions of the form

$$X(t) = e^{\lambda t}(\alpha + t\beta_1 + \cdots + t^{m-1}\beta_{m-1}),$$

where $\alpha, \beta_j \in \mathcal{C}^n$. The initial values $X(0) = \alpha$ belong to $\mathcal{F}(\mathbf{A}, \lambda)$; that is they satisfy

$$(\mathbf{A} - \lambda \mathbf{I}_n)^m \alpha = 0,$$

and there are m linearly independent choices of α satisfying this equation. For each α, the β_j are uniquely given by

$$\beta_j = \frac{(\mathbf{A} - \lambda \mathbf{I}_n)^j}{j!}\alpha, \quad j = 1, \ldots, m-1.$$

The collection of all solutions obtained by carrying out this procedure for each eigenvalue in turn yields a basis for $X' = \mathbf{A}X$.

Begin with the rather trivial example

$$\mathbf{A} = \begin{pmatrix} -2 & 0 \\ 0 & -2 \end{pmatrix}.$$

The characteristic polynomial is $p_\mathbf{A}(\lambda) = (\lambda + 2)^2$, with a root $\lambda = -2$ with multiplicity $m = 2$. There are two independent solutions of the form

$$X(t) = e^{-2t}(\alpha + t\beta),$$

which in this case are simply

$$U(t) = e^{-2t}\begin{pmatrix} 1 \\ 0 \end{pmatrix}, \quad V(t) = e^{-2t}\begin{pmatrix} 0 \\ 1 \end{pmatrix}.$$

As a second example, let

$$\mathbf{A} = \begin{pmatrix} -2 & 1 \\ 0 & -2 \end{pmatrix},$$

which was considered briefly in section 3.4.2. Here $p_\mathbf{A}(\lambda) = (\lambda + 2)^2$, with a root $\lambda = -2$ with multiplicity $m = 2$. Thus there are two solutions of the form

$$X(t) = e^{-2t}(\alpha + t\beta),$$

where

$$(\mathbf{A} + 2\mathbf{I}_2)^2\alpha = 0, \quad \beta = (\mathbf{A} + 2\mathbf{I}_2)\alpha.$$

Now

$$\mathbf{A} + 2\mathbf{I}_2 = \begin{pmatrix} 0 & 1 \\ 0 & 0 \end{pmatrix}, \quad (\mathbf{A} + 2\mathbf{I}_2)^2 = 0,$$

and hence every vector in $\mathcal{C}^2$ satisfies the equation for α. A basis for these solutions is

$$E_1 = \begin{pmatrix} 1 \\ 0 \end{pmatrix}, \quad E_2 = \begin{pmatrix} 0 \\ 1 \end{pmatrix}.$$

For $\alpha = E_1$ we find $\beta = 0$, which gives the solution

$$U(t) = e^{-2t} E_1 = e^{-2t} \begin{pmatrix} 1 \\ 0 \end{pmatrix}.$$

For $\alpha = E_2$ we obtain $\beta = E_1$, and so

$$V(t) = e^{-2t}(E_2 + t E_1) = e^{-2t} \begin{pmatrix} t \\ 1 \end{pmatrix}$$

is a solution. Thus

$$\mathbf{X}(t) = (U(t), V(t)) = \begin{pmatrix} e^{-2t} & te^{-2t} \\ 0 & e^{-2t} \end{pmatrix}, \quad t \in \mathcal{R},$$

gives a basis for the solutions of the system

$$x_1' = -2x_1 + x_2,$$

$$x_2' = -2x_2.$$

As a third example, consider

$$\mathbf{A} = \begin{pmatrix} -2 & 1 & 0 \\ 0 & -2 & 1 \\ 0 & 0 & -2 \end{pmatrix}.$$

Here $p_{\mathbf{A}}(\lambda) = (\lambda + 2)^3$ with one root $\lambda = -2$ with multiplicity $m = 3$. There are solutions of the form

$$X(t) = e^{-2t}(\alpha + t\beta_1 + t^2\beta_2),$$

with

$$(\mathbf{A} + 2\mathbf{I}_3)^3 \alpha = 0, \quad \beta_1 = (A + 2\mathbf{I}_3)\alpha, \quad \beta_2 = \frac{(\mathbf{A} + 2\mathbf{I}_3)^2}{2}\alpha.$$

In this case

$$(\mathbf{A} + 2\mathbf{I}_3) = \begin{pmatrix} 0 & 1 & 0 \\ 0 & 0 & 1 \\ 0 & 0 & 0 \end{pmatrix}, \quad (\mathbf{A} + 2\mathbf{I}_3)^2 = \begin{pmatrix} 0 & 0 & 1 \\ 0 & 0 & 0 \\ 0 & 0 & 0 \end{pmatrix}, \quad (A + 2\mathbf{I}_3)^3 = 0.$$

Every vector in $\mathcal{C}^3$ satisfies $(A + 2\mathbf{I}_3)^3\alpha = 0$, and a basis for these α is E_1, E_2, E_3. The choices for β_1, β_2 corresponding to these α are

$$\alpha = E_1 : \quad \beta_1 = 0, \quad \beta_2 = 0,$$

$$\alpha = E_2 : \quad \beta_1 = E_1, \quad \beta_2 = 0,$$

$$\alpha = E_3 : \quad \beta_1 = E_2, \quad \beta_2 = E_1/2.$$

These yield the three solutions U, V, W, where

$$U(t) = e^{-2t}E_1, \quad V(t) = e^{-2t}(E_2 + tE_1), \quad W(t) = e^{-2t}\left(E_3 + tE_2 + \frac{t^2}{2}E_1\right),$$

so that a basis is

$$\mathbf{X}(t) = (U(t), V(t), W(t)) = e^{-2t}\begin{pmatrix} 1 & t & t^2/2 \\ 0 & 1 & t \\ 0 & 0 & 1 \end{pmatrix}.$$

As a final example, let

$$\mathbf{A} = \begin{pmatrix} 2 & -1 & 1 \\ 1 & 0 & 3 \\ 0 & 0 & 2 \end{pmatrix}.$$

A calculation shows that

$$p_{\mathbf{A}}(\lambda) = \lambda^3 - 4\lambda^2 + 5\lambda - 2 = (\lambda - 1)^2(\lambda - 2),$$

and therefore there is an eigenvalue $\lambda_1 = 1$ with multiplicity $m_1 = 2$ and another eigenvalue $\lambda_2 = 2$ with multiplicity $m_2 = 1$. Corresponding to $\lambda_1 = 1$, there are two solutions of the form

$$X(t) = e^t(\alpha + t\beta),$$

where

$$(\mathbf{A} - \lambda_1 \mathbf{I}_3)^2 \alpha = 0, \quad \beta = (\mathbf{A} - \lambda_1 \mathbf{I}_3)\alpha, \quad \lambda_1 = 1.$$

Now

$$\mathbf{A} - \mathbf{I}_3 = \begin{pmatrix} 1 & -1 & 1 \\ 1 & -1 & 3 \\ 0 & 0 & 1 \end{pmatrix}, \quad (\mathbf{A} - \mathbf{I}_3)^2 = \begin{pmatrix} 0 & 0 & -1 \\ 0 & 0 & 1 \\ 0 & 0 & 1 \end{pmatrix},$$

and so two solutions of $(\mathbf{A} - \mathbf{I}_3)^2 \alpha = 0$ are E_1, E_2. The corresponding values of β are

$$\alpha = E_1 : \quad \beta = E_1 + E_2,$$
$$\alpha = E_2 : \quad \beta = -E_1 - E_2.$$

This gives two solutions:

$$U(t) = e^t[E_1 + t(E_1 + E_2)], \quad V(t) = e^t[E_2 - t(E_1 + E_2)].$$

For $\lambda_2 = 2$ there is a solution W of the form $W(t) = e^{2t}\alpha$, where $(\mathbf{A} - 2\mathbf{I}_3)\alpha = 0$. Each such α is a multiple of $E_1 - E_2 - E_3$, and this gives a solution

$$W(t) = e^{2t}(E_1 - E_2 - E_3).$$

Hence a basis $\mathbf{X}$ for the solutions to the system

$$x_1' = 2x_1 - x_2 + x_3,$$

$$x_2' = x_1 + 3x_3,$$

$$x_3' = 2x_3$$

is given by

$$\mathbf{X}(t) = (U(t), V(t), W(t)) = \begin{pmatrix} (1+t)e^t & -te^t & e^{2t} \\ te^t & (1-t)e^t & -e^{2t} \\ 0 & 0 & -e^{2t} \end{pmatrix}.$$

3.4.4   Real solutions

In case $\mathbf{A} \in M_n(\mathcal{R})$, it is convenient to consider $\mathbf{A}$ as a matrix with complex elements in order to guarantee the existence of eigenvalues $\lambda \in \mathcal{C}$ of $\mathbf{A}$. This led us to construct (in sections 3.4.1 and 3.4.2) solutions of $X' = \mathbf{A}X$ for each eigenvalue, which in general have values in $\mathcal{C}^n$. On the other hand, for $\mathbf{A} \in M_n(\mathcal{R})$ there is a basis with values in $\mathcal{R}^n$, namely that given by $\mathbf{X_A}(t) = e^{\mathbf{A}t}$. If $\mathbf{X}$ is any basis for the solutions of $X' = \mathbf{A}X$, for example, the one constructed in Theorem 3.7, then $\mathbf{X_A} = \mathbf{X}\mathbf{C}$ for some constant invertible matrix $\mathbf{C}$. Now $\mathbf{C}$ can be determined by noting that $\mathbf{X_A}(0) = \mathbf{I}_n = \mathbf{X}(0)\mathbf{C}$, so that $\mathbf{C} = \mathbf{X}^{-1}(0)$. Hence the basis $\mathbf{X_A}$, whose columns are in $\mathcal{R}^n$ if $\mathbf{A} \in M_n(\mathcal{R})$, is given by $\mathbf{X_A}(t) = \mathbf{X}(t)\mathbf{X}^{-1}(0)$.

For any $\mathbf{B} \in M_{mn}(\mathcal{C})$, whose entries are (b_{ij}), the *conjugate* of $\mathbf{B}$ is $\overline{\mathbf{B}} = (\overline{b}_{ij})$. By analogy with the complex numbers, $\mathbf{B}$ has a *real part* and an *imaginary part*,

$$\mathrm{Re}(\mathbf{B}) = \frac{1}{2}(\mathbf{B} + \overline{\mathbf{B}}), \quad \mathrm{Im}(\mathbf{B}) = \frac{1}{2i}(\mathbf{B} - \overline{\mathbf{B}}).$$

It is simple to check that $\mathbf{B} = \mathrm{Re}(\mathbf{B}) + i\,\mathrm{Im}(\mathbf{B})$, and

$$\mathrm{Re}(\mathbf{B}) = (\mathrm{Re}(b_{ij})), \quad \mathrm{Im}(\mathbf{B}) = (\mathrm{Im}(b_{ij})).$$

A matrix $\mathbf{B} \in M_{mn}(\mathcal{C})$ is said to be *real* if $\mathbf{B} \in M_{mn}(\mathcal{R})$. Thus $\mathbf{B}$ is real if and only if $\mathbf{B} = \mathrm{Re}(\mathbf{B})$ or $\mathbf{B} = \overline{\mathbf{B}}$. In particular, if $\xi \in \mathcal{C}^n$, considered as $M_{n1}(\mathcal{C})$, then $\mathrm{Re}(\xi), \mathrm{Im}(\xi) \in \mathcal{R}^n$, and $\xi = \mathrm{Re}(\xi) + i\mathrm{Im}(\xi)$.

If $\mathbf{B}$ is any function from a real interval I into $M_{mn}(\mathcal{C})$, we say $\mathbf{B}$ is real if $\mathbf{B}(t)$ is real for all $t \in I$. If $\overline{\mathbf{B}}$ is the function given by $\overline{\mathbf{B}}(t) = \overline{\mathbf{B}(t)}$, $t \in I$, then $\mathbf{B}$ is real if and only if $\mathbf{B} = \overline{\mathbf{B}}$.

Now suppose $\mathbf{A} \in M_n(\mathcal{R})$, so that $\mathbf{A}$ is real. We consider $\mathbf{A} \in M_n(\mathcal{C})$ and look at the set of all solutions $S \subset C(\mathcal{R}, \mathcal{C}^n)$ of $X' = \mathbf{A}X$. If $X(t) \in S$ then $\overline{X}(t) \in S$ for $X'(t) = \mathbf{A}X(t)$ implies $\overline{X'(t)} = \overline{\mathbf{A}X(t)} = \mathbf{A}\overline{X}(t)$. Consequently, $U = \mathrm{Re}(X)$ and $V = \mathrm{Im}(X)$ also belong to S. A solution X is real, $X = \overline{X}$, if and only if its initial value $X(0) = \xi$ is real. This follows since

$$X(t) = e^{\mathbf{A}t}\xi = e^{\mathbf{A}t}\overline{\xi} = \overline{X}(t)$$

if and only if $\xi = \overline{\xi}$.

Theorem 3.7 asserts that, for each eigenvalue λ of $\mathbf{A}$ of multiplicity m, there are m linearly independent solutions of $X' = \mathbf{A}X$ of the form

$$X(t) = e^{\lambda t}(\alpha + t\beta_1 + \cdots + t^{m-1}\beta_m),$$

where $\alpha, \beta_j \in \mathcal{C}^n$. The initial values $X(0) = \alpha$ satisfy

$$(\mathbf{A} - \lambda \mathbf{I}_n)^m \alpha = 0, \tag{3.17}$$

and the m linearly independent solutions are constructed by finding m linearly independent α satisfying (3.17) and putting

$$\beta_j = \frac{(\mathbf{A} - \lambda \mathbf{I}_m)^j}{j!}\alpha, \quad j = 1, \ldots, m-1. \tag{3.18}$$

When $\lambda \in \mathcal{R}$ the system (3.17) has a real coefficient matrix $(\mathbf{A} - \lambda \mathbf{I}_n)^m$, so that there are m linearly independent solutions α which are real (by Gaussian elimination). The corresponding solutions $X(t) \in S$ such that $X(0) = \alpha$ will then be real.

If $\lambda \in \mathcal{C}$ is not real, then $\overline{\lambda}$ is also an eigenvalue of $\mathbf{A}$ of multiplicity m. This can be seen from the fact that the characteristic polynomial $p_{\mathbf{A}}$ has real coefficients:

$$p_{\mathbf{A}}(\lambda) = \det(\lambda - \mathbf{A}) = \lambda^n + a_{n-1}\lambda^{n-1} + \cdots + a_0, \quad a_j \in \mathcal{R}.$$

Thus $\overline{p_{\mathbf{A}}(\lambda)} = p_{\mathbf{A}}(\overline{\lambda})$ shows that if λ is an eigenvalue of $\mathbf{A}$, so is $\overline{\lambda}$, and a slight extension of this argument shows that they have the same multiplicity. Moreover, $(\mathbf{A} - \lambda \mathbf{I}_n)^m \alpha = 0$ if and only if $(\mathbf{A} - \overline{\lambda}\mathbf{I}_n)^m \overline{\alpha} = 0$, so that the map $\alpha \in \mathcal{F}(\mathbf{A}, \lambda) \to \overline{\alpha} \in \mathcal{F}(\mathbf{A}, \overline{\lambda})$ is an isomorphism. Now corresponding to the eigenvalue λ there are m linearly independent solutions $X_1, \ldots, X_m \in S$ of the form
$$X_i(t) = e^{\lambda t} P_i(t), \quad P_i(0) = \alpha_i \in \mathcal{F}(\mathbf{A}, \lambda),$$

where P_i is a polynomial in t of degree less than or equal to $m-1$, and also m linearly independent $\overline{X}_1, \ldots, \overline{X}_m \in S$ of the form

$$\overline{X}_i(t) = e^{\overline{\lambda}t}\overline{P}_i(t), \quad \overline{P}_i(0) = \overline{\alpha}_i \in \mathcal{F}(\mathbf{A}, \overline{\lambda}).$$

These $2m$ solutions $X_1, \ldots, X_m, \overline{X}_1, \ldots, \overline{X}_m$ corresponding to λ form a linearly independent set, for $\alpha_1, \ldots, \alpha_m, \overline{\alpha}_1, \ldots, \overline{\alpha}_m$ is a basis for $\mathcal{F}(\mathbf{A}, \lambda) \oplus \mathcal{F}(\mathbf{A}, \overline{\lambda})$. It is not difficult to see that the $2m$ real solutions $U_1, \ldots, U_m, V_1, \ldots, V_m$, where $U_i = \operatorname{Re}(X_i)$, $V_i = \operatorname{Im}(X_i)$, also constitute a linearly independent set. When $\mathbf{A} \in M_n(\mathcal{R})$ this gives a method for constructing a real basis from that given in Theorem 3.7.

As an example, consider

$$\mathbf{A} = \begin{pmatrix} 2 & 1 \\ -1 & 2 \end{pmatrix} \in M_2(\mathcal{R}).$$

In this case $p_{\mathbf{A}}(\lambda) = \lambda^2 - 4\lambda + 5$, whose two roots are

$$\lambda_1 = 2 + i, \quad \lambda_2 = \overline{\lambda}_1 = 2 - i.$$

Eigenvectors for λ_1 and λ_2 are, respectively,

$$\alpha = \begin{pmatrix} 1 \\ i \end{pmatrix}, \quad \overline{\alpha} = \begin{pmatrix} 1 \\ -i \end{pmatrix}.$$

These yield the two solutions

$$X(t) = e^{(2+i)t} \begin{pmatrix} 1 \\ i \end{pmatrix}, \quad \overline{X}(t) = e^{(2-i)t} \begin{pmatrix} 1 \\ -i \end{pmatrix},$$

and $X, \overline{X}$ give a basis for $X' = \mathbf{A}X$. A real basis is given by $U = \mathrm{Re}(X)$, $V = \mathrm{Im}(X)$. Since

$$X(t) = e^{2t}(\cos(t) + i\sin(t)) \begin{pmatrix} 1 \\ i \end{pmatrix} = \begin{pmatrix} e^{2t}\cos(t) + ie^{2t}\sin(t) \\ -e^{2t}\sin(t) + ie^{2t}\cos(t) \end{pmatrix},$$

we see that

$$U(t) = \begin{pmatrix} e^{2t}\cos(t) \\ -e^{2t}\sin(t) \end{pmatrix}, \quad V(t) = \begin{pmatrix} e^{2t}\sin(t) \\ e^{2t}\cos(t) \end{pmatrix},$$

and a real basis $\mathbf{X} = (U, V)$ is given by

$$\mathbf{X}(t) = e^{2t} \begin{pmatrix} \cos(t) & \sin(t) \\ -\sin(t) & \cos(t) \end{pmatrix} = \mathbf{X}_{\mathbf{A}}(t).$$

3.5   The Jordan canonical form of a matrix

A more precise computation of $e^{\mathbf{A}t}$ can be made by using a more refined result from linear algebra which asserts that each $\mathbf{A} \in M_n(\mathcal{C})$ can be written as $\mathbf{A} = \mathbf{Q}\mathbf{J}\mathbf{Q}^{-1}$ for some invertible $\mathbf{Q} \in M_n(\mathcal{C})$, where $\mathbf{J}$ has a certain standard form, called a *Jordan canonical form* of $\mathbf{A}$ (see [7, 9, 10]).

Two matrices $\mathbf{A}, \mathbf{B} \in M_n(\mathcal{F})$ are said to be *similar* if there exists an invertible $\mathbf{Q} \in M_n(\mathcal{F})$ such that $\mathbf{A} = \mathbf{Q}\mathbf{B}\mathbf{Q}^{-1}$. Similar matrices have the same characteristic polynomial, since

$$p_{\mathbf{A}}(\lambda) = \det(\lambda\mathbf{I}_n - \mathbf{A}) = \det[\mathbf{Q}(\lambda\mathbf{I}_n - \mathbf{B})\mathbf{Q}^{-1}]$$

$$= \det(\mathbf{Q})\det(\lambda\mathbf{I}_n - \mathbf{B})\det(\mathbf{Q}^{-1}) = \det(\lambda\mathbf{I}_n - \mathbf{B}) = p_{\mathbf{B}}(\lambda).$$

Thus similar matrices have the same eigenvalues together with their respective multiplicities.

If $\mathbf{B} \in M_n(\mathcal{F})$ consists of blocks $\mathbf{B}_1, \ldots, \mathbf{B}_k$ of square matrices of order less than n distributed down the main diagonal, with all other elements being zero,

$$\mathbf{B} = \begin{pmatrix} \mathbf{B}_1 & 0 & 0 & \ldots & 0 \\ 0 & \mathbf{B}_2 & 0 & \ldots & 0 \\ \vdots & \vdots & \vdots & \ldots & \vdots \\ 0 & 0 & 0 & \ldots & \mathbf{B}_k \end{pmatrix},$$

we say that $\mathbf{B}$ is in *block diagonal form* and write $\mathbf{B} = \mathrm{diag}(\mathbf{B}_1, \ldots, \mathbf{B}_k)$. If all $\mathbf{B}_j$ are 1×1 matrices, then $\mathbf{B}$ is in *diagonal form*.

Jordan canonical form theorem: *Each $\mathbf{A} \in M_n(\mathcal{C})$ is similar to a matrix $\mathbf{J}$, a Jordan canonical form of $\mathbf{A}$, which is in block diagonal form*

$$\mathbf{A} = \mathbf{QJQ}^{-1}, \quad \mathbf{J} = \mathrm{diag}(\mathbf{J}_1, \ldots, \mathbf{J}_q),$$

where each $\mathbf{J}_i$ is an $r_i \times r_i$ matrix of the form

$$\mathbf{J}_i = (\lambda_i)$$

if $r_i = 1$ and

$$\mathbf{J}_i = \begin{pmatrix} \lambda_i & 1 & 0 & \ldots & 0 & 0 \\ 0 & \lambda_i & 1 & \ldots & 0 & 0 \\ \vdots & \vdots & \vdots & \ldots & \vdots & \vdots \\ 0 & 0 & 0 & \ldots & \lambda_i & 1 \\ 0 & 0 & 0 & \ldots & 0 & \lambda_i \end{pmatrix}$$

if $r_i > 1$, with all elements on the main diagonal being λ_i, all elements on the first superdiagonal being 1, and all other elements being 0.

Since $p_{\mathbf{A}}(\lambda) = p_{\mathbf{J}}(\lambda) = \det(\lambda \mathbf{I}_n - \mathbf{J})$ it follows that the λ_i in each $\mathbf{J}_i$ are eigenvalues of $\mathbf{A}$ and of $\mathbf{J}$. These eigenvalues λ_i and the integers q, r_i are uniquely determined by $\mathbf{A}$, so that $\mathbf{J}$ is unique up to a permutation of the blocks $\mathbf{J}_1, \ldots, \mathbf{J}_q$. Note that $r_1 + \cdots + r_q = n$ and that it is possible for the same eigenvalue λ to occur in several of the blocks $\mathbf{J}_i$.

From the form of $\mathbf{J}$ it is not difficult to see that the dimension of the eigenspace $\mathcal{E}(\mathbf{J}, \lambda)$ of $\mathbf{J}$ for the eigenvalue λ is just the number of blocks $\mathbf{J}_i$ containing λ along the diagonal. Since $\mathbf{J}\beta = \lambda\beta$ if and only if $\mathbf{A}(\mathbf{Q}\beta) = \lambda(\mathbf{Q}\beta)$, it follows that $\mathbf{Q} : \mathcal{E}(\mathbf{J}, \lambda) \to \mathcal{E}(\mathbf{A}, \lambda)$ is an isomorphism, and we have $\dim(\mathcal{E}(\mathbf{A}, \lambda)) = \dim(\mathcal{E}(\mathbf{J}, \lambda))$ as the number of blocks $\mathbf{J}_i$ containing λ.

The algebraic multiplicity $m = \dim(\mathcal{F}(\mathbf{J}, \lambda))$ of an eigenvalue λ of $\mathbf{J}$ is the sum of the r_i for the blocks $\mathbf{J}_i$ which contain $\lambda_i = \lambda$. Since $p_{\mathbf{A}}(\lambda) = p_{\mathbf{J}}(\lambda)$ this is also the multiplicity $m = \dim(\mathcal{F}(\mathbf{A}, \lambda))$ of the eigenvalue λ of $\mathbf{A}$. Moreover, since

$$(\mathbf{A} - \lambda \mathbf{I}_n)^m \mathbf{Q}\beta = \mathbf{Q}(\mathbf{J} - \lambda \mathbf{I}_n)^m \beta, \quad \beta \in \mathcal{C}^n,$$

it follows that $\mathbf{Q} : \mathcal{F}(\mathbf{J}, \lambda) \to \mathcal{F}(\mathbf{A}, \lambda)$ is an isomorphism.

If $\mathbf{A} = \mathbf{QJQ}^{-1}$, then the definition of $e^{\mathbf{A}t}$ gives

$$e^{\mathbf{A}t} = \sum_{p=0}^{\infty} \frac{t^p}{p!}(\mathbf{QJQ}^{-1})^p = \sum_{p=0}^{\infty} \frac{t^p}{p!}\mathbf{QJ}^p\mathbf{Q}^{-1} = \mathbf{Q}\Big(\sum_{p=0}^{\infty} \frac{t^p}{p!}\mathbf{J}^p\Big)\mathbf{Q}^{-1} = \mathbf{Q}e^{\mathbf{J}t}\mathbf{Q}^{-1},$$

since

$$(\mathbf{QJQ}^{-1})^p = (\mathbf{QJQ}^{-1})(\mathbf{QJQ}^{-1}) \cdots (\mathbf{QJQ}^{-1}) = \mathbf{QJ}^p\mathbf{Q}^{-1}.$$

Thus the computation of $e^{\mathbf{A}t}$ may be reduced to the computation of $e^{\mathbf{J}t}$, and since

$$\mathbf{J}^p = \mathrm{diag}(\mathbf{J}_1^p, \ldots, \mathbf{J}_q^p), \quad p = 0, 1, \ldots,$$

it follows that

$$e^{\mathbf{J}t} = \sum_{p=0}^{\infty} \frac{t^p}{p!}\mathbf{J}^p = \mathrm{diag}(e^{\mathbf{J}_1 t}, \ldots, e^{\mathbf{J}_q t}).$$

The $r_i \times r_i$ block $\mathbf{J}_i$ has the form

$$\mathbf{J}_i = \lambda_i \mathbf{I}_{r_i} + \mathbf{N}_i, \quad \mathbf{N}_i = \begin{pmatrix} 0 & 1 & 0 & \ldots & 0 \\ 0 & 0 & 1 & \ldots & 0 \\ \vdots & \vdots & \vdots & \ldots & \vdots \\ 0 & 0 & 0 & \ldots & 1 \\ 0 & 0 & 0 & \ldots & 0 \end{pmatrix},$$

and we see that

$$\mathbf{N}_i^2 = \begin{pmatrix} 0 & 0 & 1 & \ldots & 0 \\ 0 & 0 & 0 & \ldots & 0 \\ \vdots & \vdots & \vdots & \ldots & \vdots \\ 0 & 0 & 0 & \ldots & 1 \\ 0 & 0 & 0 & \ldots & 0 \\ 0 & 0 & 0 & \ldots & 0 \end{pmatrix}, \quad \ldots, \quad \mathbf{N}_i^{r_i-1} = \begin{pmatrix} 0 & 0 & \ldots & 0 & 1 \\ 0 & 0 & \ldots & 0 & 0 \\ \vdots & \vdots & \ldots & \vdots & \vdots \\ 0 & 0 & \ldots & 0 & 0 \\ 0 & 0 & \ldots & 0 & 0 \end{pmatrix},$$

$$\mathbf{N}_i^p = 0, \quad p \geq r_i.$$

Such a matrix is said to be *nilpotent* with *index* r_i. Since $t\lambda_i \mathbf{I}_{r_i}$ commutes with $t\mathbf{N}_i$ we have

$$e^{\mathbf{J}_i t} = e^{\lambda_i t \mathbf{I}_{r_i}} e^{\mathbf{N}_i t} = e^{\lambda_i t} e^{\mathbf{N}_i t}.$$

But

$$e^{\mathbf{N}_i t} = \mathbf{I}_{r_i} + t\mathbf{N}_i + \frac{t^2}{2!}\mathbf{N}_i^2 + \cdots + \frac{t^{r_i-1}}{(r_i-1)!}\mathbf{N}_i^{r_i-1}$$

$$= \begin{pmatrix} 1 & t & t^2/2! & \ldots & t^{r_i-1}/(r_i-1)! \\ \vdots & \vdots & \vdots & \ldots & \vdots \\ 0 & \ldots & 1 & t & t^2/2! \\ 0 & \ldots & 0 & 1 & t \\ 0 & \ldots & 0 & 0 & 1 \end{pmatrix}.$$

We summarize as follows.

Theorem 3.10: *Let* $\mathbf{A} \in M_n(\mathcal{C})$ *be such that* $\mathbf{A} = \mathbf{Q}\mathbf{J}\mathbf{Q}^{-1}$, *where* $\mathbf{J} = \mathrm{diag}(\mathbf{J}_1, \ldots, \mathbf{J}_q)$ *is a Jordan canonical form of* $\mathbf{A}$, *with* $\mathbf{J}_i$ *being* $r_i \times r_i$, $\mathbf{J}_i = (\lambda_i)$ *if* $r_i = 1$,

$$\mathbf{J}_i = \begin{pmatrix} \lambda_i & 1 & 0 & \ldots & 0 & 0 \\ 0 & \lambda_i & 1 & \ldots & 0 & 0 \\ \vdots & \vdots & \vdots & \ldots & \vdots & \vdots \\ 0 & 0 & 0 & \ldots & \lambda_i & 1 \\ 0 & 0 & 0 & \ldots & 0 & \lambda_i \end{pmatrix}$$

if $r_i > 1$. *Then*

$$e^{\mathbf{A}t} = \mathbf{Q}e^{\mathbf{J}t}\mathbf{Q}^{-1}, \quad e^{\mathbf{J}t} = \mathrm{diag}(e^{\mathbf{J}_1 t}, \ldots, e^{\mathbf{J}_q t}),$$

and

$$e^{\mathbf{J}_i t} = e^{\lambda_i t} \begin{pmatrix} 1 & t & t^2/2! & \cdots & t^{r_i-1}/(r_i-1)! \\ \vdots & \vdots & \vdots & \cdots & \vdots \\ 0 & \cdots & 1 & t & t^2/2! \\ 0 & \cdots & 0 & 1 & t \\ 0 & \cdots & 0 & 0 & 1 \end{pmatrix}.$$

As an illustration, let

$$\mathbf{J} = \begin{pmatrix} 3 & 0 & 0 & 0 & 0 \\ 0 & 2 & 1 & 0 & 0 \\ 0 & 0 & 2 & 0 & 0 \\ 0 & 0 & 0 & 2 & 1 \\ 0 & 0 & 0 & 0 & 2 \end{pmatrix} = \mathrm{diag}(\mathbf{J}_1, \mathbf{J}_2, \mathbf{J}_3),$$

where

$$\mathbf{J}_1 = (3), \quad \mathbf{J}_2 = \mathbf{J}_3 = \begin{pmatrix} 2 & 1 \\ 0 & 2 \end{pmatrix}.$$

In this case $e^{\mathbf{J}t} = \mathrm{diag}(e^{3t}, e^{\mathbf{J}_2 t}, e^{\mathbf{J}_2 t})$, and, since

$$e^{\mathbf{J}_2 t} = e^{2t} \begin{pmatrix} 1 & t \\ 0 & 1 \end{pmatrix},$$

we have

$$\mathbf{J} = \begin{pmatrix} e^{3t} & 0 & 0 & 0 & 0 \\ 0 & e^{2t} & te^{2t} & 0 & 0 \\ 0 & 0 & e^{2t} & 0 & 0 \\ 0 & 0 & 0 & e^{2t} & te^{2t} \\ 0 & 0 & 0 & 0 & e^{2t} \end{pmatrix}.$$

Suppose the columns of $\mathbf{Q}$ are $\mathbf{Q}_1, \ldots, \mathbf{Q}_n$, so that $\mathbf{Q} = (\mathbf{Q}_1, \ldots, \mathbf{Q}_n)$. If the first block of $\mathbf{J}$ is

$$\mathbf{J}_1 = \begin{pmatrix} \lambda_1 & 1 & 0 & \cdots & 0 & 0 \\ 0 & \lambda_1 & 1 & \cdots & 0 & 0 \\ \vdots & \vdots & \vdots & \cdots & \vdots & \vdots \\ 0 & 0 & 0 & \cdots & \lambda_1 & 1 \\ 0 & 0 & 0 & \cdots & 0 & \lambda_1 \end{pmatrix}, \quad r_1 \times r_1, \quad r_1 > 1,$$

then $\mathbf{AQ} = \mathbf{QJ}$ implies that

$$\mathbf{AQ}_1 = \lambda_1 \mathbf{Q}_1, \quad \mathbf{AQ}_2 = \mathbf{Q}_1 + \lambda_1 \mathbf{Q}_2, \ldots, \quad \mathbf{AQ}_{r_1} = \mathbf{Q}_{r_1-1} + \lambda_1 \mathbf{Q}_{r_1},$$

and hence

$$(\mathbf{A} - \lambda_1 \mathbf{I}_n)\mathbf{Q}_1 = 0, \quad (\mathbf{A} - \lambda_1 \mathbf{I}_n)^2 \mathbf{Q}_2 = 0, \ldots, \quad (\mathbf{A} - \lambda_1 \mathbf{I}_n)^{r_1} \mathbf{Q}_{r_1} = 0.$$

We have similar relations valid for the remaining columns of $\mathbf{Q}$.

The substitution $X = \mathbf{Q}Y$ in $X' = \mathbf{A}X$ transforms this equation into $Y' = \mathbf{Q}^{-1}\mathbf{A}\mathbf{Q}Y = \mathbf{J}Y$. A basis for this equation is $\mathbf{Y}(t) = e^{\mathbf{J}t}$, and then a basis for $X' = \mathbf{A}X$ is $\mathbf{X}(t) = \mathbf{Q}e^{\mathbf{J}t}$. Another basis is $\mathbf{X_A}(t) = e^{\mathbf{A}t} = \mathbf{Q}e^{\mathbf{J}t}\mathbf{Q}^{-1}$. The explicit structure of $e^{\mathbf{J}t}$ shows that the columns $X_1, \ldots, X_n$ of $\mathbf{X}$ are of the form $X_j(t) = e^{\lambda t}P(t)$, where λ is an eigenvalue of $\mathbf{A}$ and P is a vector polynomial. If $\mathbf{J}_1$ is as above, then the first r_1 columns of $\mathbf{X}$ are

$$X_1(t) = e^{\lambda_1 t}\mathbf{Q}_1, \quad X_2(t) = e^{\lambda_1 t}(t\mathbf{Q}_1 + \mathbf{Q}_2),$$

$$X_3(t) = e^{\lambda_1 t}\left(\frac{t^2}{2!}\mathbf{Q}_1 + t\mathbf{Q}_2 + \mathbf{Q}_3\right), \ldots, \quad X_{r_1}(t) = e^{\lambda_1 t}\left(\frac{t^{r_1-1}}{(r_1-1)!}\mathbf{Q}_1 + \cdots + \mathbf{Q}_{r_1}\right).$$

3.6   The behavior of solutions for large t

To discuss the behavior of the vector-valued function $X(t)$, it is convenient to introduce a norm. Recall that a *normed vector space* is a vector space $\mathcal{V}$ over a field $\mathcal{F}$ (which is either $\mathcal{C}$ or $\mathcal{R}$) with a *norm* $\|\ \|$ which is a function $X \in \mathcal{V} \to \|X\| \in \mathcal{R}$ such that, for all $X, Y \in \mathcal{V}$,

(a) $$\|X\| \geq 0,$$

(b) $$\|X\| = 0, \quad \text{if and only if } X = 0,$$

(c) $$\|X + Y\| \leq \|X\| + \|Y\|,$$

(d) $$\|\alpha X\| = |\alpha|\,\|X\|, \quad \alpha \in \mathcal{F}.$$

A set of vectors S in a normed vector space is said to be *bounded* if there is a positive number K such that

$$\|X\| \leq K, \quad X \in S.$$

We say that a function $X(t)$, defined on a half line (a, ∞), has *limit* Y as $t \to \infty$ if for every $\epsilon > 0$ there is a $T > 0$ such that $t > T$ implies $\|X(t) - Y\| < \epsilon$.

A variety of norms are available on $\mathcal{F}^N$. We will adopt the specific norm

$$|X| = |x_1| + \cdots + |x_n|, \quad X = (x_1, \ldots, x_n) \in \mathcal{F}^N$$

as the standard one for $\mathcal{F}^n$. This norm may also be used for $m \times n$ matrices:

$$|\mathbf{A}| = \sum_{j=1}^{n}\sum_{i=1}^{m} |a_{ij}|, \quad \mathbf{A} = (a_{ij}) \in M_{mn}(\mathcal{F}).$$

The representation $\mathbf{X}(t) = \mathbf{Q}e^{\mathbf{J}t}$ of a basis for $X' = \mathbf{A}X$ not only gives an explicit representation for solutions but also shows how solutions behave for large $|t|$. This is demonstrated by the following result.

Theorem 3.11: *Let $\lambda_1, \ldots, \lambda_k$ be the distinct eigenvalues of $\mathbf{A}$, with algebraic multiplicities $m_1, \ldots, m_k$, respectively. Then*

(i) *all solutions of $X' = \mathbf{A}X$ are bounded on $[0, \infty)$ if and only if*

$$\mathrm{Re}(\lambda_j) \le 0, \quad j = 1, \ldots, k,$$

and for those λ_j such that $\mathrm{Re}(\lambda_j) = 0$ we have $m_j = \dim(\mathcal{E}(\mathbf{A}, \lambda_j))$;

(ii) *all solutions of $X' = \mathbf{A}X$ have limit zero as $t \to \infty$ if and only if*

$$\mathrm{Re}(\lambda_j) < 0, \quad j = 1, \ldots, k.$$

Proof: The matrix $\mathbf{A}$ has a Jordan canonical form $\mathbf{J} = \mathrm{diag}(\mathbf{J}_1, \ldots, \mathbf{J}_q)$. If the conditions in (i) are met, then any block $\mathbf{J}_i$ with eigenvalue λ_i and such that $\mathrm{Re}(\lambda_i) = 0$ will be a 1×1 block. The columns $X_1(t), \ldots, X_n(t)$ of the basis $\mathbf{X}(t) = \mathbf{Q}e^{\mathbf{J}t}$ then have the form

(a) $X_l(t) = e^{\lambda_i t} P(t)$ if $\mathrm{Re}(\lambda_i) < 0$, where P is a nontrivial polynomial, or have the form

(b) $X_l(t) = e^{\lambda_i t} \alpha$ if $\mathrm{Re}(\lambda_i) = 0$, $\alpha \in \mathcal{C}^n$, $\alpha \ne 0$.

Thus in case (a)

$$|X_l(t)| = e^{\mathrm{Re}(\lambda_i)t}|P(t)| \to 0, \quad t \to \infty,$$

and in case (b)

$$|X_l(t)| = e^{\mathrm{Re}(\lambda_i)t}|\alpha| = |\alpha|, \quad t \in \mathcal{R}.$$

In both cases the solutions X_l are bounded on $[0, \infty)$. There is thus a constant $M > 0$ such that

$$|\mathbf{X}(t)| = |X_1(t)| + \cdots + |X_n(t)| \le M, \quad 0 \le t < \infty.$$

This implies that any solution $X(t)$ of $X' = \mathbf{A}X$ is bounded, for $X(t) = \mathbf{X}(t)C$, and $|\mathbf{X}(t)| \le M|C|$ on $[0, \infty)$. Conversely, if there is a λ_j such that $\mathrm{Re}(\lambda_j) > 0$, then the solution X given by

$$X(t) = e^{\lambda_j t}\alpha,$$

where $\alpha \in E(\mathbf{A}, \lambda_j)$, $\alpha \ne 0$, is such that

$$|X(t)| = e^{(\mathrm{Re}(\lambda_j)t)}|\alpha| \to \infty, \quad t \to +\infty.$$

Also if there is a λ_j for which $\mathrm{Re}(\lambda_j) = 0$ and $\dim(\mathcal{E}(A, \lambda_j)) < m_j$, then there is a Jordan block $\mathbf{J}_j$ involving λ_j which is $r_j \times r_j$, $r_j > 1$. There is then a solution of the form

$$X(t) = e^{\lambda_j t}(\alpha + t\beta),$$

where $\alpha, \beta \in \mathcal{C}^n$, $\beta \neq 0$. For $t > 0$,

$$|X(t)| = |\alpha + t\beta| = t|\beta + \alpha/t| \geq t(|\beta| - |\alpha|/t) \to \infty, \quad t \to \infty.$$

Thus (i) is proved.

As to (ii), if $\mathrm{Re}(\lambda_j) < 0$ for $j = 1, \ldots, k$, then the argument used in case (a) of (i) shows that all solutions X_l are such that $|X_l(t)| \to 0$, $t \to \infty$, and hence $|\mathbf{X}(t)| \to 0$ as $t \to \infty$, which implies that every solution $X = \mathbf{X}C$ satisfies $|X(t)| \to 0$ as $t \to \infty$. If there is a λ_j such that $\mathrm{Re}(\lambda_j) \geq 0$, there is a corresponding solution X_j such that

$$X_j(t) = e^{\lambda_j t} \alpha,$$

where $\alpha \in \mathcal{E}(\mathbf{A}, \lambda_j)$, $\alpha \neq 0$. Then if $\mathrm{Re}(\lambda_j) > 0$,

$$|X_j(t)| = e^{\mathrm{Re}(\lambda_j)t}|\alpha| \to \infty, \quad t \to +\infty,$$

and $|X_j(t)| = |\alpha| \neq 0$, if $\mathrm{Re}(\lambda_j) = 0$. In either case $|X_j(t)|$ does not have limit zero as $t \to \infty$. $\square$

3.7   Higher-order equations

An nth-order linear homogeneous equation with constant coefficients

$$x^{(n)} + a_{n-1}x^{(n-1)} + \cdots + a_0 x = 0, \quad a_j \in \mathcal{C}, \qquad (H_n)$$

has its associated first-order system

$$Y' = \mathbf{A}Y,$$

where

$$\mathbf{A} = \begin{pmatrix} 0 & 1 & 0 & 0 & \cdots & 0 \\ 0 & 0 & 1 & 0 & \cdots & 0 \\ \vdots & \vdots & \vdots & \vdots & \cdots & \vdots \\ 0 & 0 & 0 & 0 & \cdots & 1 \\ -a_0 & -a_1 & -a_2 & -a_3 & \cdots & -a_{n-1} \end{pmatrix}, \qquad (3.19)$$

which is a system with a constant coefficient matrix $\mathbf{A}$.

Theorem 3.12: *The characteristic polynomial $p_{\mathbf{A}}$ of the matrix $\mathbf{A}$ given by (3.19) is*

$$p_{\mathbf{A}}(\lambda) = \lambda^n + a_{n-1}\lambda^{n-1} + \cdots + a_0. \qquad (3.20)$$

Proof: The characteristic polynomial is

$$p_{\mathbf{A}}(\lambda) = \det(\lambda \mathbf{I}_n - \mathbf{A}) = \det \begin{pmatrix} \lambda & -1 & 0 & 0 & \cdots & 0 \\ 0 & \lambda & -1 & 0 & \cdots & 0 \\ \vdots & \vdots & \vdots & \vdots & \cdots & \vdots \\ 0 & 0 & 0 & 0 & \cdots & -1 \\ a_0 & a_1 & a_2 & a_3 & \cdots & \lambda + a_{n-1} \end{pmatrix},$$

with elements not in the last row, the main diagonal, or the first superdiagonal being zero. Suppose $\lambda \neq 0$. The determinant remains unchanged if we multiply any row by a constant and add this to another row. Multiply the first row by $-\lambda^{-1} a_0$ and add to the last row, obtaining

$$(0 \quad a_1 + \lambda^{-1} a_0 \quad a_2 \quad \ldots \quad \lambda + a_{n-1})$$

for the new last row. Then multiply the second row by $-\lambda^{-1}(a_1 + \lambda^{-1} a_0) = -\lambda^{-1} a_1 - \lambda^{-2} a_0$ and add to the last row. This gives

$$(0 \quad 0 \quad a_2 + \lambda^{-1} a_1 + \lambda^{-2} a_0 \quad a_3 \quad \ldots \quad \lambda + a_{n-1})$$

as the new last row. Continuing in this way, after $n-1$ steps, we obtain a new last row of

$$(0 \quad 0 \quad 0 \quad \ldots \quad 0 \quad q(\lambda)) ,$$

where

$$q(\lambda) = \lambda + a_{n-1} + \lambda^{-1} a_{n-2} + \cdots + \lambda^{-(n-1)} a_0 .$$

Thus

$$p_{\mathbf{A}}(\lambda) = \det \begin{pmatrix} \lambda & -1 & 0 & 0 & \ldots & 0 \\ 0 & \lambda & -1 & 0 & \ldots & 0 \\ \vdots & \vdots & \vdots & \vdots & \ldots & \vdots \\ 0 & 0 & 0 & 0 & \ldots & -1 \\ 0 & 0 & 0 & 0 & \ldots & q(\lambda) \end{pmatrix}$$

$$= \lambda^{n-1} q(\lambda) = \lambda^n + a_{n-1} \lambda^{n-1} + \cdots + a_0 .$$

This gives the result if $\lambda \neq 0$, and since the determinant of $\lambda \mathbf{I}_n - \mathbf{A}$ is a continuous function of its elements, and these elements are continuous in λ, $p_{\mathbf{A}}$ is continuous in λ. Thus $p_{\mathbf{A}}(\lambda) \to p_{\mathbf{A}}(0) = a_0$ as $\lambda \to 0$, and the result is valid for all $\lambda \in C$. $\square$

Theorem 3.12 shows that we can read off the characteristic polynomial $p_{\mathbf{A}}$ from (H_n) by replacing $x^{(k)}$ everywhere by λ^k. A Jordan canonical form $\mathbf{J} = \mathrm{diag}(\mathbf{J}_1, \ldots, \mathbf{J}_q)$ for $\mathbf{A}$ has a particularly simple form; for any eigenvalue λ of $\mathbf{A}$ there is only one block $\mathbf{J}_i$ which contains λ (see exercise 44). Instead of analyzing $\mathbf{A}$ further we study (H_n) directly.

For $x \in C^n(\mathcal{R}, \mathcal{C})$, define the operator L by

$$L(x) = x^{(n)} + a_{n-1} x^{(n-1)} + \cdots + a_0 x, \quad a_j \in \mathcal{C}.$$

This constant coefficient differential operator L is linear:

$$L(\alpha x + \beta y) = \alpha L(x) + \beta L(y)$$

for all $\alpha, \beta \in \mathcal{C}$ and all $x, y \in C^n(\mathcal{R}, \mathcal{C})$. Corresponding to L we let

$$p(\lambda) = \lambda^n + a_{n-1} \lambda^{n-1} + \cdots + a_0,$$

so that $p = p_\mathbf{A}$ is the characteristic polynomial of $\mathbf{A}$. The polynomial p is also called the *characteristic polynomial* of L, or of the equation $Lx = 0$.

To find a concrete basis for S_n, the set of all $x \in C^n(\mathcal{R}, \mathcal{C})$ satisfying $Lx = 0$, begin with the equation

$$L(e^{\lambda t}) = p(\lambda)e^{\lambda t}, \tag{3.21}$$

which is valid for all $\lambda \in \mathcal{C}$, $t \in \mathcal{R}$. If $p(\lambda_1) = 0$, then $x_1(t) = e^{\lambda_1 t}$ gives a solution of (H_n). If λ_1 is a root of p of multiplicity m_1, then

$$p(\lambda_1) = 0, \quad p'(\lambda_1) = 0, \dots, \quad p^{(m_1-1)}(\lambda_1) = 0, \quad p^{(m_1)}(\lambda_1) \neq 0.$$

This is easily seen by observing that

$$p(\lambda) = (\lambda - \lambda_1)^{m_1} q(\lambda),$$

where q is a polynomial of degree $n - m_1$ with the coefficient of λ^{n-m_1} being one, and $q(\lambda_1) \neq 0$. Using the formula

$$(fg)^{(l)} = \sum_{j=0}^{l} \binom{l}{j} f^{(l-j)} g^{(j)}, \quad \binom{l}{j} = \frac{l!}{j!(l-j)!}$$

for the lth derivative of fg, differentiation of (3.21) l times with respect to λ produces

$$\frac{\partial^l}{\partial \lambda^l} L(e^{\lambda t}) = L\left(\frac{\partial^l}{\partial \lambda^l} e^{\lambda t}\right) = \frac{\partial^l}{\partial \lambda^l} p(\lambda)(e^{\lambda t})$$

$$= \left[p^{(l)}(\lambda) + l p^{(l-1)}(\lambda)t + \frac{l(l-1)}{2!} p^{(l-2)}(\lambda)t^2 + \cdots + p(\lambda)t^l \right] e^{\lambda t}.$$

Thus

$$L(t^l e^{\lambda t}) = \left[p^{(l)}(\lambda) + l p^{(l-1)}(\lambda)t + \frac{l(l-1)}{2!} p^{(l-2)}(\lambda)t^2 + \cdots + p(\lambda)t^l \right] e^{\lambda t}, \tag{3.22}$$

and we see, by putting $l = 0, 1, \dots, m_1 - 1$, that $t^l e^{\lambda_1 t}$ is a solution of (H_n). Repeating this for each root of p gives a set of n solutions of (H_n).

Theorem 3.13: *Let $\lambda_1, \dots, \lambda_k$ be the distinct roots of p, where*

$$p(\lambda) = \lambda^n + a_{n-1}\lambda^{n-1} + \cdots + a_0,$$

and suppose λ_j has multiplicity m_j. Then the n functions x_{ij} given by

$$x_{ij}(t) = t^{i-1} e^{\lambda_j t}, \quad i = 1, \dots, m_j, \quad j = 1, \dots, k,$$

form a basis for the set of solutions S_n of (H_n) on $\mathcal{R}$.

Before completing the proof of Theorem 3.13, let us look at several examples. Consider the equation

$$x^{(3)} - 3x' + 2x = 0. \tag{3.23}$$

It has the characteristic polynomial

$$p(\lambda) = \lambda^3 - 3\lambda + 2 = (\lambda - 1)^2(\lambda + 2).$$

Thus a basis for the solutions of (3.23) is given by

$$\mathbf{X}(t) = (e^t, te^t, e^{-2t}),$$

and every solution x has the form

$$x(t) = c_1 e^t + c_2 t e^t + c_3 e^{-2t},$$

where $c_j \in \mathcal{C}$.

As a second example, consider

$$x^{(4)} - 16x = 0, \tag{3.24}$$

with the characteristic polynomial

$$p(\lambda) = \lambda^4 - 16.$$

The roots of p are $2, -2, 2i, -2i$, and hence a basis for (3.24) is given by

$$\mathbf{X}(t) = (e^{2t}, e^{-2t}, e^{2it}, e^{-2it}).$$

Every solution has the form

$$x(t) = c_1 e^{2t} + c_2 e^{-2t} + c_3 e^{2it} + c_4 e^{-2it}, \quad c_j \in \mathcal{C}.$$

A real basis is given by

$$\mathbf{Y}(t) = (e^{2t}, e^{-2t}, \cos(2t), \sin(2t)),$$

and every real solution y has the form

$$y(t) = c_1 e^{2t} + c_2 e^{-2t} + c_3 \cos(2t) + c_4 \sin(2t), \quad c_j \in \mathcal{R}.$$

Proof of Theorem 3.13: Since the $x_{ij} \in S_n$, all that remains is to show the linear independence of these solutions. Suppose that

$$\sum_{j=1}^{k} \sum_{i=1}^{m_j} c_{ij} x_{ij} = 0, \quad c_{ij} \in \mathcal{C}.$$

Then

$$e^{\lambda_1 t} p_1(t) + \cdots + e^{\lambda_k t} p_k(t) = 0, \quad t \in R,$$

where p_j is a polynomial of degree at most $m_j - 1$,

$$p_j(t) = \sum_{i=1}^{m_j} c_{ij} t^{i-1}.$$

Suppose that some polynomial $p_j(t)$ is not identically zero. In particular if $deg(p_J) = d$ then $c_{d,J} \neq 0$. Employing the notation

$$Df = f',$$

we observe that

$$(D - \lambda_k)e^{\lambda_j t}p_j(t) = (\lambda_j - \lambda_k)e^{\lambda_j t}p_j(t) + e^{\lambda_j t}p_j'(t),$$

with the degree of p_j' lower than that of p_j as long as $deg(p_j) > 0$, and $p_j' = 0$ if $deg(p_j) = 0$. By choosing any $j \neq J$ and repeatedly performing the operations $D - \lambda_j$, each summand with a factor $e^{\lambda_j t}$ can be removed. After annihilating all terms with $j \neq J$ the remaining expression has the form

$$\left[\prod_{j \neq J} (\lambda_J - \lambda_j)^{d_j} \right] e^{\lambda_J t}p_J(t) + e^{\lambda_J t}Q(t) = 0, \quad t \in \mathcal{R},$$

where $deg(Q) < deg(p_J)$, or $Q = 0$ if $deg(p_J) = 0$. This implies that $c_{d,J} = 0$, a contradiction. □

3.8   Exercises

1. Compute $e^{\mathbf{A}t}$, where $\mathbf{A}$ is given by

(a)
$$\mathbf{A} = \begin{pmatrix} -2 & 0 \\ 0 & 1 \end{pmatrix},$$

(b)
$$\mathbf{A} = \begin{pmatrix} 0 & -2 \\ 2 & 0 \end{pmatrix},$$

(c)
$$\mathbf{A} = \begin{pmatrix} a & 0 \\ 0 & b \end{pmatrix}, \quad a, b \in \mathcal{C},$$

(d)
$$\mathbf{A} = \begin{pmatrix} 0 & a \\ -a & 0 \end{pmatrix}, \quad a \in \mathcal{R}.$$

2. Compute $e^{\mathbf{A}t}$, when

(a)
$$\mathbf{A} = \begin{pmatrix} -2 & 0 \\ 1 & -2 \end{pmatrix},$$

(Hint: Find the basis $\mathbf{X_A}$ of $X' = \mathbf{A}X$ satisfying $\mathbf{X_A}(0) = \mathbf{I}_2$.)

(b)
$$\mathbf{A} = \begin{pmatrix} -1 & 1 \\ -1 & -1 \end{pmatrix}.$$

3. Find the solution X of the system

$$x_1' = -2x_1,$$

$$x_2' = x_1 - 2x_2$$

satisfying

$$X(0) = \begin{pmatrix} 2 \\ -3 \end{pmatrix}.$$

(Hint: See exercise 2(a).)

4. Find that solution X of the system

$$x_1' = -x_1 + x_2,$$

$$x_2' = -x_1 - x_2$$

satisfying

$$X(1) = \begin{pmatrix} -1 \\ 1 \end{pmatrix}.$$

(Hint: See exercise 2(b).)

5. If $\mathbf{X}$ is any basis for the system (3.1) on $\mathcal{R}$, show that $\mathbf{X}$, restricted to I, is a basis for (3.1) on the interval $I \subset \mathcal{R}$. In particular, $\mathbf{X_A}(t) = e^{\mathbf{A}t}$ gives a basis for (3.1) on any interval $I \subset \mathcal{R}$.

6. Compute $e^{\mathbf{A}t}$ for the following $\mathbf{A}$:

(a) $\qquad\qquad \mathbf{A} = \begin{pmatrix} a & b \\ -b & a \end{pmatrix}, \quad a, b \in \mathcal{R},$

(Hint: $\mathbf{A} = a\mathbf{I}_2 + b\mathbf{J}$.)

(b) $\qquad\qquad \mathbf{A} = \begin{pmatrix} a & 1 \\ 0 & a \end{pmatrix}, \quad a \in \mathcal{C},$

(c) $\qquad\qquad \mathbf{A} = \begin{pmatrix} a & 1 & 0 \\ 0 & a & 1 \\ 0 & 0 & a \end{pmatrix}, \quad a \in \mathcal{C}.$

7. Let $\mathbf{A} \in M_n(\mathcal{C})$, with $\mathbf{A} = (a_{ij})$. If $\mathbf{A}^T = (a_{ji})$ is the *transpose* of $\mathbf{A}$, $\overline{\mathbf{A}} = (\overline{a}_{ij})$ is the *complex conjugate* of $\mathbf{A}$, and $\mathbf{A}^* = \overline{\mathbf{A}}^T = (\overline{a}_{ji})$ is the *adjoint* of $\mathbf{A}$, show that

(a) $\qquad\qquad \exp(\mathbf{A}^T) = (\exp(\mathbf{A}))^T,$

(b) $\qquad\qquad \exp(\overline{\mathbf{A}}) = \overline{\exp(\mathbf{A})},$

(c) $\qquad\qquad \exp(\mathbf{A}^*) = (\exp(\mathbf{A}))^*,$

(d) If $\mathbf{A}^T = -\mathbf{A}$ ($\mathbf{A}$ is *skew symmetric*), show that $(\exp(\mathbf{A}))^T = \exp(-\mathbf{A}) = (\exp(\mathbf{A}))^{-1}$.

(e) If $\mathbf{A}^* = -\mathbf{A}$ ($\mathbf{A}$ is *skew hermitian*), show that $(\exp(\mathbf{A}))^* = \exp(-\mathbf{A}) = (\exp(\mathbf{A}))^{-1}$.

(f) If $\operatorname{tr}(\mathbf{A}) = 0$ show that $\det(e^{\mathbf{A}}) = 1$.

8. Suppose $\mathbf{A} \in M_n(\mathcal{C})$ and $\mathbf{A}^* = -\mathbf{A}$. Show that every solution X of $X' = \mathbf{A}X$ satisfies

$$\|X(t)\|_2 = \text{constant}, \quad t \in \mathcal{R}.$$

(Hint: $X(t) = e^{\mathbf{A}t}\xi$, where $\xi = X(0)$ and $\|X(t)\|_2^2 = X^*(t)X(t)$.)

9. Although $\mathbf{AB} = \mathbf{BA}$ implies $e^{\mathbf{A}+\mathbf{B}} = e^{\mathbf{A}}e^{\mathbf{B}}$ for $\mathbf{A}, \mathbf{B} \in M_n(\mathcal{F})$, it is not true that $e^{\mathbf{A}+\mathbf{B}} = e^{\mathbf{A}}e^{\mathbf{B}}$ implies $\mathbf{AB} = \mathbf{BA}$. Give an example of matrices $\mathbf{A}, \mathbf{B}$ such that $e^{\mathbf{A}+\mathbf{B}} = e^{\mathbf{A}}e^{\mathbf{B}}$, but $\mathbf{AB} \neq \mathbf{BA}$.

10. Compute the solution X of the system

$$x_1' = 3x_1 + x_2 + 1,$$

$$x_2' = -x_1 + 3x_2$$

satisfying

$$\begin{pmatrix} x_1(0) \\ x_2(0) \end{pmatrix} = E_2 = \begin{pmatrix} 0 \\ 1 \end{pmatrix}.$$

(Hint: See exercise 6(a).)

11. Find that solution X of

$$x_1' = 2x_1 + x_2 - 1,$$

$$x_2' = 2x_2 + 2$$

satisfying $X(1) = E_1$. (Hint: See exercise 6(b).)

12. Let

$$\mathbf{A} = \begin{pmatrix} 3 & 2 \\ 4 & 1 \end{pmatrix}.$$

(a) Compute the characteristic polynomial $p_{\mathbf{A}}$ of $\mathbf{A}$.

(b) Compute the eigenvalues λ_1, λ_2 of $\mathbf{A}$.

(c) Compute eigenvectors α_1, α_2 for λ_1, λ_2, respectively.

13. Find a basis $\mathbf{X} = (u, v)$ for the solutions of the system:

$$x_1' = 3x_1 + 2x_2,$$

$$x_2' = 4x_1 + x_2.$$

(Hint: Use exercise 12.)

14. Let

$$\mathbf{A} = \begin{pmatrix} 3 & -2 \\ 1 & 1 \end{pmatrix}$$

(a) Compute the characteristic polynomial $p_{\mathbf{A}}$ of $\mathbf{A}$.

(b) Compute the eigenvalues λ_1, λ_2 of $\mathbf{A}$.

(c) Compute eigenvectors α_1, α_2 for λ_1, λ_2, respectively.

15. Find a basis $\mathbf{X} = (u, v)$ for the solutions of the system:

$$x_1' = 3x_1 - 2x_2,$$

$$x_2' = x_1 + x_2.$$

(Hint: Use exercise 14.)

16. (a) Find all solutions of the system

$$x_1' = -2x_1 + x_2,$$

$$x_2' = -2x_2.$$

(b) If

$$\mathbf{A} = \begin{pmatrix} -2 & 1 \\ 0 & -2 \end{pmatrix},$$

compute $e^{\mathbf{A}t}$.

17. Find a basis $\mathbf{X} = (u, v)$ for the system

$$x_1' = ix_1 + 4x_2,$$

$$x_2' = (i - 1)x_1 + (4 - i)x_2.$$

18. We have shown that two bases for the system

$$X' = \mathbf{A}X, \quad \mathbf{A} = \begin{pmatrix} 0 & 1 \\ -1 & 0 \end{pmatrix}$$

are given by

$$\mathbf{X}(t) = \begin{pmatrix} e^{it} & ie^{-it} \\ ie^{it} & e^{-it} \end{pmatrix}, \quad \mathbf{X_A}(t) = \begin{pmatrix} \cos(t) & \sin(t) \\ -\sin(t) & \cos(t) \end{pmatrix}.$$

Thus $\mathbf{X}(t) = \mathbf{X_A}(t)\mathbf{C}$ for some invertible constant matrix $\mathbf{C}$. Find $\mathbf{C}$.

19. Find a basis $\mathbf{X} = (u, v)$ for the system

$$x_1' = -2x_1 + 3x_2 + x_3,$$

$$x_2' = x_2 + x_3,$$

$$x_3' = -3x_1 + 4x_2 + x_3.$$

20. Let $\lambda_1, \ldots, \lambda_k$ be distinct eigenvalues for $\mathbf{A} \in M_n(\mathcal{C})$, with eigenvectors $\alpha_1, \ldots, \alpha_k$, respectively. Show that $\alpha_1, \ldots, \alpha_k$ is a linearly independent set as follows. Suppose that

$$c_1\alpha_1 + \cdots + c_k\alpha_k = 0.$$

Then

$$0 = (\mathbf{A} - \lambda_k)(c_1\alpha_1 + \cdots + c_k\alpha_k) = c_1(\lambda_1 - \lambda_k)\alpha_1 + \cdots + (\lambda_k - \lambda_k)c_k\alpha_k$$

$$= c_1(\lambda_1 - \lambda_k)\alpha_1 + \cdots + (\lambda_{k-1} - \lambda_k)c_{k-1}\alpha_{k-1}.$$

Next apply $\mathbf{A} - \lambda_{k-1}$, etc.

21. Find a basis for the solutions of the system $X' = \mathbf{A}X$ if $\mathbf{A}$ is given by

(a) $\mathbf{A} = \begin{pmatrix} -1 & 1 \\ -5 & 3 \end{pmatrix}$, (b) $\mathbf{A} = \begin{pmatrix} 4 & 1 \\ -1 & 2 \end{pmatrix}$, (c) $\mathbf{A} = \begin{pmatrix} 3i & 2 \\ 2 & -i \end{pmatrix}$.

22. If

$$\mathbf{A} = \begin{pmatrix} 4 & 1 \\ -1 & 2 \end{pmatrix},$$

compute $e^{\mathbf{A}t}$. (Hint: See exercise 21(b).)

23. Find a basis for the solutions of $X' = \mathbf{A}X$ if $\mathbf{A}$ is given by

(a) $\mathbf{A} = \begin{pmatrix} 3 & 0 & 0 \\ 1 & 3 & 0 \\ 0 & 0 & 3 \end{pmatrix}$, (b) $\mathbf{A} = \begin{pmatrix} 3 & -1 & 1 \\ -5 & 1 & 7 \\ 1 & -1 & 3 \end{pmatrix}$,

(c) $\mathbf{A} = \begin{pmatrix} 2 & 1 & 2 \\ -1 & 0 & -2 \\ 0 & 0 & 1 \end{pmatrix}$.

24. Consider the general homogeneous system of two equations:

$$x_1' = ax_1 + bx_2, \tag{3.25}$$

$$x_2' = cx_1 + dx_2,$$

where $a, b, c, d \in \mathcal{C}$. The *discriminant* of

$$\mathbf{A} = \begin{pmatrix} a & b \\ c & d \end{pmatrix}$$

is $\delta = (a - d)^2 + 4bc$.

(a) Show that if $\delta \neq 0$ then $\mathbf{A}$ has distinct eigenvalues λ_1, λ_2. Compute these and find a basis X for (3.25).

(b) Suppose $\delta = 0$. Show that $\mathbf{A}$ has one eigenvalue λ with multiplicity 2. If $\delta = 0$ and $b = c = 0$, show that there are two linearly independent eigenvectors for λ. Find a basis for (3.25) in this case. If $\delta = 0$ and $b \neq 0$ or $c \neq 0$, show that there is only one linearly independent eigenvector for λ. Find a basis for (3.25) in this case.

25. For $\mathbf{A} \in M_n(\mathcal{C})$ consider the nonhomogeneous system

$$X' = \mathbf{A}X + B(t). \tag{NH}$$

(a) Let $B(t) = e^{\mu t}\beta$, where $\mu \in \mathcal{C}$ is not an eigenvalue of $\mathbf{A}$, and $\beta \in \mathcal{C}^n$. Show that (NH) has a solution Y of the form

$$Y(t) = e^{\mu t}\gamma, \quad \gamma \in \mathcal{C}^n.$$

(Hint: Show that $\gamma = -(\mathbf{A} - \mu \mathbf{I}_n)^{-1}\beta$ works.)

(b) Let $B(t) = \beta_0 + \beta_1 t + \cdots + \beta_r t^r$, $\beta_r \neq 0$, where the $\beta_j \in \mathcal{C}^n$, and suppose 0 is not an eigenvalue of $\mathbf{A}$. Show that (NH) has a solution Y of the form

$$Y(t) = \gamma_0 + \gamma_1 t + \cdots + \gamma_r t^r, \quad \gamma_r \neq 0,$$

as follows. Note that $B^{(r+1)}(t) = 0$ for all $t \in \mathcal{R}$. Show that

$$Y(t) = -[\mathbf{A}^{-1}b(t) + \mathbf{A}^{-2}b'(t) + \cdots + \mathbf{A}^{-r}b^{(r-1)}(t) + \mathbf{A}^{-r-1}b^{(r)}(t)]$$

works, and $\gamma_r = -\mathbf{A}^{-1}\beta_r$.

(c) Let $B(t) = e^{\mu t}P(t)$, where

$$P(t) = \beta_0 + \beta_1 t + \cdots + \beta_r t^r, \quad \beta_r \neq 0,$$

and μ is not an eigenvalue of $\mathbf{A}$. Show that (NH) has a solution Y of the form $Y(t) = e^{\mu t}Q(t)$, where

$$Q(t) = \gamma_0 + \gamma_1 t + \cdots + \gamma_r t^r, \quad \gamma_r \neq 0.$$

(Hint: Let $X(t) = e^{\mu t}Z(t)$ in (NH), and reduce to the case $\mu = 0$ considered in (b).)

26. Use the results of exercise 25 to find particular solutions Y for each of the following systems:

(a)
$$x_1' = 2x_1 - 3x_2 + 1,$$
$$x_2' = x_1 - 2x_2 - 1;$$

(b)
$$x_1' = ix_1 + x_2 + e^{it},$$
$$x_2' = -x_1 + ix_2 - 2e^{it};$$

(c)
$$x_1' = x_1 + 2x_2 + 1 + t,$$
$$x_2' = 3x_1 + 4x_2 - 2 - 3t;$$

(d)
$$x_1' = -2x_1 + t,$$
$$x_2' = x_1 - 2x_2 - t^2;$$

(e)
$$x_1' = x_1 + 2x_2 + te^{-t},$$

$$x_2' = 3x_1 + 4x_2 - e^{-t}.$$

27. Find a real basis for the solutions of $X' = \mathbf{A}X$ if $\mathbf{A}$ is given by

$$\text{(a)} \quad \mathbf{A} = \begin{pmatrix} 3 & -1 \\ 1 & 3 \end{pmatrix}, \quad \text{(b)} \quad \mathbf{A} = \begin{pmatrix} -1 & 1 \\ -5 & 3 \end{pmatrix}.$$

28. Find a real basis for the solutions of $X' = \mathbf{A}X$, where

$$\mathbf{A} = \begin{pmatrix} 0 & 0 & 2 \\ -1 & 0 & 1 \\ 0 & -1 & 2 \end{pmatrix}.$$

29. Find all real solutions of the following systems:

(a)
$$x_1' = 2ix_1 + x_2 - 2ix_3,$$

$$x_2' = -x_3,$$

$$x_3' = x_2.$$

(b)
$$x_1' = 3ix_1 + 2x_2,$$

$$x_2' = 2x_1 - ix_2.$$

30. Let p be a polynomial with real coefficients

$$p(\lambda) = \lambda^n + a_{n-1}\lambda^{n-1} + \cdots + a_0, \quad a_j \in \mathcal{R}.$$

Show that if λ is a complex root of p which is not real and λ has multiplicity m, then $\bar{\lambda}$ is a root of p with multiplicity m.

31. Use the fundamental theorem of algebra to show that if $p(z)$ is a polynomial with degree $n \geq 1$, then

$$p(z) = a_n(z - r_1)^{m_1} \cdots (z - r_k)^{m_k},$$

where $m_1 + \cdots + m_k = n$. If the roots r_j are distinct in the above factorization, show that each polynomial $p(z)$ has a unique set of roots r_j and multiplicities m_j.

32. Let $\mathbf{A} \in M_n(\mathcal{R})$ and let λ be a nonreal complex eigenvalue of $\mathbf{A}$ with multiplicity m. There are m linearly independent solutions $X_1, \ldots, X_m$ of $X' = \mathbf{A}X$ of the form

$$X_i(t) = e^{\lambda t} P_i(t), \quad P_i(0) = \alpha_i \in \mathcal{F}(A, \lambda),$$

where P_i is a polynomial, $\deg(P_i) \leq m - 1$. Show that if $U_i = \mathrm{Re}(X_i)$, $V_i = \mathrm{Im}(X_i)$, then

$$U_1, \ldots, U_m, V_1, \ldots, V_m$$

is a linearly independent set.

33. Find a Jordan canonical form $\mathbf{J}$ for $\mathbf{A}$ if $\mathbf{A}$ is given by

(a) $\mathbf{A} = \begin{pmatrix} 1 & -1 \\ 1 & 1 \end{pmatrix}$,   (b) $\mathbf{A} = \begin{pmatrix} -1 & 1 \\ -5 & 3 \end{pmatrix}$,   (c) $\mathbf{A} = \begin{pmatrix} 4 & 1 \\ -1 & 2 \end{pmatrix}$.

34. (a) For each $\mathbf{A}$ in exercise 33, find an invertible $\mathbf{Q} \in M_2(\mathcal{C})$ such that $\mathbf{AQ} = \mathbf{QJ}$ (i.e., $\mathbf{A} = \mathbf{QJQ}^{-1}$).
(b) Compute $e^{\mathbf{J}t}$ and $e^{\mathbf{A}t}$ for each $\mathbf{A}$ in exercise 33.

35. Let
$$\mathbf{A}_1 = \begin{pmatrix} 1 & -1 \\ 1 & 1 \end{pmatrix}, \quad \mathbf{A}_2 = \begin{pmatrix} -1 & 1 \\ -5 & 3 \end{pmatrix}.$$
Show that there is an invertible $\mathbf{P}$ such that $\mathbf{A}_2 = \mathbf{PA}_1\mathbf{P}^{-1}$, and compute such a $\mathbf{P}$. Thus $\exp(\mathbf{A}_2 t) = \mathbf{P} \exp(\mathbf{A}_1 t)\mathbf{P}^{-1}$. (Hint: see exercises 33 and 34.)

36. (a) Find a Jordan canonical form $\mathbf{J}$ for $\mathbf{A}$ if $\mathbf{A}$ is given by

(i) $\mathbf{A} = \begin{pmatrix} 3 & -1 & 1 \\ -5 & 1 & 7 \\ 1 & -1 & 3 \end{pmatrix}$,   (ii) $\mathbf{A} = \begin{pmatrix} 2 & 1 & 2 \\ -1 & 0 & -2 \\ 0 & 0 & 1 \end{pmatrix}$.

(b) Compute $e^{\mathbf{J}t}$ and $e^{\mathbf{A}t}$ for each $\mathbf{A}$ in (a).

37. Let
$$\mathbf{A} = \begin{pmatrix} 0 & 1 \\ -a_0 & -a_1 \end{pmatrix}, \quad a_0, a_1 \in \mathcal{C}.$$

(a) Show that $p_{\mathbf{A}}(\lambda) = \lambda^2 + a_1\lambda + a_0$.
(b) If λ is an eigenvalue for $\mathbf{A}$, show that $\dim(\mathcal{E}(\mathbf{A}, \lambda)) = 1$.
(c) Let λ_1, λ_2 be the eigenvalues of $\mathbf{A}$. Compute a Jordan canonical form for $\mathbf{A}$ in the two cases

(i) $\lambda_1 \neq \lambda_2$,   (ii) $\lambda_1 = \lambda_2$.

38. Let
$$\mathbf{A} = \begin{pmatrix} 0 & 1 & 0 \\ 0 & 0 & 1 \\ -a_0 & -a_1 & -a_2 \end{pmatrix}, \quad a_0, a_1, a_2 \in \mathcal{C}.$$

(a) Compute $p_{\mathbf{A}}$.
(b) If λ is an eigenvalue for $\mathbf{A}$, show that $\dim(\mathcal{E}(\mathbf{A}, \lambda)) = 1$.
(c) Let λ_1, λ_2, λ_3 be the eigenvalues of $\mathbf{A}$. Compute a Jordan canonical form for $\mathbf{A}$ in the three cases

(i) $\lambda_1, \lambda_2, \lambda_3$ are distinct,

(ii) $\lambda_1 = \lambda_2, \quad \lambda_1 \neq \lambda_3,$

(iii) $\lambda_1 = \lambda_2 = \lambda_3.$

39. Let $\mathbf{A} \in M_n(\mathcal{C})$ and let $\lambda_1, \ldots, \lambda_n$ be the n eigenvalues of $\mathbf{A}$, not necessarily distinct. Show that

(a)
$$\det(\mathbf{A}) = \lambda_1 \cdots \lambda_n,$$

(b)
$$\text{tr}(\mathbf{A}) = \lambda_1 + \cdots + \lambda_n.$$

(Hint: $\mathbf{A} = \mathbf{QJQ}^{-1}$, $p_\mathbf{A} = p_\mathbf{J}$. Show that the coefficient of λ^{n-1} in $p_\mathbf{A}$ is $a_{n-1} = -\text{tr}(\mathbf{A})$.)

40. For $\mathbf{A} \in M_n(\mathcal{C})$ consider the nonhomogeneous system

$$X' = \mathbf{A}X + B(t). \tag{NH}$$

(a) Suppose B is a polynomial of degree r,

$$B(t) = \beta_0 + \beta_1 t + \cdots + \beta_r t^r, \quad \beta_j \in \mathcal{C}^n, \quad \beta_r \neq 0.$$

If 0 is an eigenvalue of $\mathbf{A}$ of multiplicity m, show that (NH) has a solution X which is a polynomial, $\deg(X) \leq r + m$,

$$X(t) = \gamma_0 + \gamma_1 t + \cdots + \gamma_{r+m} t^{r+m}, \quad \gamma_j \in \mathcal{C}^n,$$

as follows. A Jordan canonical form $\mathbf{J}$ for $\mathbf{A}$ can be written as $\mathbf{J} = \text{diag}(\mathbf{J}_0, \mathbf{J}_1)$ where $\mathbf{J}_0$ is $m \times m$ and contains all blocks with the eigenvalue 0 along the diagonal. Thus

$$\mathbf{J}_0 = \begin{pmatrix} 0 & \delta_{1,2} & 0 & \ldots & 0 & 0 \\ 0 & 0 & \delta_{2,3} & \ldots & 0 & 0 \\ \vdots & \vdots & \vdots & \ldots & \vdots & \vdots \\ 0 & 0 & 0 & \ldots & 0 & \delta_{m-1,m} \\ 0 & 0 & 0 & \ldots & 0 & 0 \end{pmatrix}, \quad \delta_{i,i+1} = 1 \text{ or } 0,$$

and $\mathbf{J}_1$ is an invertible $(n-m) \times (n-m)$ matrix . Let $\mathbf{A} = \mathbf{QJQ}^{-1}$ and put $X(t) = \mathbf{Q}Y(t)$ in (NH). Then Y satisfies

$$Y' = \mathbf{J}Y + C(t), \quad C(t) = \mathbf{Q}^{-1}B(t), \tag{3.26}$$

and C is a polynomial of degree r. Let

$$Y = \begin{pmatrix} U \\ V \end{pmatrix}, \quad C = \begin{pmatrix} P \\ Q \end{pmatrix},$$

where U, P are the first m components of Y, C, respectively. Then (3.26) splits into two systems

(i)
$$U' = \mathbf{J}_0 U + P(t),$$

(ii)
$$V' = \mathbf{J}_1 V + Q(t).$$

Show that (i) has a polynomial solution U, $deg(U) \le r+m$. Use exercise 25(b) to show (ii) has a polynomial solution V, $deg(V) \le r$. Thus (3.26), and hence (NH), has a polynomial solution of degree $\le r + m$.

(b) Let $B(t) = e^{\mu t}P(t)$, where μ is an eigenvalue of $\mathbf{A}$ of multiplicity m and P is a polynomial, $deg(P) = r$. Show that (NH) has a solution X of the form $X(t) = e^{\mu t}Q(t)$, where Q is a polynomial, $deg(Q) \le r + m$. (Hint: Let $X(t) = e^{\mu t}Z(t)$ in (NH), and reduce to the case $\mu = 0$ considered in (a).)

41. Find a basis $X = (u, v)$ for the solutions of the following equations:

(a) $x'' - 4x = 0$, (b) $x'' + 9x = 0$, (c) $x'' - 2ix' - x = 0$.

42. Find all solutions of the following equations:

(a) $x''' - 8x = 0$, (b) $x''' - 3x' - 2x = 0$, (c) $x^{(4)} + 16x = 0$.

43. Let $\mathbf{X} = (x_1, \ldots, x_n)$ be any basis for the solutions S_n of (H_n) on $\mathcal{R}$. Show that $\mathbf{X}$, when restricted to I, is a basis for the solutions of (H_n) on the interval $I \subset \mathcal{R}$.

44. Consider the matrix $\mathbf{A} \in M_n(\mathcal{C})$ given by (3.19).

(a) If λ is an eigenvalue of $\mathbf{A}$, show that $\dim(\mathcal{E}(\mathbf{A}, \lambda)) = 1$. Try showing that every eigenvector α of $\mathbf{A}$ is a nonzero constant multiple of

$$E_1 + \lambda E_2 + \lambda^2 E_3 + \cdots + \lambda^{n-1}E_n.$$

(b) If $\lambda_1, \ldots, \lambda_k$ are the distinct eigenvalues of $\mathbf{A}$, with $m_1, \ldots, m_k$, their respective multiplicities, show that a Jordan canonical form for $\mathbf{A}$ is

$$\mathbf{J} = \mathrm{diag}(\mathbf{J}_1, \ldots, \mathbf{J}_k),$$

where $\mathbf{J}_i$ is an $m_i \times m_i$ block

$$\mathbf{J}_i = \begin{pmatrix} \lambda_i & 1 & 0 & \cdots & 0 & 0 \\ 0 & \lambda_i & 1 & \cdots & 0 & 0 \\ \vdots & \vdots & \vdots & \cdots & \vdots & \vdots \\ 0 & 0 & 0 & \cdots & \lambda_i & 1 \\ 0 & 0 & 0 & \cdots & 0 & \lambda_i \end{pmatrix}.$$

45. Consider the equation (H_n) on $\mathcal{R}$ with the characteristic polynomial p.

(a) Show that all solutions of (H_n) are bounded on $[0, \infty)$ if and only if
  (i) $\mathrm{Re}(\lambda) \le 0$ for each root λ of p of multiplicity 1,
  (ii) $\mathrm{Re}(\lambda) < 0$ for each root λ of p of multiplicity greater than 1.

(b) Show that all solutions of (H_n) have limit zero as $t \to +\infty$ if and only if $\mathrm{Re}(\lambda) < 0$ for each root of p.

46. Find a real basis for the solutions of

(a) $x'' + x = 0$, (b) $x^{(4)} - x = 0$, (c) $x^{(4)} + x = 0$.

47. Determine all real solutions of

$$\text{(a)} \quad x''' + ix'' + x' + ix = 0, \quad \text{(b)} \quad x'' - 2ix' - x = 0.$$

48. Find that solution x of

$$x''' + x = 0$$

satisfying $\tilde{x}(0) = E_3$, $(x(0) = 0, x'(0) = 0, x''(0) = 1)$.

49. Suppose that L is the constant coefficient operator given by

$$Lx = x^{(n)} + a_{n-1}x^{(n-1)} + \cdots + a_0 x, \quad a_j \in \mathcal{C}.$$

In addition to the variation of parameters method, there is an alternate approach to solving the nonhomogeneous equation $Lx = b(t)$, the *method of undetermined coefficients*, in case $b(t)$ satisfies a constant coefficient homogeneous differential equation

$$Mb = b^{(m)} + c_{m-1}b^{(m-1)} + \cdots + c_0 b = 0, \quad c_j \in \mathcal{C}.$$

(a) Show that any solution $x(t)$ of $Lx = b$ is also a solution of the homogeneous equation $MLx = 0$.

(b) Use Theorem 3.13 to describe the structure of solutions to $Lx = b$.

(c) Consider the problem of finding x satisfying

$$Lx = x'' - x = 2e^t.$$

Now $b(t) = 2e^t$ satisfies $Mb = 0$, where

$$My = y' - y.$$

Show that

$$ML(y) = M(Ly) = y^{(3)} - y^{(2)} - y' + y = 0.$$

(d) Find the characteristic polynomial of ML.

(e) By inserting the function

$$x(t) = c_1 e^{-t} + c_2 e^t + c_3 t e^t$$

into the equation

$$Lx = x'' - x = 2e^t,$$

show that the general solution is given by

$$x(t) = te^t + c_1 e^t + c_2 e^{-t}, \quad c_1, c_2 \in \mathcal{C}.$$

50. Use the method of undetermined coefficients (exercise 49) to find all solutions of the following equations:

(a) $x'' - 4x = e^t,$

(b) $$x'' - 4x = e^{2t},$$

(c) $$x'' + x = \sin(t),$$

(d) $$x'' + 16x = t^2 e^{4t}.$$

51. Suppose that L and M are the constant coefficient operators given by

$$Lx = x^{(n)} + a_{n-1}x^{(n-1)} + \cdots + a_0 x, \quad a_j \in \mathcal{C},$$

$$Mx = x^{(m)} + c_{m-1}x^{(m-1)} + \cdots + c_0 x, \quad c_j \in \mathcal{C},$$

with characteristic polynomials

$$p(\lambda) = \lambda^n + a_{n-1}\lambda^{n-1} + \cdots + a_0 \lambda, \quad a_j \in \mathcal{C}$$

and

$$q(\lambda) = \lambda^m + c_{m-1}\lambda^{n-1} + \cdots + c_0 \lambda, \quad c_j \in \mathcal{C},$$

respectively. Let $b(t)$ satisfy $Mb = 0$.

Suppose further that all roots of both p and q have multiplicity one and p and q have no roots in common. Show that there is a unique solution $y(t)$ of $Ly = b$ such that $My = 0$. Find a formula for y, given that

$$b(t) = \sum_{j=1}^{m} c_j e^{\mu_j t},$$

where the μ_j are the distinct roots of q.

Chapter 4

Periodic Coefficients

4.1 Introduction

Human fascination with periodic phenomena governed by differential equations, such as the periodic waxing and waning of the moon, extends back into prehistory. In this chapter we consider linear systems of differential equations whose coefficients are periodic functions. This class of equations is much larger than the systems with constant coefficients, and in general explicit descriptions of the solutions will not be obtained. Nonetheless, the periodicity does impose some structure on solutions, a structure which can be described using a bit of linear algebra.

Before launching into the theoretical development, we describe two models which use linear systems with periodic coefficients. The first model is a modification of the toxic discharge problem in Chapter 1 that is pictured in Figure 1.1 and led to equations (1.1) and (1.2). That case considered the build up of toxic material in a system of lakes after a discharge of toxic material upstream.

In the modified model some of the water flowing in the channel between Lakes 1 and 2 is removed from the system for crop irrigation. The rate at which water is removed is not constant. Each morning at the same time a valve is opened, and water is drained from the channel. The valve is closed later, again at the same time each day. Thus the rate at which water is removed from the channel is described by a function $w(t)$ which is periodic in time. If time is measured in days, then $w(t + 1) = w(t)$.

When the derivation of the equations describing the lake contamination is reexamined, we observe first that the behavior in Lake 1 is unchanged:

$$\frac{d\tau_1}{dt} = r(t) - \frac{i_1}{v_1}\tau_1.$$

However, the rate i_2 at which water enters Lake 2 is no longer constant but may now be represented by $l_1 - w(t)$, where l_1 is the constant rate at which water flows from Lake 1 into the channel. The second equation thus becomes

$$\frac{d\tau_2}{dt} = \frac{l_1 - w(t)}{v_1}\tau_1 - \frac{o_2}{v_2}\tau_2.$$

In the matrix formulation the nonhomogeneous system of equations has assumed the form

$$X' = \mathbf{A}(t)X + B(t),$$

where $\mathbf{A}(t)$ is no longer constant but is periodic with period 1,

$$\mathbf{A}(t+1) = \mathbf{A}(t).$$

Since the variation of parameters formula reduces the problem of solving a nonhomogeneous system to that of solving the associated homogeneous system, the fundamental problem is the analysis of

$$X' = \mathbf{A}(t)X, \quad \mathbf{A}(t+1) = \mathbf{A}(t).$$

A simple but essential observation initiates the study of such systems: shifting the solutions of the homogeneous system by a period is a linear transformation on a finite-dimensional vector space. Linear algebra thus offers a key with which to open our investigation.

Periodicity can also enter linear systems through the coefficient $B(t)$. A mechanical example, the mass and spring system considered in section 1.3, leads to the equation

$$mx'' + cx' + kx = f(t).$$

Recall that in this model the coefficients m, c, and k are constant functions, which are thus periodic with every period. When the external forcing function $f(t)$ is periodic,

$$f(t + \omega) = f(t), \quad \omega > 0,$$

the whole equation is periodic with period ω. A natural question in this setting is whether the equation has solutions with period ω. This problem is taken up after consideration of the homogeneous system with periodic coefficients.

4.2   Floquet's theorem

Consider the system

$$X' = \mathbf{A}(t)X, \tag{4.1}$$

where $\mathbf{A} \in C(\mathcal{R}, M_n(\mathcal{C}))$, and $\mathbf{A}$ is periodic with period $\omega > 0$,

$$\mathbf{A}(t + \omega) = \mathbf{A}(t), \quad t \in \mathcal{R}.$$

A basic observation about such a system is that if $X(t)$ satisfies (4.1), then so does $Y(t) = X(t + \omega)$, since

$$Y'(t) = X'(t + \omega) = \mathbf{A}(t + \omega)X(t + \omega) = \mathbf{A}(t)Y(t), \quad t \in \mathcal{R}.$$

Thus if $\mathbf{X}$ is a basis for the solutions of (4.1) then so is $\mathbf{Y}$, where $\mathbf{Y}(t) = \mathbf{X}(t+\omega)$. This fact leads to a fundamental representation result for a basis $\mathbf{X}$.

Theorem 4.1 (Floquet): *Each basis $\mathbf{X}$ for the solutions of* (4.1) *can be represented as*

$$\mathbf{X}(t) = \mathbf{P}(t)e^{\mathbf{R}t}, \quad t \in \mathcal{R}, \tag{4.2}$$

where $\mathbf{P} \in C^1(\mathcal{R}, M_n(\mathcal{C}))$, $\mathbf{P}(t)$ is invertible, $\mathbf{P}(t + \omega) = \mathbf{P}(t)$ for all $t \in \mathcal{R}$, and $\mathbf{R} \in M_n(\mathcal{C})$.

Proof: Since $\mathbf{Y}(t) = \mathbf{X}(t + \omega)$ gives a basis for the solutions of (4.1), $\mathbf{X}(t + \omega) = \mathbf{X}(t)\mathbf{C}$ for some invertible $\mathbf{C} \in M_n(\mathcal{C})$. Theorem 4.2 below shows that for every invertible $\mathbf{C}$ there exists a matrix $\mathbf{R} \in M_n(\mathcal{C})$ such that $e^{\mathbf{R}\omega} = \mathbf{C}$; that is, $\mathbf{R}\omega$ is a logarithm of $\mathbf{C}$. It follows that

$$\mathbf{X}(t + \omega) = \mathbf{X}(t)e^{\mathbf{R}\omega}, \quad t \in \mathcal{R}. \tag{4.3}$$

If $\mathbf{P}(t) = \mathbf{X}(t)e^{-\mathbf{R}t}$ we see that for all $t \in \mathcal{R}$,

$$\mathbf{P}(t + \omega) = \mathbf{X}(t + \omega)e^{-\mathbf{R}(t+\omega)} = \mathbf{X}(t)e^{\mathbf{R}\omega}e^{-\mathbf{R}(t+\omega)} = \mathbf{X}(t)e^{-\mathbf{R}t} = \mathbf{P}(t).$$

Since both $\mathbf{X}(t)$ and $e^{-\mathbf{R}t}$ are invertible, so is $\mathbf{P}(t)$, and $\mathbf{P} \in C^1(\mathcal{R}, M_n(\mathcal{C}))$. $\square$

Since a basis $\mathbf{X}$ of solutions for (4.1) satisfies $\mathbf{X}' = \mathbf{A}\mathbf{X}$, substitution of $\mathbf{X}(t) = \mathbf{P}(t)e^{\mathbf{R}t}$ leads to

$$\mathbf{P}'(t)e^{\mathbf{R}t} + \mathbf{P}(t)\mathbf{R}e^{\mathbf{R}t} = \mathbf{A}(t)\mathbf{P}(t)e^{\mathbf{R}t},$$

or $\mathbf{P}' + \mathbf{P}\mathbf{R} = \mathbf{A}\mathbf{P}$. Putting $X = \mathbf{P}Y$ in (4.1) then gives

$$\mathbf{P}Y' + \mathbf{P}'Y = \mathbf{A}\mathbf{P}Y = \mathbf{P}'Y + \mathbf{P}\mathbf{R}Y,$$

so that

$$Y' = \mathbf{R}Y. \tag{4.4}$$

Thus the substitution $X = \mathbf{P}Y$ transforms the system (4.1) with a periodic coefficient matrix $\mathbf{A}$ to the constant coefficient system (4.4). In fact the map $Y \to X = \mathbf{P}Y$ is an isomorphism between the set of solutions of (4.4) and the set $\mathcal{S}$ of solutions of (4.1).

4.3   The logarithm of an invertible matrix

A matrix $\mathbf{B} \in M_n(\mathcal{C})$ such that $e^{\mathbf{B}} = \mathbf{C}$ is said to be a *logarithm* of $\mathbf{C}$. Theorem 4.1 made use of the fact that every invertible matrix $\mathbf{C} \in M_n(\mathcal{C})$ has a logarithm, which is the next result. The proof makes use of some power series techniques which are discussed more fully in Chapter 5.

Theorem 4.2: *If $\mathbf{C} \in M_n(\mathcal{C})$ is invertible, there exists a $\mathbf{B} \in M_n(\mathcal{C})$ such that $e^{\mathbf{B}} = \mathbf{C}$.*

Proof: Let $\mathbf{J} = \mathbf{Q}^{-1}\mathbf{C}\mathbf{Q}$ be a Jordan canonical form for $\mathbf{C}$. If there is a $\mathbf{K}$ such that $e^{\mathbf{K}} = \mathbf{J}$, then it will follow from the power series definition of the matrix exponential (section 3.1) that $\mathbf{C} = \mathbf{QJQ}^{-1} = e^{\mathbf{B}}$, where $\mathbf{B} = \mathbf{QKQ}^{-1}$. A further reduction is possible because the matrix $\mathbf{J}$ has block diagonal form, $\mathbf{J} = \mathrm{diag}(\mathbf{J}_1, \ldots, \mathbf{J}_q)$. If K_j satisfies $e^{\mathbf{K}_j} = \mathbf{J}_j$ and if $\mathbf{K} = \mathrm{diag}(\mathbf{K}_1, \ldots, \mathbf{K}_q)$, then

$$e^{\mathbf{K}} = \mathrm{diag}(e^{\mathbf{K}_1}, \ldots, e^{\mathbf{K}_q}) = \mathrm{diag}(\mathbf{J}_1, \ldots \mathbf{J}_q) = \mathbf{J}.$$

Since $\mathbf{C}$ is invertible, none of its eigenvalues μ_j is 0. Therefore it suffices to find a $\mathbf{K}$ such that $e^{\mathbf{K}} = \mathbf{J}$ when $\mathbf{J}$ is the 1×1 matrix

(i)
$$\mathbf{J} = (\mu), \quad \mu \neq 0$$

or when $\mathbf{J}$ is the $r \times r$ matrix

(ii)
$$\mathbf{J} = \begin{pmatrix} \mu & 1 & 0 & \ldots & 0 & 0 \\ 0 & \mu & 1 & \ldots & 0 & 0 \\ \vdots & \vdots & \vdots & \ldots & \vdots & \vdots \\ 0 & 0 & 0 & \ldots & \mu & 1 \\ 0 & 0 & 0 & \ldots & 0 & \mu \end{pmatrix} = \mu \mathbf{I}_r + \mathbf{N}, \quad r > 1, \quad \mu \neq 0.$$

In case (i) choose $\mathbf{K}$ to be any value of $\log(\mu)$. For example, if $\mu = |\mu| e^{i\theta}$, $-\pi < \theta \leq \pi$, then we choose $\log(\mu) = \log(|\mu|) + i\theta$, the principle value of $\log(\mu)$.

In case (ii), write

$$\mathbf{J} = \mu(\mathbf{I}_r + \mathbf{N}/\mu),$$

and note that $(\mathbf{N}/\mu)^r = 0$. We will find a matrix $l(\mathbf{N}/\mu)$ such that

$$e^{l(\mathbf{N}/\mu)} = \mathbf{I}_r + \mathbf{N}/\mu, \tag{4.5}$$

and then the matrix

$$\mathbf{K} = \log(\mu)\mathbf{I}_r + l(\mathbf{N}/\mu)$$

will be such that $e^{\mathbf{K}} = \mathbf{J}$. The determination of $l(\mathbf{N}/\mu)$ is based on the power series

$$l(z) = \log(1 + z) = z - z^2/2 + z^3/3 - \cdots = \sum_{k=1}^{\infty} (-1)^{k+1} z^k / k,$$

which is absolutely convergent for $|z| < 1$, $z \in C$. Since

$$1 + z = \exp(l(z)) = 1 + (z - z^2/2 + \cdots) + \frac{1}{2!}(z - z^2/2 + \cdots)^2 + \cdots, \tag{4.6}$$

the rearrangement of the terms on the right obtained by collecting ascending powers of z must agree with $1 + z$; that is, the coefficients of the first two powers of z sum to 1, while the remainder sum to zero.

Since $(\mathbf{N}/\mu)^r = 0$, the sum

$$l(\mathbf{N}/\mu) = (\mathbf{N}/\mu) - (\mathbf{N}/\mu)^2/2 + (\mathbf{N}/\mu)^3/3 - \cdots = \sum_{k=1}^{\infty} (-1)^{k+1} (\mathbf{N}/\mu)^k / k$$

is a finite sum, hence is convergent. Now compute $e^{l(z)}$ as in (4.6), with the same rearrangement of terms. Since the coefficients after rearrangement agree with those in (4.6), (4.5) is obtained. $\square$

Equation (4.3) shows that

$$\mathbf{C} = e^{\mathbf{R}\omega} = \mathbf{X}^{-1}(0)\mathbf{X}(\omega),$$

so $\mathbf{R}$ can be computed once we know $\mathbf{X}$ at 0 and ω. Also the periodicity of $\mathbf{P}$ means it is uniquely determined by its value on $[0, \omega]$. Thus a basis $\mathbf{X}$ is determined on all of $\mathcal{R}$ by (4.2) once it is known on $[0, \omega]$.

4.4   Multipliers

The matrices $\mathbf{C}$ and $\mathbf{R}$ depend on the basis $\mathbf{X}$. If $\mathbf{X}_1$ is another basis for $\mathcal{S}$, the set of solutions of (4.1), then $\mathbf{X}_1 = \mathbf{X}\mathbf{T}$ for some invertible $\mathbf{T} \in M_n(\mathcal{C})$, and if

$$\mathbf{X}_1(t + \omega) = \mathbf{X}_1(t)\mathbf{C}_1, \quad t \in \mathcal{R},$$

then

$$\mathbf{C}_1 = \mathbf{X}_1^{-1}(0)\mathbf{X}_1(\omega) = \mathbf{T}^{-1}\mathbf{X}^{-1}(0)\mathbf{X}(\omega)\mathbf{T} = \mathbf{T}^{-1}\mathbf{C}\mathbf{T},$$

so that $\mathbf{C}_1$ is similar to $\mathbf{C}$. Thus the eigenvalues of $\mathbf{C}$, together with their multiplicities, depend only on $\mathbf{A}(t)$ and are independent of the basis chosen. The eigenvalues μ_j of $\mathbf{C}$ are called the *multipliers* of (4.1). The name is suggested by the following property of the multipliers.

Theorem 4.3: *The system* (4.1) *has a nontrivial solution X satisfying*

$$X(t + \omega) = \mu X(t), \quad t \in \mathcal{R}, \tag{4.7}$$

if and only if μ is a multiplier for (4.1). *In particular,* (4.1) *has a nontrivial solution of period ω if and only if $\mu = 1$ is a multiplier for* (4.1).

Proof: If $\mathbf{X}$ is a basis for (4.1), then the nontrivial solution $X = \mathbf{X}\xi,\ \xi \neq 0$, satisfies (4.7) if and only if

$$X(t + \omega) = \mathbf{X}(t + \omega)\xi = \mathbf{X}(t)\mathbf{C}\xi = \mu X(t) = \mu\mathbf{X}(t)\xi,$$

or $\mathbf{C}\xi = \mu\xi$. $\square$

Now let $\mathbf{X}$ be the basis for (4.1) satisfying $\mathbf{X}(0) = \mathbf{I}_n$. Then $\mathbf{C} = e^{\mathbf{R}\omega} = \mathbf{X}(\omega)$, and the multipliers for (4.1) are the eigenvalues of $\mathbf{X}(\omega)$. Let $\mathcal{S}(\mu)$ be the set of all $X \in \mathcal{S}$ satisfying (4.7). The structure of $\mathcal{S}(\mu)$ is given in the following result.

Theorem 4.4: *For each $\mu \in \mathcal{C}$,*
(i) $\mathcal{S}(\mu)$ *is a vector subspace of $\mathcal{S}$,*
(ii) $\dim \mathcal{S}(\mu) = \dim \mathcal{E}(\mathbf{X}(\omega), \mu) = \dim N(\mu\mathbf{I}_n - \mathbf{X}(\omega))$.
Proof: If $X, Y \in \mathcal{S}(\mu)$ and $\alpha, \beta \in \mathcal{C}$, then $\alpha X + \beta Y \in \mathcal{S}$ and

$$(\alpha X + \beta Y)(t + \omega) = \alpha X(t + \omega) + \beta Y(t + \omega)$$

$$= \alpha\mu X(t) + \beta\mu Y(t) = \mu(\alpha X + \beta Y)(t), \quad t \in \mathcal{R},$$

so that $\alpha X + \beta Y \in \mathcal{S}(\mu)$, proving (i). As to (ii), we note that the proof of Theorem 4.3 shows that the map

$$S : \xi \in \mathcal{E}(\mathbf{X}(\omega), \mu) \to X = \mathbf{X}\xi$$

takes the eigenspace of $\mathbf{X}(\omega)$ for μ onto $\mathcal{S}(\mu)$. It is clearly linear, for $\alpha, \beta \in \mathcal{C}$ and $\xi, \eta \in \mathcal{E}(\mathbf{X}(\omega), \mu)$,

$$S(\alpha\xi + \beta\eta) = \mathbf{X}(\alpha\xi + \beta\eta) = \alpha\mathbf{X}\xi + \beta\mathbf{X}\eta = \alpha S(\xi) + \beta S(\eta),$$

and it is one to one since $\mathbf{X}$ is invertible; that is, $S(\xi) = S(\eta)$, or $\mathbf{X}\xi = \mathbf{X}\eta$, if and only if $\xi = \eta$. Thus S is an isomorphism of $\mathcal{E}(\mathbf{X}(\omega), \mu)$ onto $\mathcal{S}(\mu)$, and (ii) follows, since two finite-dimensional vector spaces are isomorphic if and only if they have the same dimension. $\square$

If $\mu_1, \ldots, \mu_n$ is an enumeration of all the multipliers, where each multiplier is repeated as many times as its multiplicity, then (see Chapter 3, exercise 39(a))

$$\mu_1 \cdots \mu_n = \det(\mathbf{X}(\omega)).$$

Since the Wronskian $W_{\mathbf{X}} = \det(\mathbf{X})$ satisfies (Theorem 2.6)

$$\det(\mathbf{X}(\omega)) = \det(\mathbf{X}(0)) \exp\left[\int_0^\omega \operatorname{tr}(\mathbf{A}(s)) \, ds\right],$$

it follows that

$$\mu_1 \cdots \mu_n = \exp\left[\int_0^\omega \operatorname{tr}(\mathbf{A}(s)) \, ds\right]. \tag{4.8}$$

In particular, none of the μ_j is zero, and if $n-1$ of them are known, the remaining one can be obtained from (4.8).

4.5   The behavior of solutions for large t

We continue using the basis $\mathbf{X}$ for (4.1) such that $\mathbf{X}(0) = \mathbf{I}_n$. The eigenvalues of a matrix $\mathbf{R}$ such that $\mathbf{X}(\omega) = e^{\mathbf{R}\omega}$ are called the *characteristic exponents* for (4.1). Let $\mathbf{J}$ be a Jordan canonical form for $\mathbf{R}$, so that $\mathbf{J} = \mathbf{Q}^{-1}\mathbf{R}\mathbf{Q}$ for some invertible $\mathbf{Q}$ and

$$\mathbf{J} = \operatorname{diag}(\mathbf{J}_1, \ldots, \mathbf{J}_q),$$

with the $\mathbf{J}_i$ being $r_i \times r_i$, $\mathbf{J}_i = (\lambda_i)$ if $r_i = 1$, and

$$\mathbf{J}_i = \begin{pmatrix} \lambda_i & 1 & 0 & \ldots & 0 & 0 \\ 0 & \lambda_i & 1 & \ldots & 0 & 0 \\ \vdots & \vdots & \vdots & \ldots & \vdots & \vdots \\ 0 & 0 & 0 & \ldots & \lambda_i & 1 \\ 0 & 0 & 0 & \ldots & 0 & \lambda_i \end{pmatrix}, \quad r_i > 1.$$

Then $\mathbf{X}(\omega) = e^{\mathbf{R}\omega} = \exp(\mathbf{QJQ}^{-1}\omega) = \mathbf{Q}\exp(\mathbf{J}\omega)\mathbf{Q}^{-1}$, and since

$$\exp(\mathbf{J}\omega) = \operatorname{diag}(\exp(\mathbf{J}_1\omega), \ldots, \exp(\mathbf{J}_q\omega)),$$

the eigenvalues of $\mathbf{X}(\omega)$, that is the multipliers for (4.1), are the same as the eigenvalues of $\exp(\mathbf{J}\omega)$, which are given by $\exp(\lambda_j\omega)$. Thus

$$\mu_j = \exp(\lambda_j\omega), \quad j = 1, \ldots, n,$$

for some enumeration of the μ_j and λ_j. If $\mathbf{R}_1$ is another matrix such that $\exp(\mathbf{R}_1\omega) = \mathbf{X}(\omega)$, with eigenvalues ν_j, then

$$\mu_j = \exp(\nu_j\omega), \quad j = 1, \ldots, n,$$

and it follows that

$$\lambda_j = \nu_j + (2\pi i k)/\omega$$

for some integer k. Thus the characteristic exponents are not uniquely determined by (4.1), but their real parts are since

$$|\mu_j| = \exp(\operatorname{Re}(\lambda_j)\omega), \quad \operatorname{Re}(\lambda_j) = \frac{1}{\omega}\log|\mu_j|, \quad j = 1, \ldots, n.$$

From the representation (4.2) for $\mathbf{X}$ and a Jordan canonical form $\mathbf{J} = \mathbf{Q}^{-1}\mathbf{RQ}$ for $\mathbf{R}$, we can obtain a basis $\mathbf{X}_1$, where

$$\mathbf{X}_1(t) = \mathbf{X}(t)\mathbf{Q} = \mathbf{P}_1(t)e^{\mathbf{J}t}, \quad \mathbf{P}_1(t) = \mathbf{P}(t)\mathbf{Q}, \quad \mathbf{P}_1(t+\omega) = \mathbf{P}_1(t).$$

From the explicit structure of $e^{\mathbf{J}t}$ as given in Theorem 3.10, we see that the columns $X_1, \ldots, X_n$ of $\mathbf{X}_1$ have the form

$$X_j(t) = e^{\lambda t}P(t),$$

where λ is an eigenvalue of $\mathbf{R}$ and P is a vector polynomial with coefficients which are periodic functions of period ω. For example, if $\mathbf{P}_1 = (P_1, \ldots, P_n)$ and the first block of $\mathbf{J}$ is the $r_1 \times r_1$ matrix

$$\mathbf{J}_1 = \begin{pmatrix} \lambda_1 & 1 & 0 & \cdots & 0 & 0 \\ 0 & \lambda_1 & 1 & \cdots & 0 & 0 \\ \vdots & \vdots & \vdots & \cdots & \vdots & \vdots \\ 0 & 0 & 0 & \cdots & \lambda_1 & 1 \\ 0 & 0 & 0 & \cdots & 0 & \lambda_1 \end{pmatrix}, \quad r_1 > 1,$$

then

$$X_1(t) = \exp(\lambda_1 t)P_1(t),$$

$$X_2(t) = \exp(\lambda_1 t)[tP_1(t) + P_2(t)],$$

$$X_3(t) = \exp(\lambda_1 t)\left[\frac{t^2}{2!}P_1(t) + tP_2(t) + P_3(t)\right],$$

$$\vdots$$

$$X_{r_1}(t) = \exp(\lambda_1 t)\left[\frac{t^{r_1-1}}{(r_1-1)!}P_1(t) + \cdots + P_{r_1}(t)\right].$$

From this representation of $\mathbf{X}_1(t) = \mathbf{P}_1(t)e^{\mathbf{J}t}$ we have the following analogue of Theorem 3.11. The proof is entirely similar once we note that $|\mu_j| = \exp(\text{Re}(\lambda_j)\omega)$.

Theorem 4.5: *Let $\mu_1, \ldots, \mu_k$ be the distinct multipliers for (4.1), with multiplicities $m_1, \ldots, m_k$, respectively. Then*

(i) all solutions of (4.1) are bounded on $[0, \infty)$ if and only if $|\mu_j| \leq 1$, $j = 1, \ldots, k$, and for those μ_j such that $|\mu_j| = 1$ we have $m_j = \dim(\mathcal{E}(\mathbf{X}(\omega), \mu_j))$;

(ii) all solutions of (4.1) tend to zero as $t \to \infty$ if and only if $|\mu_j| < 1$, $j = 1, \ldots, k$.

4.6 First-order nonhomogeneous systems

Consider a nonhomogeneous system

$$X' = \mathbf{A}(t)X + B(t), \tag{4.9}$$

where $\mathbf{A} \in C(\mathcal{R}, M_n(\mathcal{C}))$, $B \in C(\mathcal{R}, \mathcal{C}^n)$, and $\mathbf{A}, B$ are periodic with period $\omega > 0$,

$$\mathbf{A}(t+\omega) = \mathbf{A}(t), \quad B(t+\omega) = B(t), \quad t \in \mathcal{R}.$$

It is interesting to determine when (4.9) has periodic solutions of period ω, and what the structure of the set of all such solutions is. Let $\mathcal{S}_p$ be the set of all solutions of $X' = \mathbf{A}(t)X$ which are periodic with period ω, and let $\mathcal{S}_p(B)$ denote the set of all solutions of (4.9) which have period ω.

Theorem 4.6: *The set $\mathcal{S}_p(B)$ is an affine space,*

$$\mathcal{S}_p(B) = V + \mathcal{S}_p = \{X | X = V + U, U \in \mathcal{S}_p\},$$

where V is a particular solution in $\mathcal{S}_p(B)$.

Proof: If $V \in \mathcal{S}_p(B)$ and $U \in \mathcal{S}_p$, then $X = V + U$ is a solution of (4.9) and it clearly is periodic of period ω, so that $X \in \mathcal{S}_p(B)$. Thus $V + \mathcal{S}_p \subset \mathcal{S}_p(B)$. Conversely, if $X, V \in \mathcal{S}_p(B)$, then $U = X - V$ satisfies $X' = \mathbf{A}(t)X$ and has period ω. Hence $U \in \mathcal{S}_p$, and we have $\mathcal{S}_p(B) \subset V + \mathcal{S}_p$. $\square$

As an example, consider the equation of period $\omega = 2\pi$,

$$x' = -\cos(t)x + \cos(t),$$

with $n = 1$, $\mathbf{A}(t) = -\cos(t)$, $B(t) = \cos(t)$. The solution x satisfying $x(0) = \xi$ is given by

$$x(t) = 1 - \exp(-\sin(t)) + \exp(-\sin(t))\xi, \quad \xi \in \mathcal{C},$$

and this is periodic of period 2π for every $\xi \in \mathcal{C}$. In this case $x = v + u$, where

$$v(t) = 1 - \exp(-\sin(t)), \quad u(t) = \exp(-\sin(t))\xi.$$

On the other hand, consider

$$x' = -\cos(t)x + 1,$$

with $\mathbf{A}(t) = -\cos(t)$, $B(t) = 1$, having period 2π. Here the solution x such that $x(0) = \xi$ is given by

$$x(t) = \exp(-\sin(t)) \int_0^t \exp(\sin(s)) \, ds + \exp(-\sin(t))\xi, \quad \xi \in \mathcal{C},$$

and

$$x(2\pi) - x(0) = \int_0^{2\pi} \exp(\sin(s)) \, ds > 0.$$

Therefore this equation has no solution of period 2π for any $\xi \in \mathcal{C}$.

Thus a central problem is to determine when $\mathcal{S}_p(B)$ is nonempty. Just as in the homogeneous case, if X is a solution of (4.9) then so is $Y(t) = X(t+\omega)$, for

$$Y'(t) = X'(t + \omega) = \mathbf{A}(t + \omega)X(t + \omega) + B(t + \omega) = \mathbf{A}(t)Y(t) + B(t).$$

Thus a solution X of (4.9) has period ω if and only if $X(t) = Y(t) = X(t+\omega)$ for all $t \in \mathcal{R}$, and this is true if and only if $X(0) = X(\omega)$. If $\mathbf{X}$ is the basis for the solutions of $X' = \mathbf{A}(t)X$ such that $\mathbf{X}(0) = \mathbf{I}_n$, the variation of parameters formula gives the solution X of (4.9) satisfying $X(0) = \xi$ as

$$X(t) = \mathbf{X}(t)\xi + \mathbf{X}(t) \int_0^t \mathbf{X}^{-1}(s)B(s) \, ds, \quad t \in \mathcal{R}.$$

Thus $X(0) = X(\omega)$, that is, $X \in \mathcal{S}_p(B)$, if and only if

$$[\mathbf{I}_n - \mathbf{X}(\omega)]\xi = \mathbf{X}(\omega) \int_0^\omega \mathbf{X}^{-1}(s)B(s) \, ds. \tag{4.10}$$

If $\mathbf{I}_n - \mathbf{X}(\omega)$ is invertible, then clearly

$$\xi = [\mathbf{I}_n - \mathbf{X}(\omega)]^{-1}\mathbf{X}(\omega) \int_0^\omega \mathbf{X}^{-1}(s)B(s) \, ds$$

is the unique $\xi = X(0)$ which yields a periodic solution X of (4.9) of period ω. Now $\mathbf{I}_n - \mathbf{X}(\omega)$ is invertible if and only if $\mu = 1$ is not a multiplier for $X' = \mathbf{A}(t)X$, that is, the homogeneous system has no nontrivial solution of period ω. We have shown the following result.

Theorem 4.7: *There exists an $X \in \mathcal{S}_p(B)$ if and only if there is a $\xi \in \mathcal{C}^n$ satisfying (4.10). If $\dim(\mathcal{S}_p) = 0$ there exists a unique $X \in \mathcal{S}_p(B)$ for every $B \in C(\mathcal{R}, \mathcal{C}^n)$ of period ω.*

The analysis of the case when $\dim(\mathcal{S}_p) > 0$ depends on a result for linear equations

$$\mathbf{T}\xi = \eta, \quad \mathbf{T} \in M_n(\mathcal{C}), \quad \xi, \eta \in \mathcal{C}^n.$$

Note that (4.10) has this form, where

$$\mathbf{T} = \mathbf{I}_n - \mathbf{X}(\omega), \quad \eta = \mathbf{X}(\omega) \int_0^\omega \mathbf{X}^{-1}(s) B(s) \, ds.$$

This result relates the solvability of $\mathbf{T}\xi = \eta$ to the solutions ζ of the adjoint homogeneous equation $\mathbf{T}^*\zeta = 0$. Recall that the adjoint $\mathbf{T}^*$ of $\mathbf{T}$ is the conjugate transpose of $\mathbf{T} = (t_{ij})$, so that $\mathbf{T}^* = (\bar{t}_{ji})$.

The inner product (ξ, η) of $\xi, \eta \in \mathcal{C}^n$ is given by $(\xi, \eta) = \eta^*\xi$. If $S \subset \mathcal{C}^n$, its *orthogonal complement* $S^\perp$ is the set

$$S^\perp = \{\eta \in \mathcal{C}^n | (\xi, \eta) = \eta^*\xi = 0, \quad \xi \in S\}.$$

The set $S^\perp$ is a vector subspace of $\mathcal{C}^n$, and if S is a subspace of $\mathcal{C}^n$, then

$$\mathcal{C}^n = S \oplus S^\perp, \tag{4.11}$$

a direct sum, so that every $\zeta \in \mathcal{C}^n$ can be written uniquely as $\zeta = \xi + \eta$, where $\xi \in S$, $\eta \in S^\perp$. Moreover, we have $(S^\perp)^\perp = S$. A consequence of (4.11) is that if S is a subspace of $\mathcal{C}^n$, then

$$\dim(S) + \dim(S^\perp) = n. \tag{4.12}$$

Theorem 4.8: *If* $\mathbf{T} \in M_n(\mathcal{C})$, *then*

(i) $$Ran(\mathbf{T}) = (N(\mathbf{T}^*))^\perp,$$

(ii) $$\dim(N(\mathbf{T})) = \dim(N(\mathbf{T}^*)).$$

Proof: For every $\xi, \zeta \in \mathcal{C}^n$ we have

$$(\mathbf{T}\xi, \zeta) = \zeta^*\mathbf{T}\xi = (\mathbf{T}^*\zeta)^*\xi = (\xi, \mathbf{T}^*\zeta).$$

From this it follows that $\zeta \in (Ran(\mathbf{T}))^\perp$, that is, $(\mathbf{T}\xi, \zeta) = 0$ for all $\xi \in \mathcal{C}^n$ if and only if $(\xi, \mathbf{T}^*\zeta) = 0$ for all $\xi \in \mathcal{C}^n$ or $\mathbf{T}^*\zeta = 0$, that is, $\zeta \in N(\mathbf{T}^*)$. Thus $(Ran(\mathbf{T}))^\perp = N(\mathbf{T}^*)$, and hence $Ran(\mathbf{T}) = ((Ran(\mathbf{T}))^\perp)^\perp = (N(\mathbf{T}^*))^\perp$, which is (i).

From (i) and (4.12), with $S = N(\mathbf{T}^*)$, we have

$$\dim(N(\mathbf{T}^*)) + \dim(Ran(\mathbf{T})) = n.$$

But we know that

$$\dim(N(\mathbf{T})) + \dim(Ran(\mathbf{T})) = n,$$

and therefore (ii) is valid. □

Applying Theorem 4.8 (i) to $\mathbf{T} = \mathbf{I}_n - \mathbf{X}(\omega)$ in (4.10), we see that there is an $X \in \mathcal{S}_p(B)$ if and only if

$$\zeta^* \mathbf{X}(\omega) \int_0^\omega \mathbf{X}^{-1}(s) B(s) \, ds = 0 \qquad (4.13)$$

for every $\zeta \in \mathcal{C}^n$ such that

$$[\mathbf{I}_n - \mathbf{X}^*(\omega)]\zeta = 0. \qquad (4.14)$$

This condition can be interpreted in terms of the solutions $\mathcal{S}^*$ of the *adjoint homogeneous differential equation*

$$Y' = -\mathbf{A}^*(t)Y. \qquad (4.15)$$

The basis $\mathbf{Y}$ for $\mathcal{S}^*$ such that $\mathbf{Y}(0) = \mathbf{I}_n$ is just $\mathbf{Y} = (\mathbf{X}^*)^{-1}$, where $\mathbf{X}$ is the basis for the solutions $\mathcal{S}$ of $X' = \mathbf{A}(t)X$ satisfying $\mathbf{X}(0) = \mathbf{I}_n$. This follows since (see Chapter 2, exercise 13)

$$\mathbf{Y}' = -(\mathbf{X}^*)^{-1}(\mathbf{X}^*)'(\mathbf{X}^*)^{-1} = -(\mathbf{X}^*)^{-1}\mathbf{X}^*\mathbf{A}^*(\mathbf{X}^*)^{-1} = -\mathbf{A}^*\mathbf{Y},$$

$$\mathbf{Y}(0) = (\mathbf{X}^*)^{-1}(0) = \mathbf{I}_n.$$

Let $\mathcal{S}_p^*$ denote the set of all $Y \in \mathcal{S}^*$ which have period ω. The equality

$$\mathbf{I}_n - \mathbf{X}^*(\omega) = -\mathbf{X}^*(\omega)[\mathbf{I}_n - \mathbf{Y}(\omega)] \qquad (4.16)$$

shows that $\zeta \in \mathcal{C}^n$ satisfies (4.14) if and only if

$$[\mathbf{I}_n - \mathbf{Y}(\omega)]\zeta = 0,$$

and this is true if and only if $Y = \mathbf{Y}\zeta \in \mathcal{S}_p^*$. Now $\mathbf{X}^*(\omega)\zeta = \zeta$ implies $\zeta^* \mathbf{X}(\omega) = \zeta^*$, and then (4.13) becomes

$$\zeta^* \int_0^\omega \mathbf{X}^{-1}(s) B(s) \, ds = \int_0^\omega Y^*(s) B(s) \, ds = 0$$

for every $Y = \mathbf{Y}\zeta \in \mathcal{S}_p^*$. Moreover, the equality (4.16) and Theorem 4.8 applied to $\mathbf{I}_n - \mathbf{X}(\omega)$ imply that

$$\dim(\mathcal{S}_p^*) = \dim(N(\mathbf{I}_n - \mathbf{Y}(\omega))) = \dim(N(\mathbf{I}_n - \mathbf{X}^*(\omega)))$$

$$= \dim(N(\mathbf{I}_n - \mathbf{X}(\omega))) = \dim(\mathcal{S}_p).$$

We have now proved the following theorem.

Theorem 4.9: *Given* $B \in C(\mathcal{R}, \mathcal{C}^n)$ *which is periodic of period* ω, *there exists an* $X \in \mathcal{S}_p(B)$ *if and only if*

$$\int_0^\omega Y^*(s) B(s) \, ds = 0$$

for every $Y \in \mathcal{S}_p^*$. We have

$$\dim(\mathcal{S}_p^*) = \dim(\mathcal{S}_p),$$

and hence there is a unique $X \in \mathcal{S}_p(B)$ if and only if $\dim(\mathcal{S}_p) = 0$.

As an application of Theorem 4.9 we have the following interesting result.

Theorem 4.10: *There exists an $X \in \mathcal{S}_p(B)$ if and only if a solution X of (4.9) which is bounded on $[0, \infty)$ exists. Equivalently, either $\mathcal{S}_p(B)$ is nonempty or every solution of (4.9) is unbounded on $[0, \infty)$.*

Proof: Clearly, if $X \in \mathcal{S}_p(B)$, then X is bounded on $\mathcal{R}$, and hence on $[0, \infty)$. Conversely, suppose there is a solution X of (4.9) which is bounded by $M > 0$,

$$|X(t)| \leq M, \quad t \in [0, \infty). \tag{4.17}$$

The variation of parameters formula gives

$$X(t) = \mathbf{X}(t)X(0) + \mathbf{X}(t) \int_0^t \mathbf{X}^{-1}(s)B(s) \, ds, \quad t \in \mathcal{R},$$

where $\mathbf{X}$ is the basis for $X' = \mathbf{A}(t)X$ such that $\mathbf{X}(0) = \mathbf{I}_n$. Thus

$$X(\omega) = \mathbf{X}(\omega)X(0) + \eta, \quad \eta = \mathbf{X}(\omega) \int_0^\omega \mathbf{X}^{-1}(s)B(s) \, ds.$$

Since $Y(t) = X(t + \omega)$ is also a solution of (4.9), we have

$$Y(\omega) = \mathbf{X}(\omega)Y(0) + \eta$$

or

$$X(2\omega) = \mathbf{X}(\omega)X(\omega) + \eta = \mathbf{X}^2(\omega)X(0) + [\mathbf{I}_n + \mathbf{X}(\omega)]\eta.$$

A simple induction then shows that for each $k = 1, 2, \ldots$,

$$X(k\omega) = \mathbf{X}^k(\omega)X(0) + [\mathbf{I}_n + \mathbf{X}(\omega) + \cdots + \mathbf{X}^{k-1}(\omega)]\eta. \tag{4.18}$$

Suppose $\mathcal{S}_p(B) = \emptyset$. Then there exists a $\zeta \in \mathcal{C}^n$ satisfying (4.14), $\zeta = \mathbf{X}^*(\omega)\zeta$ such that (4.13) is not valid, that is, $\zeta^*\eta \neq 0$. This is equivalent to the existence of a $Y \in \mathcal{S}_p^*$ such that

$$\int_0^\omega Y^*(s)B(s) \, ds \neq 0.$$

Now $\zeta = \mathbf{X}^*(\omega)\zeta$ implies

$$\zeta^* = \zeta^*\mathbf{X}^2(\omega) = \cdots = \zeta^*\mathbf{X}^k(\omega), \quad k = 1, 2, \ldots,$$

and multiplying (4.18) by ζ^* results in

$$\zeta^*X(k\omega) = \zeta^*X(0) + k\zeta^*\eta, \quad k = 1, 2, \ldots.$$

From (4.17) it follows that

$$k|\zeta^*\eta| = |\zeta^*X(k\omega) - \zeta^*X(0)| \leq 2M|\zeta|, \quad k = 1, 2, \ldots,$$

but this is clearly false since $\zeta^*\eta \neq 0$. Hence $\mathcal{S}_p(B)$ is not empty. $\square$

4.7 Second-order homogeneous equations

Important applications of the results in sections 4.1–4.4 occur in the case of second-order equations. For example, the *Mathieu equation*

$$x'' + (a + b\cos(2t))x = 0,$$

where $a, b \in \mathcal{R}$, arises in the study of wave motion associated with elliptic membranes and elliptic cylinders. The more general *Hill's equation*

$$x'' + a(t)x = 0,$$

where $a(t)$ is real valued and $a(t+\omega) = a(t)$ for some $\omega > 0$, arises in the study of the motion of the moon. In that case $a(t)$ is an even function of period π and is often written as a series

$$a(t) = b_0 + \sum_{k=1}^{\infty} b_k \cos(2kt).$$

Let us first interpret our results in the case of the equation

$$x'' + a_1(t)x' + a_0(t)x = 0, \quad a_j \in C(\mathcal{R}, \mathcal{C}), \tag{4.19}$$

$$a_j(t + \omega) = a_j(t), \quad j = 0, 1,$$

for some $\omega > 0$. The system
$$y' = \mathbf{A}(t)y \tag{4.20}$$

associated with (4.19) has a coefficient matrix $\mathbf{A}$ given by

$$\mathbf{A} = \begin{pmatrix} 0 & 1 \\ -a_0 & -a_1 \end{pmatrix},$$

and $\mathbf{X} = (x_1, x_2)$ is a basis for (4.19) if and only if

$$\tilde{\mathbf{X}} = (\tilde{x}_1, \tilde{x}_2) = \begin{pmatrix} x_1 & x_2 \\ x_1' & x_2' \end{pmatrix}$$

is a basis for (4.20). Let $\mathbf{X}$ be the basis for (4.19) such that $\tilde{\mathbf{X}}(0) = \mathbf{I}_2$. Then the multipliers for (4.20) are the eigenvalues μ_1, μ_2 of $\tilde{\mathbf{X}}(\omega)$, that is, the μ satisfying

$$\det[\mu \mathbf{I}_2 - \tilde{\mathbf{X}}(\omega)] = \mu^2 - 2a\mu + b = 0,$$

$$2a = \operatorname{tr}(\tilde{\mathbf{X}}(\omega)) = x_1(\omega) + x_2'(\omega) = \mu_1 + \mu_2,$$

$$b = \det(\tilde{\mathbf{X}}(\omega)) = \mu_1\mu_2.$$

Note that the Wronskian $W_{\mathbf{X}} = \det(\tilde{\mathbf{X}})$ satisfies

$$\det(\tilde{\mathbf{X}}(\omega)) = \exp\left(-\int_0^\omega a_1(s)\, ds\right) = \mu_1\mu_2,$$

so that if the average of a_1 over a period is zero,

$$\frac{1}{\omega} \int_0^\omega a_1(s) \, ds = 0, \tag{4.21}$$

then $\det(\tilde{\mathbf{X}}(\omega)) = \mu_1 \mu_2 = 1$, and the equation for the μ_j becomes

$$\mu^2 - 2a\mu + 1 = 0, \quad 2a = x_1(\omega) + x_2'(\omega). \tag{4.22}$$

Let us assume (4.21), so that $\mu_2 = 1/\mu_1$, and also assume that the a_j are real valued. Then $\tilde{\mathbf{X}}$ is real and a in (4.22) is real. If $\text{Im}(\mu_1) \neq 0$, then $\mu_2 = \bar{\mu}_1$ and $|\mu_1| = |\mu_2| = 1$, $\text{Re}(\mu_1) = \text{Re}(\mu_2) = a$. Solving (4.22), we may choose

$$\mu_1 = a + \sqrt{a^2 - 1}, \quad \mu_2 = a - \sqrt{a^2 - 1}.$$

There are three cases: (i) $a^2 > 1$, (ii) $a^2 < 1$, (iii) $a^2 = 1$.

In case (i), μ_1 and μ_2 are real and distinct and $\mu_2 = 1/\mu_1$. There is a basis $\mathbf{U} = (u_1, u_2)$ for the solutions of (4.19) having the form

$$u_1(t) = e^{\lambda t} p_1(t), \quad u_2(t) = e^{-\lambda t} p_2(t), \tag{4.23}$$

$$e^{\lambda \omega} = \mu_1, \quad p_j(t + \omega) = p_j(t), \quad j = 1, 2.$$

If $a > 0$ we have $0 < \mu_2 < 1 < \mu_1$ so that $-\text{Re}(\lambda) < 0 < \text{Re}(\lambda)$. Now

$$|u_1(t)| = e^{\text{Re}(\lambda)t} |p_1(t)|, \quad |u_2(t)| = e^{-\text{Re}(\lambda)t} |p_2(t)|,$$

and, since the p_j are not identically zero, there are $\tau_j \in [0, \omega]$ such that $p_j(\tau_j) \neq 0$. Then

$$|u_1(\tau_1 + m\omega)| = |u_1(\tau_1)| e^{\text{Re}(\lambda)m\omega}, \quad |u_2(\tau_2 + m\omega)| = |u_2(\tau_2)| e^{-\text{Re}(\lambda)m\omega}$$

for every integer m. Thus

$$|u_1(\tau_1 + m\omega)| \to \infty, \quad m \to \infty, \quad \text{and} \quad |u_2(\tau_2 + m\omega)| \to \infty, \quad m \to -\infty,$$

and so u_1 is unbounded as $t \to +\infty$, whereas u_2 is unbounded as $t \to -\infty$. Also we have

$$|u_1(t)| \to 0, \quad t \to -\infty, \quad \text{and} \quad |u_2(t)| \to 0, \quad t \to +\infty.$$

If $a < 0$ we have $0 < \mu_1 < 1 < \mu_2$ and the roles of u_1, u_2 are reversed. Thus every nontrivial solution $x = c_1 u_1 + c_2 u_2$ of (4.19), where c_1, c_2 are not both 0, is unbounded on $\mathcal{R}$.

In case (ii), μ_1 and μ_2 are distinct nonreal complex numbers of magnitude one,

$$\mu_1 = e^{i\theta\omega}, \quad \mu_2 = e^{-i\theta\omega}, \quad \theta \in \mathcal{R}, \quad \theta \neq k\pi.$$

There is now a basis $\mathbf{U} = (u_1, u_2)$ of the form (4.23) above, where we can take $\lambda = i\theta$. Since

$$|u_j(t)| = |p_j(t)|, \quad j = 1, 2,$$

all solutions x of (4.19) are bounded on $\mathcal{R}$. In fact $\tilde{\mathbf{U}} = (\tilde{u}_1, \tilde{u}_2)$ is bounded on $\mathcal{R}$.

For case (iii) there is one real multiplier $\mu_1 = a = \pm 1$ of multiplicity 2. If $\mu_1 = 1$ there exists a nontrivial solution u_1 such that $\tilde{u}_1(t + \omega) = \tilde{u}_1(t)$ for all $t \in \mathcal{R}$, so that u_1 is periodic of period ω. A second basis element u_2 may or may not be periodic of period ω. We have u_2 of period ω if and only if $\tilde{\mathbf{U}}(t + \omega) = \tilde{\mathbf{U}}(t)$, $t \in \mathcal{R}$, and this is true if and only if $\tilde{\mathbf{U}}(\omega) = \tilde{\mathbf{U}}(0)$, or

$$\tilde{\mathbf{X}}(\omega) = \tilde{\mathbf{U}}(\omega)\tilde{\mathbf{U}}^{-1}(0) = \mathbf{I}_2.$$

If $\mu_1 = -1$ there is a nontrivial solution u_1 satisfying $\tilde{u}_1(t + \omega) = -\tilde{u}_1(t)$ for all $t \in \mathcal{R}$, and then $\tilde{u}_1(t + 2\omega) = -\tilde{u}_1(t + \omega) = \tilde{u}_1(t)$, so that u_1 has period 2ω. A second basis element u_2 satisfies $\tilde{u}_2(t + \omega) = -\tilde{u}_2(t)$, if and only if $\tilde{\mathbf{U}}(\omega) = -\tilde{\mathbf{U}}(0)$, or

$$\tilde{\mathbf{X}}(\omega) = \tilde{\mathbf{U}}(\omega)\tilde{\mathbf{U}}^{-1}(0) = -\mathbf{I}_2.$$

In this case all solutions have period 2ω.

4.8   Second-order nonhomogeneous equations

Finally, consider the nonhomogeneous equation

$$x'' + a_1(t)x' + a_0(t)x = b(t), \quad a_j, b \in C(\mathcal{R}, \mathcal{C}), \tag{4.24}$$

$$a_j(t + \omega) = a_j(t), \quad b(t + \omega) = b(t)$$

for some $\omega > 0$. The set $\mathcal{S}_{2p}(b)$ of all solutions of (4.24) which have period ω is an affine space

$$\mathcal{S}_{2p}(b) = v + \mathcal{S}_{2p} = \{x \mid x = v + u, \quad u \in \mathcal{S}_{2p}\},$$

where v is a particular solution in $\mathcal{S}_{2p}(b)$ and $\mathcal{S}_{2p}$ is the set of all solutions of the homogeneous equation (4.19) of period ω. The system associated with (4.24) is

$$Y' = \mathbf{A}(t)Y + B(t), \tag{4.25}$$

where

$$\mathbf{A} = \begin{pmatrix} 0 & 1 \\ -a_0 & -a_1 \end{pmatrix}, \quad B = \begin{pmatrix} 0 \\ b \end{pmatrix}.$$

$\mathbf{A}$ and B are clearly periodic with period ω, and a solution x of (4.24) has period ω if and only if $Y = \tilde{x}$ is a solution of (4.25) of period ω. Theorem 4.7 then implies that if the homogeneous equation (4.19) has no solution of period ω other than the trivial one, then (4.24) has a unique solution of period ω for every $b \in C(\mathcal{R}, \mathcal{C})$ which is periodic with period ω.

In the general case, when (4.19) may have solutions of period ω, Theorem 4.9 implies that (4.24) has a solution of period ω if and only if

$$\int_0^\omega Z^*(s)B(s)\ ds = 0$$

for every solution Z of the adjoint homogeneous equation

$$Z' = -\mathbf{A}^*(t)Z$$

which has period ω. If

$$Z = \begin{pmatrix} z_1 \\ z_2 \end{pmatrix},$$

this says that

$$\int_0^\omega \bar{z}_2(s)b(s)\ ds = 0$$

for every Z satisfying

$$z_1' = \bar{a}_0 z_2, \tag{4.26}$$

$$z_2' = -z_1 + \bar{a}_1 z_2$$

which has period ω. For such a Z, $y = z_2$ has period ω, $y' - \bar{a}_1 y$ is differentiable, and

$$(y' - \bar{a}_1 y)' + \bar{a}_0 y = 0. \tag{4.27}$$

Conversely, if $y = z_2$ has period ω and satisfies (4.27) and $z_1 = \bar{a}_1 y - y'$, then Z will be a solution of (4.26) of period ω. The equation (4.27) is called the *adjoint equation* to (4.19). Thus (4.24) has a solution of period ω if and only if

$$\int_0^\omega \bar{y}(s)b(s)\ ds = 0$$

for every solution y of the adjoint homogeneous equation (4.27) which has period ω.

4.9 Notes

A classical discussion of Mathieu's and Hill's equations can be found in [27, pp. 404–428]. Hill's equation is also the subject of a short and accessible book [16]. A short historical commentary on Hill's work can be found in [12, pp. 730–732].

In the 1970s there was a series of remarkable works related to Hill's equation. The first surprise was a close relationship between Hill's equation and a nonlinear partial differential equation. Subsequent research revealed an extremely rich and subtle theory for certain problems related to Hill's equation. References to this recent work can be found in [16, pp. iii–iv].

4.10 Exercises

1. Consider the equation $x' = a(t)x$, where $a \in C(\mathcal{R}, \mathcal{C})$ is periodic with period $\omega > 0$.

(a) Find the solution x satisfying $x(0) = 1$.

(b) Find the constant c such that

$$x(t + \omega) = cx(t), \quad t \in \mathcal{R}.$$

(c) What condition must a satisfy in order that there exist a nontrivial solution (i) of period ω, (ii) of period 2ω ?

(d) In case a is real valued, what are the answers to (c)(i) and (c)(ii)?

(e) If $a(t) = a$, a constant, what must this constant be in order that a nontrivial solution of period 2ω exist ?

2. Consider the nonhomogeneous equation

$$x' = a(t)x + b(t), \tag{4.28}$$

where $a, b \in C(\mathcal{R}, \mathcal{C})$ are periodic with period $\omega > 0$.

(a) Show that a solution x of (4.28) is periodic of period ω if and only if $x(\omega) = x(0)$.

(b) Show that there is a unique solution of period ω if there is no nontrivial solution of the homogeneous equation $x' = a(t)x$ of period ω.

(c) Suppose there is a nontrivial periodic solution of the homogeneous equation of period ω. Show that (4.28) has periodic solutions of period ω if and only if

$$\int_0^\omega e^{-\alpha(t)} b(t) \, dt = 0,$$

where

$$\alpha(t) = \int_0^t a(s) \, ds.$$

(d) Find all solutions of period 2π for the equations

(i)
$$x' = 3x + \cos(t),$$

(ii)
$$x' = \cos(t)x + \sin(2t),$$

(iii)
$$x' = \cos(t)x + 2.$$

3. (a) If $\mathbf{A} \in C(\mathcal{R}, M_n(\mathcal{C}))$, show that $\mathbf{A}$ is periodic of period $\omega \neq 0$, $\omega \in \mathcal{R}$,

$$\mathbf{A}(t + \omega) = \mathbf{A}(t), \quad t \in \mathcal{R},$$

if and only if $\mathbf{A}$ is periodic with period $-\omega$,

$$\mathbf{A}(t - \omega) = \mathbf{A}(t), \quad t \in \mathcal{R}.$$

Thus there is no loss of generality in assuming that $\mathbf{A}$ has period $\omega > 0$ in equation (4.1).

(b) If $\mathbf{A} \in C(\mathcal{R}, M_n(\mathcal{C}))$ is periodic of period $\omega > 0$, show that $\mathbf{A}$ is also periodic with period $k\omega$ for any $k = 2, 3, \ldots$.

4. Consider the equation

$$x'' + \mu x = 0, \quad \mu \in \mathcal{R}.$$

For what values of μ will there exist a nontrivial solution x satisfying $\tilde{x}(0) = \tilde{x}(2\pi)$? Find all solutions for each such μ.

5. Consider the system $X' = \mathbf{A}(t)X$, where $\mathbf{A}$ has period 2π and is given by

$$\mathbf{A}(t) = \begin{pmatrix} -2\sin^2(t) & 1 - 2\sin(t)\cos(t) \\ -1 - \sin(t)\cos(t) & -2\cos^2(t) \end{pmatrix}.$$

(a) Verify that the basis $\mathbf{X}$ satisfying $\mathbf{X}(0) = \mathbf{I}_2$ is given by

$$\mathbf{X}(t) = \begin{pmatrix} \cos(t) & e^{-2t}\sin(t) \\ -\sin(t) & e^{-2t}\cos(t) \end{pmatrix}.$$

(b) Compute $\mathbf{X}(2\pi)$ and find a matrix $\mathbf{R}$ such that $\mathbf{X}(2\pi) = e^{2\pi\mathbf{R}}$. Then write $\mathbf{X}$ as $\mathbf{X}(t) = \mathbf{P}(t)e^{\mathbf{R}t}$, where $\mathbf{P}(t + 2\pi) = \mathbf{P}(t)$, $t \in \mathcal{R}$.

(c) Find all solutions X of $X' = \mathbf{A}(t)X$ such that

$$X(t + 2\pi) = \mu X(t), \quad t \in \mathcal{R},$$

for some $\mu \in \mathcal{R}$. In particular, find all solutions which are periodic with period 2π.

6. Compute a logarithm $\mathbf{B}$ for the following $\mathbf{C} \in M_2(\mathcal{R})$, that is, a matrix $\mathbf{B}$ satisfying $e^{\mathbf{B}} = \mathbf{C}$.

(a)
$$\mathbf{C} = \begin{pmatrix} 3 & 0 \\ 0 & -2 \end{pmatrix},$$

(b)
$$\mathbf{C} = \begin{pmatrix} -1 & 1 \\ 0 & -1 \end{pmatrix},$$

(c)
$$\mathbf{C} = \begin{pmatrix} 1 & \sqrt{3} \\ -\sqrt{3} & 1 \end{pmatrix}.$$

7. Show that if there exists a nontrivial solution of

$$x^{(n)} + a_{n-1}x^{(n-1)} + \cdots + a_0 x = 0, \quad a_j \in \mathcal{R},$$

satisfying $\tilde{x}(0) = \tilde{x}(1)$, then there exists a nontrivial real solution which is periodic with period 1.

8. The toxic waste problem from the introduction to this chapter leads to a homogeneous system

$$\frac{d\tau_1}{dt} = a_{11}\tau_1,$$

$$\frac{d\tau_2}{dt} = a_{21}(t)\tau_1 + a_{22}\tau_2,$$

where a_{11} and a_{22} are constant and $a_{21}(t)$ is a nonconstant function with least period 1. If $\mathbf{X}(t)$ is a basis for this system satisfying $\mathbf{X}(0) = \mathbf{I}$, find $\mathbf{X}(1)$.

9. Let $\mathbf{X}$ be the basis for (4.1) such that $\mathbf{X}(0) = \mathbf{I}_n$, and let $\mu_1, \ldots, \mu_n$ be the multipliers for (4.1), where each is repeated as many times as its multiplicity.

(a) Show that

$$|\mu_1 \cdots \mu_n| = \exp\left(\operatorname{Re} \int_0^\omega \operatorname{tr}(\mathbf{A}(s))\, ds\right).$$

(b) If $\operatorname{Re} \int_0^\omega \operatorname{tr}(\mathbf{A}(s))\, ds < 0$, show that there is at least one μ_j such that $|\mu_j| < 1$.

(c) If $\operatorname{Re} \int_0^\omega \operatorname{tr}(\mathbf{A}(s))\, ds > 0$, show that there is at least one μ_j such that $|\mu_j| > 1$.

10. Let μ be a multiplier for (4.1), and let X be a nontrivial solution satisfying $X(t + \omega) = \mu X(t)$.

(a) Show that

$$X(t + k\omega) = \mu^k X(t), \quad t \in \mathcal{R},$$

for each $k = 1, 2, 3, \ldots$.

(b) If $|\mu| < 1$ show that $|X(s)| \to 0$ as $s \to \infty$. (Hint: If $s \geq 0$ then s/ω lies between two consecutive integers, $k \leq s/\omega < k+1$, $k = 0, 1, \ldots$, and hence $s = t + k\omega$, $0 < t < \omega$. Use (a).)

(c) If $|\mu| > 1$ show that X is unbounded on $[0, \infty)$.

(d) If $|\mu| = 1$ show that

$$0 < m \leq |X(s)| \leq M, \quad s \in \mathcal{R},$$

where

$$m = \inf\{|X(t)|\ |0 \leq t \leq \omega\},$$

$$M = \sup\{|X(t)|\ |0 \leq t \leq \omega\}.$$

11. (a) If $\mathbf{X}$ is the basis for (4.1) satisfying $\mathbf{X}(0) = \mathbf{I}_n$, show that $\mathbf{X}(t + k\omega) = \mathbf{X}(t)\mathbf{X}^k(\omega)$, $k = 1, 2, 3, \ldots$.

(b) Show that $|\mathbf{X}(s)| \to 0$, $s \to \infty$, if and only if $|\mathbf{X}^k(\omega)| \to 0$, $k \to \infty$.

(c) Show that $|\mathbf{X}(s)|$ is bounded on $[0, \infty)$ if and only if $|\mathbf{X}^k(\omega)|$ is bounded for $k = 1, 2, 3, \ldots$.

12. Suppose that $\epsilon_1, \epsilon_2 > 0$ and $\epsilon_1 + \epsilon_2 < 1$. Define an $n \times n$ matrix-valued function $\mathbf{A}(t)$ to be the constant matrix $\mathbf{A}_1$ on intervals of the form

$(K - \epsilon_1, K + \epsilon_2)$, where $K = 0, \pm 1, \pm 2, \ldots$, and another constant matrix $\mathbf{A}_2$ on intervals of the form $(K + \epsilon_2, K + 1 - \epsilon_1)$. Notice that $\mathbf{A}(t+1) = \mathbf{A}(t)$.

The differential equation

$$X' = \mathbf{A}(t)X$$

does not have a continuous coefficient matrix $\mathbf{A}(t)$, but we may consider as solutions those continuous functions $X(t)$ which satisfy the equation on intervals where $\mathbf{A}(t)$ is continuous.

(a) Using the existence and uniqueness theorem for linear systems, show that the initial value problem

$$X' = \mathbf{A}(t)X, \quad X(t_0) = \xi$$

has a unique solution for any $t_0 \in \mathcal{R}$ and $\xi \in \mathcal{F}^n$.

(b) If $\mathbf{X}(t)$ is the basis satisfying $\mathbf{X}(0) = I$, describe $\mathbf{X}(1)$. (Hint: What is $\mathbf{X}(\epsilon_2)$?)

(c) Describe the relevance of this problem to the toxic discharge problem from the introduction to this chapter.

13. Consider the system

$$X' = \mathbf{A}X, \quad \mathbf{A} \in M_n(\mathcal{C}), \quad t \in \mathcal{R}. \tag{4.29}$$

The coefficient function $\mathbf{A}(t) = \mathbf{A}$ has period ω for any $\omega > 0$.

(a) Show that (4.29) has a nontrivial solution of period $\omega > 0$ if and only if $\mathbf{A}$ has an eigenvalue λ of the form $\lambda = 2\pi i k/\omega$, $k = 0, \pm 1, \pm 2, \ldots$. (Hint: The eigenvalues μ of $\mathbf{E} = e^{\mathbf{A}\omega}$ have the form $\mu = e^{\lambda\omega}$, where λ is an eigenvalue of $\mathbf{A}$.)

(b) If $\lambda = 2\pi i k/\omega$, $k = 0, \pm 1, \pm 2, \ldots$, is an eigenvalue of $\mathbf{A}$, show that (4.29) has a nontrivial solution of period ω of the form

$$X(t) = e^{\lambda t}\xi, \quad \xi \in \mathcal{C}^n.$$

(c) Show that (4.29) has a nontrivial solution X satisfying

$$X(t + \omega) = \mu X(t), \quad t \in \mathcal{R}, \tag{4.30}$$

for some

$$\mu = |\mu|e^{i\theta}, \quad 0 \le \theta < 2\pi,$$

if and only if $\mathbf{A}$ has an eigenvalue λ of the form

$$\lambda = \frac{1}{\omega}[\log(|\mu|) + i\theta + 2\pi ik], \quad k = 0, \pm 1, \pm 2, \ldots. \tag{4.31}$$

(d) If λ is given by (4.31) show that (4.29) has a nontrivial solution satisfying (4.30) of the form

$$X(t) = e^{\lambda t}\xi, \quad \xi \in \mathcal{C}^n.$$

14. Suppose X is a nontrivial solution of (4.1) satisfying

$$X(t + \omega) = \mu X(t), \quad t \in \mathcal{R},$$

for some $\mu \in \mathcal{C}$. Show that

$$X(t) = e^{\lambda t} P(t),$$

where $P(t + \omega) = P(t)$, $t \in \mathcal{R}$, and $e^{\lambda \omega} = \mu$.

15. Show that if (4.9) (with B not the zero function) has a solution X satisfying

$$X(t + \omega) = \mu X(t), \quad t \in \mathcal{R},$$

then $\mu = 1$.

16. Consider the nonhomogeneous equation (4.9) when $\mathbf{A}(t) = \mathbf{A}$, a constant matrix.

(a) Show that $X' = \mathbf{A}X + B(t)$ has a unique solution of period ω if and only if $\mathbf{A}$ has no eigenvalue of the form $\lambda = 2\pi ik/\omega$, $k = 0, \pm 1 \pm 2, \dots$.

(b) Show that $X' = \mathbf{A}X + B(t)$ has a solution of period ω if and only if

$$\zeta^* \int_0^\omega e^{-\mathbf{A}s} B(s) \, ds = 0,$$

for every $\zeta \in \mathcal{C}^n$ satisfying $e^{\mathbf{A}^* \omega} \zeta = \zeta$.

17. Let $C_p(\mathcal{R}, \mathcal{C}^n)$ be the set of all continuous functions $f : \mathcal{R} \to \mathcal{C}^n$ having period $\omega > 0$, and define

$$(f, g) = \int_0^\omega g^*(s) f(s) \, ds, \quad f, g \in C_p(\mathcal{R}, \mathcal{C}^n).$$

(a) Show that $(,)$ is an inner product on $C_p(\mathcal{R}, \mathcal{C}^n)$.

(b) Let $\mathbf{A} \in C(\mathcal{R}, M_n(\mathcal{C}))$ have period ω, and put

$$L(X) = X' - \mathbf{A}X$$

for each $X \in C_p^1(\mathcal{R}, \mathcal{C}^n)$, the set of all $X \in C^1(\mathcal{R}, \mathcal{C}^n)$, having period ω, and define

$$L^*(Y) = -Y' - \mathbf{A}^* Y$$

for $Y \in C_p^1(\mathcal{R}, \mathcal{C}^n)$. Thus L, L^* are operators taking $C_p^1(\mathcal{R}, \mathcal{C}^n)$ into $C_p(\mathcal{R}, \mathcal{C}^n)$. Define the *range* $Ran(L)$ and *nullspace* $N(L)$ by

$$Ran(L) = \{B \in C_p(\mathcal{R}, \mathcal{C}^n) | L(X) = B \text{ for some } X \in C_p^1(\mathcal{R}, \mathcal{C}^n)\},$$

$$N(L) = \{X \in C_p^1(\mathcal{R}, \mathcal{C}^n) | L(X) = 0\},$$

and similarly for $Ran(L^*), N(L^*)$. With these definitions show that Theorem 4.9 can be stated as

(i) $Ran(L) = (N(L^*))^\perp,$

(ii) $$\dim(N(L)) = \dim(N(L^*)).$$

This is a direct analogue of Theorem 4.8.

18. From Theorem 4.7 (or Theorem 4.9) we know that if $\dim(S_p) = 0$ then there is a unique $X \in S_p(B)$ for every $B \in C(\mathcal{R}, \mathcal{C}^n)$ of period ω. Show that if there is an $X \in S_p(B)$ for every $B \in C_p(\mathcal{R}, \mathcal{C}^n)$, then $\dim(S_p) = 0$. (In the terminology of exercise 13, $Ran(L) = C_p(\mathcal{R}, \mathcal{C}^n)$ if and only if $\dim(S_p) = 0$.)

19. Consider the homogeneous system

$$X' = \mathbf{A}(t)X, \quad t \in \mathcal{R},$$

where $\mathbf{A} \in C(\mathcal{R}, M_n(\mathcal{C}))$ has period $\omega > 0$, and let $S(\mu)$ denote the set of all solutions X satisfying

$$X(t + \omega) = \mu X(t), \quad t \in \mathcal{R}.$$

Similarly, let $S^*(\nu)$ denote the set of all solutions Y of the adjoint equation

$$Y' = -\mathbf{A}^*(t)Y, \quad t \in \mathcal{R}.$$

(a) Show that if $\mu \neq 0$ then

$$\dim(S^*(1/\overline{\mu})) = \dim(S(\mu)).$$

(Hint: Show that

$$[\mu \mathbf{I}_n - \mathbf{X}(\omega)]^* = -\overline{\mu}\mathbf{X}^*(\omega)[1/\overline{\mu}\mathbf{I}_n - \mathbf{Y}(\omega)],$$

where $\mathbf{X}$ is the basis of $X' = \mathbf{A}(t)X$ such that $\mathbf{X}(0) = \mathbf{I}_n$ and $\mathbf{Y} = (\mathbf{X}^*)^{-1}$.)

(b) Suppose $\mathbf{A}$ is real, $\mathbf{A}(t) = \overline{\mathbf{A}}(t)$, $t \in \mathcal{R}$. Show that

$$\dim(S^*(1/\mu)) = \dim(S^*(1/\overline{\mu})) = \dim(S(\mu)) = \dim(S(\overline{\mu}))$$

for each $\mu \neq 0$.

20. Show the following extension of formula (4.18):

$$X(t + k\omega) = X(t) + \mathbf{X}(t)[\mathbf{I}_n + \mathbf{X}(\omega) + \cdots + \mathbf{X}^{k-1}(\omega)]C, \quad t \in \mathcal{R},$$

where X satisfies $X' = \mathbf{A}(t)X + B(t)$ and $C = X(\omega) - X(0) = \mathbf{X}(\omega)X(0) + \eta$. (Putting $t = 0$ gives (4.18).)

21. Consider the nonhomogeneous equation

$$X' = \mathbf{A}(t)X + B(t), \tag{4.32}$$

where $\mathbf{A} \in C(\mathcal{R}, M_n(\mathcal{C}))$ and $B \in C(\mathcal{R}, \mathcal{C}^n)$ have period $\omega > 0$, together with the boundary condition

$$X(0) - X(\omega) = \beta, \quad \beta \in \mathcal{C}^n. \tag{4.33}$$

Let $S(B, \beta)$ denote the set of all solutions X of (4.32) satisfying (4.33); thus $S(B, 0) = S_p(B)$.

(a) Show that $S(B, \beta)$ is an affine space,

$$S(B, \beta) = V + S_p,$$

where V is a particular solution of (4.32) satisfying (4.33).

(b) Show that there exists an $X \in S(B, \beta)$ if and only if

$$\int_0^\omega Y^*(s) B(s) \, ds + Y^*(0)\beta = 0$$

for every $Y \in S_p^*$. (Hint: A solution X of (4.32) satisfies (4.33) if and only if there exists a $\xi \in \mathcal{C}^n$ satisfying

$$[\mathbf{I}_n - \mathbf{X}(\omega)]\xi = \eta + \beta,$$

where

$$\eta = \mathbf{X}(\omega) \int_0^\omega \mathbf{X}^{-1}(s) B(s) \, ds$$

and $\mathbf{X}$ is the basis for $X' = \mathbf{A}(t) X$ such that $\mathbf{X}(0) = \mathbf{I}_n$.)

22. The following equations have coefficients with period 2π.

(i)
$$x'' + x = 0,$$

(ii)
$$(2 + \cos(t)) x'' + \cos(t) x = 0.$$

For both of these

(a) find the basis $\mathbf{X} = (u, v)$ satisfying $\tilde{\mathbf{X}}(0) = \mathbf{I}_2$. (Hint: For (ii), $x(t) = 2 + \cos(t)$ is a solution. Try a second solution of the form $y = ux$.)

(b) What are the multipliers?

(c) Determine a basis for the set of all solutions of period 2π.

23. Consider the equation (4.19) with real-valued coefficients a_1, a_0 with the condition (4.21). In case (iii), when $\mu = -1$ is the multiplier, we showed that if there is a basis $\mathbf{U} = (u_1, u_2)$ satisfying

$$\tilde{\mathbf{U}}(t + \omega) = -\tilde{\mathbf{U}}(t), \quad t \in \mathcal{R}, \tag{4.34}$$

then $\mathbf{U}$ is a basis satisfying

$$\tilde{\mathbf{U}}(t + 2\omega) = \tilde{\mathbf{U}}(t), \quad t \in \mathcal{R}. \tag{4.35}$$

Show that if (4.19) has a basis $\mathbf{U}$ satisfying (4.35) (with $\mu = -1$ being the only multiplier), then $\tilde{\mathbf{U}}$ satisfies (4.34), as follows. If $\mathbf{X}$ is the basis satisfying $\tilde{\mathbf{X}}(0) = \mathbf{I}_2$, then a Jordan form $\mathbf{J}$ for $\tilde{\mathbf{X}}(\omega)$ has one of the two forms

(i)
$$\mathbf{J} = -\mathbf{I}_2,$$

(ii)
$$\mathbf{J} = -\mathbf{I}_2 + N, \quad N = \begin{pmatrix} 0 & 1 \\ 0 & 0 \end{pmatrix},$$

and $\tilde{\mathbf{X}}(2\omega) = (\tilde{\mathbf{X}}(\omega))^2 = \tilde{\mathbf{X}}(0) = \mathbf{I}_2$.

24. Consider the general linear equation (4.19) with coefficients a_0, a_1 having period $\omega > 0$ and where we assume that $a_1 \in C^1(\mathcal{R}, \mathcal{C})$, $a_0 \in C(\mathcal{R}, \mathcal{C})$. Show that x satisfies (4.19) if and only if

$$y(t) = x(t) \exp\left(\frac{1}{2}\int_0^t a_1(s)\, ds\right)$$

satisfies Hill's equation

$$y'' + a(t)y = 0, \quad a = a_0 - a_1'/2 - a_1^2/4.$$

25. Consider Hill's equation

$$x'' + a(t)x = 0, \quad t \in \mathcal{R}, \tag{4.36}$$

where $a \in C(\mathcal{R}, \mathcal{R})$, $a(t + \omega) = a(t)$ for some $\omega > 0$, and $a(t) \leq 0$ for $t \in \mathcal{R}$.

(a) If u is the solution of (4.36) satisfying $u(0) = 1$, $u'(0) = 1$, show that u is real.

(b) Show that $u(t) > 0$ for all $t \geq 0$ as follows. If not, let t_1 be the smallest $t > 0$ such that $u(t_1) = 0$. Show that $u'(t_1) < 0$. Then

$$0 \leq \int_0^{t_1} u''(s)\, ds = u'(t_1) - 1,$$

which is a contradiction.

(c) Show that $u(t) \geq 1 + t$ for $t \geq 0$, so that $u(t) \to +\infty$ as $t \to +\infty$. (Hint: $u''(t) = -a(t)u(t) \geq 0$ for $t \geq 0$.)

(d) Conclude that the multipliers μ_1, μ_2 for (4.36) are real and, if equal ($\mu_1 = \mu_2 = \pm 1$), there is no basis of (4.36) consisting entirely of periodic solutions.

26. Let $u \in C^2(I, \mathcal{R})$, $I = [a, b]$, and suppose $u(a) = u(b) = 0$, $u(t) > 0$ for $t \in (a, b)$. Show that

$$(b - a)\int_a^b \frac{|u''(s)|}{u(s)}\, ds > 4.$$

One approach is to let

$$\max\{u(t)|a \leq t \leq b\} = u(c) > 0, \quad a < c < b.$$

The mean value theorem implies there are σ, τ satisfying

$$u(c) - u(a) = u'(\sigma)(c - a), \quad a < \sigma < c,$$

$$u(b) - u(c) = u'(\tau)(b - a), \quad c < \tau < b.$$

Then

$$\frac{1}{c-a} + \frac{1}{b-c} = \frac{1}{u(c)}\left[\frac{u(c)-u(a)}{c-a} - \frac{u(b)-u(c)}{b-c}\right]$$

$$= \frac{u'(\sigma)-u'(\tau)}{u(c)} \le \frac{1}{u(c)}\int_\sigma^\tau |u''(s)|\,ds < \int_a^b \frac{|u''(s)|}{u(s)}\,ds.$$

If $\alpha = c - a$, $\beta = b - c$, then $(\alpha + \beta)^2 \ge 4\alpha\beta$ implies

$$\frac{4}{(b-a)} \le 1/\alpha + 1/\beta < \int_a^b \frac{|u''(s)|}{u(s)}\,ds.$$

27. Consider Hill's equation

$$x'' + a(t)x = 0, \tag{4.37}$$

where $a \in C(\mathcal{R},\mathcal{R})$, $a(t+\omega) = a(t)$, $t \in \mathcal{R}$ for some $\omega > 0$. Suppose μ is a real multiplier for this equation, and u is a nontrivial solution satisfying

$$u(t+\omega) = \mu u(t), \quad t \in \mathcal{R}.$$

Show that we may assume that u is real.

28. Assume that in Hill's equation (4.37) the following hold: (i) a is not the zero function,

$$\text{(ii)} \quad \int_0^\omega a(s)\,ds \ge 0, \quad \text{(iii)} \quad \omega\int_0^\omega |a(s)|\,ds \le 4.$$

(a) Show that any nontrivial real solution u satisfying

$$u(t+\omega) = \mu u(t), \quad t \in \mathcal{R}, \tag{4.38}$$

for some $\mu \in \mathcal{R}$ is such that $u(t) = 0$ for some $t \in [0,\omega]$. (Hint: Suppose $u(t) > 0$ for $t \in [0,\omega]$. Then

$$0 = \int_0^\omega \frac{u''(s)}{u(s)}\,ds + \int_0^\omega a(s)\,ds = \int_0^\omega \left[\frac{u'(s)}{u(s)}\right]^2\,ds + \int_0^\omega a(s)\,ds,$$

which contradicts (i) and (ii).)

(b) Show that all solutions of (4.37) are bounded on $\mathcal{R}$, together with their first derivatives.

To establish (b), show that (i)–(iii) imply that the multipliers μ_1, μ_2 for (4.37) are not real. If they are real, there is a nontrivial real solution u satisfying (4.38) for some $\mu \in \mathcal{R}$. By (a) there are points $a < b$ such that $u(a) = u(b) = 0$, $b - a \le \omega$, and $u(t) \ne 0$, $t \in (a,b)$. Then by exercise 26 and (iii),

$$4 < (b-a)\int_a^b \frac{|u''(s)|}{u(s)}\,ds \le \omega\int_0^\omega |a(s)|\,ds \le 4,$$

which is a contradiction. Thus the multipliers μ_1, μ_2 are distinct nonreal complex numbers ($\mu_2 = \bar{\mu}_1$) of magnitude 1.

Chapter 5

Analytic Coefficients

5.1 Introduction

Linear systems of ordinary differential equations with constant coefficients have the great advantage of being explicitly solvable. (This sweeping assertion ignores the nontrivial problem of solving the polynomial equations for the eigenvalues.) To significantly extend the class of equations whose solutions can be "exhibited," the class of functions in which solutions are found must also be extended. A fruitful approach is to consider functions with power series representations.

An example of a power series representation for a solution arose in the case of constant coefficient systems

$$X' = \mathbf{A}X, \quad X(0) = \xi,$$

where $\mathbf{A}$ was a constant $n \times n$ matrix and $\xi \in \mathcal{F}^n$. In that case a formal argument led to the representation of the solution $X(t)$ as

$$X(t) = \sum_{k=0}^{\infty} \frac{\mathbf{A}^k t^k}{k!} \xi, \quad t \in \mathcal{R}.$$

The argument was formal in the sense that we neither described precisely what was meant by this sum of infinitely many terms nor considered when such a series might be integrated or differentiated term by term.

The next two sections of this chapter present the basic material on convergence of sequences and series which is needed to justify the earlier formal calculations. These ideas are then applied to the more general case of first-order linear systems

$$X' = \mathbf{A}(t)X + B(t), \tag{5.1}$$

where the components of the matrix-valued function $\mathbf{A}$ and the vector-valued function B can be written as convergent power series in some open interval about a point τ. In this case it is not difficult to show that the solution $X(t)$, with initial condition $X(\tau) = \xi$, may also be represented as a convergent power series. Moreover, the power series coefficients of the solution may be explicitly calculated from the power series coefficients of $\mathbf{A}$ and B.

Solutions of many of the important differential equations arising in applications can be expressed using power series and their generalizations. After the results for general systems and nth order equations are established, one of these examples, the *Legendre equation*, is considered. This equation arises in the study of wave phenomenon or heat conduction for spherical bodies. As with many other examples, the general theory only hints at the rich structure possessed by particular equations.

5.2   Convergence

5.2.1   Normed vector spaces

Let us begin the discussion of convergence of series by recalling how these questions are treated in calculus. The problem is to make sense of the sum

$$\sum_{k=0}^{\infty} c_k, \quad c_k \in \mathcal{F}, \tag{5.2}$$

in which infinitely many of the c_k may be nonzero. The behavior of this series is understood by considering the sequence of partial sums s_m, where the mth partial sum is the sum of the first m terms,

$$s_m = \sum_{k=0}^{m-1} c_k.$$

In general if $a_m \in \mathcal{F}$, the sequence $\{a_m\}$ *converges to a limit* L, or

$$\lim_{m \to \infty} a_m = L,$$

if for every $\epsilon > 0$ there is an integer M such that $m > M$ implies $|a_m - L| < \epsilon$. The series

$$\sum_{k=0}^{\infty} c_k, \quad c_k \in \mathcal{F},$$

converges to (the sum) s if its sequence $\{s_m\}$ of partial sums converges to s, i.e.,

$$\lim_{m \to \infty} s_m = s.$$

For applications to differential equations it is important to consider series whose terms c_k may be vectors in $\mathcal{F}^n$, matrices, or vector- or matrix-valued functions. Consider for example the power series for the matrix exponential function

$$\exp(\mathbf{A}t) = \sum_{k=0}^{\infty} \frac{\mathbf{A}^k t^k}{k!}, \quad t \in \mathcal{R}, \quad \mathbf{A} \in M_n(\mathcal{F}).$$

If t is fixed then the summands are the $n \times n$ matrices $\mathbf{A}^k t^k/k!$. For some purposes, however, it will be important to simultaneously consider all t in some interval I. In this context the summands are the functions $t \to \mathbf{A}^k t^k/k!$, and the partial sums are the matrix polynomial functions

$$s_m(t) = \sum_{k=0}^{m-1} \frac{\mathbf{A}^k t^k}{k!}, \quad t \in I.$$

In order to define convergence of such series, we need a generalization of the absolute value which can serve to indicate the size of a vector, a matrix, or a function.

With this goal in mind, a *normed vector space* is a vector space $\mathcal{V}$ over a field $\mathcal{F}$ (which is either $\mathcal{C}$ or $\mathcal{R}$) with a *norm* $\| \ \|$, which is a function $X \in \mathcal{V} \to \|X\| \in \mathcal{R}$ such that, for all $X, Y \in \mathcal{V}$,

(a) $$\|X\| \geq 0,$$

(b) $$\|X\| = 0, \quad \text{if and only if } X = 0,$$

(c) $$\|X + Y\| \leq \|X\| + \|Y\|,$$

(d) $$\|\alpha X\| = |\alpha| \, \|X\|, \quad \alpha \in \mathcal{F}.$$

The property (c) implies that

$$\left| \|X\| - \|Y\| \right| \leq \|X + (-Y)\| \leq \|X\| + \|Y\|.$$

Important norms on $X = (x_1, \ldots, x_n) \in \mathcal{F}^n$ are given by

$$\|X\|_1 = |x_1| + \cdots + |x_n|,$$

$$\|X\|_2 = (|x_1|^2 + \cdots + |x_n|^2)^{1/2},$$

$$\|X\|_\infty = \sup\{|x_1|, \ldots, |x_n|\}.$$

It is not difficult to see that

$$\|X\|_2 \leq \|X\|_1 \leq \sqrt{n} \, \|X\|_2,$$

$$\|X\|_\infty \leq \|X\|_1 \leq n \, \|X\|_\infty,$$

and

$$\|X\|_\infty \leq \|X\|_2 \leq \sqrt{n} \, \|X\|_\infty.$$

It turns out that any two norms $\| \ \|$, $\| \ \|_+$ on a finite-dimensional vector space $\mathcal{V}$ are *equivalent* in the sense that there exist two positive numbers m, M such that

$$m\|X\| \leq \|X\|_+ \leq M\|X\|, \quad X \in \mathcal{V}. \tag{5.3}$$

For this reason, many of the results which involve a norm on $\mathcal{V}$ are independent of which norm is used. We will adopt the first of the above norms as the standard one for $\mathcal{F}^n$ and denote it simply by $|\ |$,

$$|X| = \|X\|_1 = |x_1| + \cdots + |x_n|, \quad X = (x_1, \ldots, x_n) \in \mathcal{F}^N.$$

A *metric space* is a set $\mathcal{V}$ with a *distance function*, or *metric*, which is a function

$$d : (X, Y) \in \mathcal{V} \times \mathcal{V} \to d(X, Y) \in \mathcal{R}$$

such that for all $X, Y, Z \in \mathcal{V}$,

(a) $$d(X, Y) \geq 0,$$

(b) $$d(X, Y) = 0 \quad \text{if and only if} \quad X = Y,$$

(c) $$d(X, Y) = d(Y, X),$$

(d) $$d(X, Z) \leq d(X, Y) + d(Y, Z).$$

The inequality (d) is called the *triangle inequality*. From the properties of a norm it is immediate that every normed vector space $\mathcal{V}$ is a metric space with a metric d given by

$$d(X, Y) = \|X - Y\|, \quad X, Y \in \mathcal{V}. \tag{5.4}$$

In particular, $\mathcal{R}^n$ and $\mathcal{C}^n$ are metric spaces with

$$d(X, Y) = |X - Y| = |x_1 - y_1| + \cdots + |x_n - y_n|. \tag{5.5}$$

If $\mathcal{V}$ is a metric space with a metric d, a sequence $X_k \in \mathcal{V}$, $k \in \{0, 1, 2, \ldots\}$ *converges* if there is an $X \in \mathcal{V}$ such that

$$d(X_k, X) \to 0, \quad \text{as} \quad k \to \infty.$$

In this case $\{X_k\}$ has a *limit* X, which is also written

$$\lim_{k \to \infty} X_k = X, \quad \text{or} \quad X_k \to X \quad \text{as} \quad k \to \infty.$$

In case $\mathcal{V}$ is a normed vector space with $d(X, Y) = \|X - Y\|$, the sequence $\{X_k\}$ converges to X, or

$$\lim_{k \to \infty} X_k = X$$

if $X \in \mathcal{V}$ and

$$\|X_k - X\| \to 0.$$

Some of the usual properties for limits hold in a normed vector space. If $X_k \to X$, $Y_k \to Y$ as $k \to \infty$, and $\alpha \in \mathcal{F}$, then

$$X_k + Y_k \to X + Y, \quad k \to \infty,$$

$$\alpha X_k \to \alpha X, \quad k \to \infty.$$

A *Cauchy sequence* $\{X_k\}$ is one such that

$$d(X_k, X_l) \to 0, \quad \text{as} \quad k, l \to \infty.$$

That is, $\{X_k\}$ is a Cauchy sequence if for every $\epsilon > 0$ there is an integer M such that $k, l > M$ implies $d(X_k, X_l) < \epsilon$. A metric space is *complete* if every Cauchy sequence $\{X_k\}$, $X_k \in \mathcal{V}$ has a limit $X \in \mathcal{V}$. A *Banach space* is a normed vector space $\mathcal{V}$ which is complete in the metric (5.4).

The field $\mathcal{F}$, where $\mathcal{F}$ is $\mathcal{C}$ or $\mathcal{R}$, is a normed vector space over $\mathcal{F}$ with the norm of $x \in \mathcal{F}$ given by the absolute value $|x|$. It is a basic result from analysis (see [17, p. 50] or [22, pp. 46–47]) that this space is complete, so that $\mathcal{F}$ is a Banach space.

Using the completeness of $\mathcal{F}$ it is not difficult to see that $\mathcal{F}^n$ is a Banach space; that is, it is complete with the metric (5.5). Suppose $X_k \in \mathcal{F}^n$ with

$$X_k = (x_{1,k}, \ldots, x_{n,k}), \quad k = 1, 2, 3, \ldots.$$

If $\{X_k\}$ is a Cauchy sequence, then

$$|X_k - X_l| = |x_{1,k} - x_{1,l}| + \cdots + |x_{n,k} - x_{n,l}| \to 0, \quad k, l \to \infty.$$

This implies that each of the component sequences satisfies

$$|x_{j,k} - x_{j,l}| \to 0, \quad k, l \to \infty, \quad j = 1, \ldots, n.$$

Thus each $\{x_{j,k}\}$ is a Cauchy sequence in $\mathcal{F}$ and so has a limit x_j. Consequently, if $X = (x_1, \ldots, x_n)$, then $X_k \to X$, and so $\mathcal{F}^n$ is complete.

Define a norm $|\ |$ on $M_{mn}(\mathcal{F})$ by

$$|\mathbf{A}| = \sum_{j=1}^{n} \sum_{i=1}^{m} |a_{ij}|, \quad \mathbf{A} = (a_{ij}) \in M_{mn}(\mathcal{F}). \tag{5.6}$$

Using the previous argument, $M_{mn}(\mathcal{F})$ is a Banach space with this norm. The norm of a product of matrices satisfies

$$|\mathbf{AB}| \le |\mathbf{A}||\mathbf{B}|. \tag{5.7}$$

To see this, let $\mathbf{A} = (a_{ij})$, $\mathbf{B} = (b_{jk})$. If $\mathbf{C} = (c_{ik}) = \mathbf{AB}$, then

$$|\mathbf{AB}| = |\mathbf{C}| = \sum_{k=1}^{p} \sum_{i=1}^{m} |c_{ik}| = \sum_{k=1}^{p} \sum_{i=1}^{m} \left| \sum_{j=1}^{n} a_{ij} b_{jk} \right|$$

$$\leq \sum_{k=1}^{p} \sum_{i=1}^{m} \sum_{j=1}^{n} |a_{ij}||b_{jk}|$$

$$\leq \left(\sum_{j=1}^{n} \sum_{i=1}^{m} |a_{ij}| \right) \left(\sum_{l=1}^{n} \sum_{k=1}^{m} |b_{lk}| \right) = |\mathbf{A}||\mathbf{B}|.$$

Notice that the norm (5.6) reduces to our earlier choice for the norm on $\mathcal{F}^n$ in the case of $M_{n1}(\mathcal{F}) = \mathcal{F}^n$, and the property (5.7) shows that

$$|\mathbf{A}X| \leq |\mathbf{A}||X|.$$

Having considered norms on $\mathcal{F}^n$ and $M_{mn}(\mathcal{F})$, let us return to the problem of convergence of infinite series. Suppose that $\{X_k\}$ and S belong to the normed vector space $\mathcal{V}$. Then we say that

$$\sum_{k=0}^{\infty} X_k$$

converges to $S \in \mathcal{V}$, or

$$S = \sum_{k=0}^{\infty} X_k,$$

if the sequence of partial sums

$$S_m = \sum_{k=0}^{m-1} X_k$$

converges to S,

$$\lim_{m \to \infty} \|S_m - S\| = 0.$$

If the normed vector space $\mathcal{V}$ is complete with respect to its norm $\| \; \|$, that is, $\mathcal{V}$ is a Banach space, then convergence of a series in $\mathcal{V}$ can often be established by showing the convergence of a related series of nonnegative real numbers. Say that the series

$$\sum_{k=0}^{\infty} X_k, \quad X_k \in \mathcal{V},$$

converges absolutely if

$$\sum_{k=0}^{\infty} \|X_k\|$$

converges.

Theorem 5.1: *If $\{X_k\}$ is a sequence from a Banach space $\mathcal{V}$ and the series*

$$\sum_{k=0}^{\infty} X_k$$

converges absolutely, then it converges.

Proof: Pick an $\epsilon > 0$ and let $\delta = \epsilon/2$. The assumption that

$$\sum_{k=0}^{\infty} X_k$$

converges absolutely means

$$\sigma_m = \sum_{j=0}^{m-1} \|X_j\| \to \sigma,$$

or there is an M such that $m > M$ implies

$$|\sigma_m - \sigma| < \delta.$$

Suppose that $l \geq k > M$. Then

$$\delta \geq |\sigma_l - \sigma| = |\sigma_k + \sum_{j=k}^{l-1} \|X_j\| - \sigma|.$$

Since $k > M$ we also have $|\sigma_k - \sigma| < \delta$, and so

$$\sum_{j=k}^{l-1} \|X_j\| < 2\delta = \epsilon.$$

Looking at the partial sums

$$S_k = \sum_{j=0}^{k-1} X_k,$$

we have

$$\|S_l - S_k\| = \left\|\sum_{j=0}^{l-1} X_j - \sum_{j=0}^{k-1} X_j\right\|.$$

Then

$$\|S_l - S_k\| = \left\|\sum_{j=k}^{l-1} X_j\right\| \leq \sum_{j=k}^{l-1} \|X_j\| < \epsilon,$$

and so $l \geq k > M$ implies

$$\|S_l - S_k\| < \epsilon,$$

or the sequence of partial sums $\{S_k\}$ is a Cauchy sequence.

Since $\{S_k\}$ is a Cauchy sequence in a Banach space $\mathcal{V}$, it converges to some limit $S \in \mathcal{V}$, and so the series

$$\sum_{k=0}^{\infty} X_k$$

converges. $\square$

For an example of an absolutely convergent series, consider the exponential function

$$\exp(\mathbf{A}t) = \sum_{k=0}^{\infty} \frac{\mathbf{A}^k t^k}{k!}, \quad t \in \mathcal{R}.$$

In this series each summand is an $n \times n$ matrix.

The inequality (5.7) and a simple induction argument imply that for any square matrix $\mathbf{B}$,

$$|\mathbf{B}^k| \leq |\mathbf{B}|^k,$$

and so

$$|\frac{\mathbf{A}^k t^k}{k!}| \leq \frac{|\mathbf{A}|^k |t|^k}{k!}.$$

The terms on the right are the summands for the series $\exp(|\mathbf{A}||t|)$, which converges for all values of $|\mathbf{A}|$ and $|t|$, and so the matrix exponential series is always convergent.

The idea of rearranging the terms in a series was used in (4.6) to show that every nonsingular $n \times n$ matrix has a logarithm. Suppose that $p(k)$ is a permutation of the nonnegative integers, that is, a one-to-one function from $\{0, 1, 2, \ldots\}$ onto $\{0, 1, 2, \ldots\}$. A series

$$\sum_{k=0}^{\infty} X_{p(k)}$$

is then called a *rearrangement* of the series

$$\sum_{k=0}^{\infty} X_k.$$

Suppose the series $\sum_{k=0}^{\infty} X_k$ converges absolutely, and let $\epsilon > 0$. Informally, if the series converges to S, then there is an N such that the sum of the first N terms is within ϵ of S and the sum of the norms of any finite collection of the remaining terms is less than ϵ. If this series is rearranged, we merely take M large enough so that the terms $X_0, \ldots, X_{N-1}$ appear as terms in the sum $\sum_{k=0}^{M-1} X_{p(k)}$. Then if $L \geq M$ the difference $|\sum_{k=0}^{L-1} X_{p(k)} - S|$ is no more than 2ϵ, so the rearranged series converges to the same value as the original series. The reader is asked to make this argument precise, and thus prove the following theorem, in exercise 5.

Theorem 5.2: *If $\{X_k\}$ is a sequence from a Banach space $\mathcal{V}$ and the series*

$$\sum_{k=0}^{\infty} X_k$$

converges absolutely, then any rearrangement converges to the same sum.

5.2.2   Normed vector spaces of functions

As noted above, the series defining the exponential function

$$\exp(\mathbf{A}t) = \sum_{k=0}^{\infty} \frac{\mathbf{A}^k t^k}{k!}, \quad t \in \mathcal{R}, \quad \mathbf{A} \in M_n(\mathcal{F}),$$

converges for every value of t. The fact that the series provides a well-defined function is not sufficient for applications to differential equations. We would like to know that this function is differentiable and integrable and that the derivative or integral can be computed by termwise differentiation or integration. This goal requires more refined information about the convergence of the functions defined by the partial sums

$$S_m(t) = \sum_{k=0}^{m-1} X_k(t)$$

to the limit function $S(t)$. This section provides the groundwork for such an analysis.

If $I \subset \mathcal{R}$ is an interval, the set $C(I, \mathcal{F})$ of all continuous functions is a vector space over $\mathcal{F}$, since $f + g$ and αf are in $C(I, \mathcal{F})$ if $f, g \in C(I, \mathcal{F})$, $\alpha \in \mathcal{F}$. If $I = [a, b]$ is a compact interval, then every $f \in C(I, \mathcal{F})$ is bounded. This allows us to define a norm on $C(I, \mathcal{F})$, the *sup norm*,

$$\|f\|_\infty = \sup\{|f(t)| \mid t \in I\}. \tag{5.8}$$

Since $|f|$ assumes its maximum on I, there is a point $t_0 \in I$ such that $|f(t_0)| = \|f\|_\infty$.

It is not difficult to verify that $\|\ \|_\infty$ is a norm. Clearly, $\|f\|_\infty \geq 0$, and $\|f\|_\infty = 0$ if and only if $f = 0$. If $f, g \in C(I, \mathcal{F})$, then

$$|(f + g)(t)| = |f(t) + g(t)| \leq |f(t)| + |g(t)| \leq \|f\|_\infty + \|g\|_\infty,$$

so

$$\|f + g\|_\infty \leq \|f\|_\infty + \|g\|_\infty.$$

Finally, if $\alpha \in \mathcal{F}$, $f \in C(I, \mathcal{F})$, then $|(\alpha f)(t)| = |\alpha||f(t)|$, and hence

$$\|\alpha f\|_\infty = \sup\{|(\alpha f)(t)| \mid t \in I\} = |\alpha| \sup\{|f(t)| \mid t \in I\} = |\alpha|\ \|f\|_\infty.$$

Thus if I is compact, $C(I, \mathcal{F})$ is a normed vector space with the norm given in (5.8), and it is a metric space with the metric

$$d(f, g) = \|f - g\|_\infty. \tag{5.9}$$

If $f_k, f \in C(I, \mathcal{F})$ and

$$\|f_k - f\|_\infty \to 0 \quad \text{as} \quad k \to \infty, \tag{5.10}$$

then given any $\epsilon > 0$ there is a positive integer N such that $\|f_k - f\|_\infty < \epsilon$ for $k > N$, and (5.10) implies that

$$|f_k(t) - f(t)| < \epsilon, \quad k > N, \quad t \in I. \tag{5.11}$$

Hence $f_k \to f$ *uniformly* on I; that is, the N in (5.11) is independent of $t \in I$. Conversely, if (5.11) is valid, then

$$\|f_k - f\|_\infty = \sup\{|f_k(t) - f(t)| \mid t \in I\} < \epsilon, \quad k > N,$$

and therefore (5.10) is valid. Thus convergence in the metric (5.9) is the same as *uniform convergence* on I.

It is a basic fact from analysis (see Theorem 7.1 or [17, p. 271]) that $C(I, \mathcal{F})$ is complete in the metric (5.9), and hence $C(I, \mathcal{F})$ with the norm $\| \ \|_\infty$ is a Banach space.

There are other useful norms for $C(I, \mathcal{F})$; among these are $\| \ \|_1$ and $\| \ \|_2$, where

$$\|f\|_1 = \int_a^b |f(t)| dt,$$

$$\|f\|_2 = \left(\int_a^b |f(t)|^2 dt \right)^{1/2}.$$

Although $C(I, \mathcal{F})$ is a metric space with either of the two metrics $d_1(f, g) = \|f - g\|_1$ or $d_2(f, g) = \|f - g\|_2$, it is not a complete space with these metrics. A Cauchy sequence in either of these metrics may have a limit which is not continuous on I (see exercise 9).

If $\mathbf{A} = (a_{ij})$ is a matrix-valued function which is defined on an interval I, except perhaps at $c \in I$, we say $\mathbf{A}$ has $L = (L_{ij}) \in M_{mn}(\mathcal{F})$ as a *limit* at c and write

$$\lim_{t \to c} \mathbf{A}(t) = L, \quad \text{or} \quad \mathbf{A}(t) \to L \quad \text{as} \quad t \to c$$

if $|\mathbf{A}(t) - L| \to 0$ as $0 < |t - c| \to 0$. Since

$$|a_{ij}(t) - L_{ij}| \le |\mathbf{A}(t) - L| = \sum_{j=1}^n \sum_{i=1}^m |a_{ij}(t) - L_{ij}|,$$

it follows that $\mathbf{A}(t) \to L$ as $t \to c$ if and only if $a_{ij}(t) \to L_{ij}$ for all $i = 1, \ldots, m$, and $j = 1, \ldots, n$. Thus the usual rules for limits hold. If $\mathbf{A}, \mathbf{B} : I \setminus \{c\} \to M_{mn}(\mathcal{F})$ are such that $\mathbf{A}(t) \to L$, $\mathbf{B}(t) \to M$, as $t \to c$, then

$$(\mathbf{A} + \mathbf{B})(t) \to L + M, \quad \text{as} \quad t \to c,$$

$$(\alpha \mathbf{A})(t) \to \alpha L.$$

Moreover, if $\mathbf{C} : I \setminus \{c\} \to M_{np}(\mathcal{F})$ and $\mathbf{C}(t) \to N$ as $t \to c$, then $(\mathbf{AC})(t) \to LN$ as $t \to c$.

We say $\mathbf{A} \in F(I, M_{mn}(\mathcal{F}))$ is *continuous* at $c \in I$ if $\mathbf{A}(t) \to \mathbf{A}(c)$ as $t \to c$. Clearly $\mathbf{A} = (a_{ij})$ is continuous at c if and only if each a_{ij} is continuous at c. We say $\mathbf{A}$ is continuous on I if it is continuous at each $c \in I$. The set of all $\mathbf{A} \in F(I, M_{mn}(\mathcal{F}))$ which are continuous on I is denoted by $C(I, M_{mn}(\mathcal{F}))$. It is easy to check that $\mathbf{A}+\mathbf{B}$ and $\alpha\mathbf{A}$ are in $C(I, M_{mn}(\mathcal{F}))$ if $\mathbf{A}, \mathbf{B} \in C(I, M_{mn}(\mathcal{F}))$ and $\alpha \in \mathcal{F}$. Hence $C(I, M_{mn}(\mathcal{F}))$ is a vector space over $\mathcal{F}$. Also if $\mathbf{C} \in C(I, M_{np}(\mathcal{F}))$, then $\mathbf{A}\mathbf{C} \in C(I, M_{mp}(\mathcal{F}))$.

The properties of $C(I, \mathcal{F})$ now carry over to the more general function space $C(I, M_{mn}(\mathcal{F}))$. Let $I = [a, b]$ be a compact interval. Then if $\mathbf{A} = (a_{ij}) \in C(I, M_{mn}(\mathcal{F}))$, the boundedness of each a_{ij} on I implies that there is a constant $M > 0$ such that for $t \in I$

$$|\mathbf{A}(t)| = \sum_{j=1}^{n} \sum_{i=1}^{m} |a_{ij}(t)| \leq M,$$

and a norm $\| \ \|_\infty$ can be defined on $C(I, M_{mn}(\mathcal{F}))$ by

$$\|\mathbf{A}\|_\infty = \sup\{|\mathbf{A}(t)| \ \big| \ t \in I\}, \quad \mathbf{A} \in C(I, M_{mn}(\mathcal{F})).$$

The proof that $\| \ \|_\infty$ is a norm follows just as in the case $C(I, \mathcal{F})$. Note that

$$\|a_{ij}\|_\infty \leq \|\mathbf{A}\|_\infty \leq \sum_{j=1}^{n} \sum_{i=1}^{m} \|a_{ij}\|_\infty,$$

since

$$|a_{ij}| \leq |\mathbf{A}(t)| \leq \|\mathbf{A}\|_\infty, \quad t \in I,$$

and

$$|\mathbf{A}(t)| \leq \sum_{j=1}^{n} \sum_{i=1}^{m} \|a_{ij}\|_\infty, \quad t \in I.$$

Now $C(I, M_{mn}(\mathcal{F}))$ with the metric d given by

$$d(\mathbf{A}, \mathbf{B}) = \|\mathbf{A} - \mathbf{B}\|_\infty, \quad \mathbf{A}, \mathbf{B} \in C(I, M_{mn}(\mathcal{F})), \tag{5.12}$$

is a metric space. A particular case is $C(I, \mathcal{F}^n)$. This is a normed vector space with the norm $\| \ \|_\infty$ given by

$$\|F\|_\infty = \sup\{|F(t)| \ \big| \ t \in I\},$$

where

$$F = \begin{pmatrix} f_1 \\ \vdots \\ f_n \end{pmatrix}, \quad |F(t)| = |f_1(t)| + \cdots + |f_n(t)|.$$

Hence $C(I, \mathcal{F}^n)$ is a metric space with the metric d given by

$$d(F, G) = \|F - G\|_\infty, \quad F, G \in C(I, \mathcal{F}^n).$$

In Chapter 7 we will show the $C(I, M_{mn}(\mathcal{F}))$, and in particular $C(I, \mathcal{F}^n)$, is complete in the metric (5.12), so that with the norm $\| \ \|_\infty$ it is a Banach space.

Integration of matrix-valued functions was introduced in section 2.2.3. The important inequality satisfied by the integral is

$$\left| \int_a^b \mathbf{A}(t) \, dt \right| \leq \int_a^b |\mathbf{A}(t)| \, dt.$$

This follows easily from the result for $m = n = 1$:

$$\left| \int_a^b \mathbf{A}(t) \, dt \right| = \sum_{j=1}^n \sum_{i=1}^m \left| \int_a^b a_{ij}(t) \, dt \right|$$

$$\leq \sum_{j=1}^n \sum_{i=1}^m \int_a^b |a_{ij}(t)| \, dt = \int_a^b \sum_{j=1}^n \sum_{i=1}^m |a_{ij}(t)| \, dt = \int_a^b |\mathbf{A}(t)| \, dt.$$

If $\mathbf{A} \in C(I, M_{mn}(\mathcal{F}))$ and $|I| = b - a$ is the *length* of I, then since $|\mathbf{A}(t)| \leq \|\mathbf{A}\|_\infty$ for all $t \in I$ we have

$$\left| \int_a^b \mathbf{A}(t) \, dt \right| \leq \|\mathbf{A}\|_\infty |I|.$$

As an example, if

$$\mathbf{A}(t) = \begin{pmatrix} t^2 & t - it^2 \\ -t & it^2 \end{pmatrix}, \quad 0 \leq t \leq 1,$$

then

$$\mathbf{A}'(t) = \begin{pmatrix} 2t & 1 - 2it \\ -1 & 2it \end{pmatrix}, \quad \int_0^1 \mathbf{A}(t) \, dt = \begin{pmatrix} 1/3 & 1/2 - i/3 \\ -1/2 & 1/3 \end{pmatrix},$$

$$\left| \int_0^1 \mathbf{A}(t) \, dt \right| = \frac{7 + \sqrt{13}}{6}, \quad |\mathbf{A}(t)| = t + 2t^2 + t\sqrt{1 + t^2},$$

$$\|\mathbf{A}\|_\infty = 3 + \sqrt{2} = |\mathbf{A}(1)|, \quad \int_0^1 |\mathbf{A}(t)| \, dt = \frac{5 + 4\sqrt{2}}{6}.$$

5.3　Analytic functions

Our goal for this chapter was to begin studying differential equations with power series solutions. Having treated some of the preliminary material from analysis, we are now ready to consider functions which are defined by convergent power series. A complex-valued function f defined on some open interval I is said to be *analytic* at $\tau \in I$ if on some interval J of radius ρ centered at τ,

$$J = \{t \mid |t - \tau| < \rho\}, \quad \rho > 0,$$

the function f can be represented by a convergent power series in powers of $t - \tau$,

$$f(t) = \sum_{k=0}^{\infty} a_k(t - \tau)^k, \quad |t - \tau| < \rho, \quad a_k \in \mathcal{C}. \tag{5.13}$$

The function f is analytic on I if it is analytic at each $\tau \in I$. If f is defined by (5.13), where the series is convergent for $|t - \tau| < \rho$, it can be shown that f has a convergent power series expansion about any center τ_0 as long as $|\tau_0 - \tau| < \rho$; that is, the function is analytic on the interval

$$I = \{t \in \mathcal{R} |\; |t - \tau| < \rho\}.$$

To begin the study, it will be helpful to recall two basic results in the theory of series. The first of these is the comparison test, which says that if $|a_k| \leq c_k$ and the series (of nonnegative terms)

$$\sum_{k=0}^{\infty} c_k$$

converges, then the series

$$\sum_{k=0}^{\infty} a_k$$

converges absolutely. A second important test for convergence of series is the ratio test. This test says that if $a_k \in \mathcal{C}$ and

$$|a_{k+1}/a_k| \to r, \quad k \to \infty,$$

then the series converges absolutely (and hence converges) if $r < 1$, and it diverges if $r > 1$.

Our first important result about functions defined by power series describes the set of points where the series converges.

Theorem 5.3: *Suppose that the power series*

$$f(t) = \sum_{k=0}^{\infty} a_k(t - \tau)^k, \quad a_k \in \mathcal{C},$$

converges at a point $t_0 \neq \tau$. Then the series converges absolutely for all t satisfying $|t - \tau| < |t_0 - \tau|$. The convergence is uniform on any compact subinterval $[a, b] \subset \{|t - \tau| < |t_0 - \tau|\}$.

Proof: Since the series

$$\sum_{k=0}^{\infty} a_k(t_0 - \tau)^k$$

is convergent, its terms must tend to zero,

$$|a_k(t_0 - \tau)^k| = |a_k||t_0 - \tau|^k \to 0, \quad k \to \infty.$$

In particular the sequence $|a_k||t_0 - \tau|^k$ is bounded by some positive number M.

Now for $|t - \tau| < |t_0 - \tau|$ we consider the series

$$\sum_{k=0}^{\infty} a_k(t - \tau)^k = \sum_{k=0}^{\infty} a_k(t_0 - \tau)^k \frac{(t - \tau)^k}{(t_0 - \tau)^k}.$$

The terms for the series on the right satisfy

$$\left| a_k(t_0 - \tau)^k \frac{(t - \tau)^k}{(t_0 - \tau)^k} \right| \leq M \left(\frac{|t - \tau|}{|t_0 - \tau|} \right)^k.$$

Since $r = |t - \tau|/|t_0 - \tau| < 1$, this series converges absolutely by comparison with the geometric series

$$\sum_{k=0}^{\infty} Mr^k.$$

On a compact subinterval $[a, b] \subset \{|t - \tau| < |t_0 - \tau|\}$ there is an $R < 1$ such that $|t - \tau|/|t_0 - \tau| < R$ for all $t \in [a, b]$, and so the comparison with a geometric series may be made simultaneously for all such t. □

Suppose that a series with nonnegative coefficients

$$\sum_{k=0}^{\infty} d_k(t - \tau)^k, \quad d_k \geq 0,$$

is convergent for $|t - \tau| < \rho$. If

$$|a_k| \leq d_k, \quad k = 0, 1, 2, \ldots,$$

then the series (5.13) is also convergent for $|t - \tau| < \rho$. To show this, let $|t - \tau| = r < \rho$, and note that

$$|a_k(t - \tau)^k| = |a_k|r^k \leq d_k r^k.$$

Since

$$\sum_{k=0}^{\infty} d_k r^k$$

is convergent, the comparison test implies that (5.13) is absolutely convergent, and hence convergent.

Next we want to show that if $f(t)$ is defined by a power series (5.13), with center τ, which is convergent on the open interval $I = \{|t - \tau| < \rho\}$, then f has all derivatives on this interval I, and these derivatives may be computed by differentiating the series term by term. A similar result holds for termwise integration. These results are established by using a pair of theorems from analysis [17, pp. 295–297].

Theorem 5.4: *Suppose that $\{g_n(t)\}$ is a sequence of complex-valued continuous functions defined on the compact interval $[a,b]$. If $g_n \to g$ uniformly on $[a,b]$, then g is continuous on $[a,b]$ and*

$$\lim_{n\to\infty} \int_a^b g_n(t)\ dt = \int_a^b g(t)\ dt.$$

Theorem 5.5: *Suppose that $\{g_n(t)\}$ is a sequence of complex-valued, continuously differentiable functions defined on the open interval (a,b). If $g_n \to g$ pointwise, and $g_n' \to h$ uniformly on (a,b), then g is differentiable on (a,b) and $g' = h$.*

Theorem 5.4 implies that a convergent power series may be integrated term by term.

Theorem 5.6: *Suppose that*

$$f(t) = \sum_{k=0}^{\infty} a_k(t-\tau)^k, \quad a_k \in \mathcal{C},$$

the series convergent on the open interval $I = \{|t-\tau| < \rho\}$. If $[a,b] \subset I$, then f is continuous on $[a,b]$ and

$$\int_a^b f(t)\ dt = \sum_{k=0}^{\infty} a_k \int_a^b (t-\tau)^k\ dt.$$

Proof: Apply Theorem 5.4 when the functions $g_n(t)$ are the partial sums

$$g_n(t) = \sum_{k=0}^{n-1} a_k(t-\tau)^k.$$

Since these functions are polynomials, they are continuous. The sequence converges uniformly to $f(t)$ on the compact subinterval $[a,b]$ by Theorem 5.3. $\square$

Theorem 5.5 can be used to show that a convergent power series may be differentiated term by term.

Theorem 5.7: *Suppose that*

$$f(t) = \sum_{k=0}^{\infty} a_k(t-\tau)^k, \quad a_k \in \mathcal{C},$$

the series convergent on the open interval $I = \{|t-\tau| < \rho\}$. Then f is differentiable on I and

$$f'(t) = \sum_{k=1}^{\infty} k a_k(t-\tau)^{k-1} = \sum_{k=0}^{\infty} (k+1)a_{k+1}(t-\tau)^k.$$

The power series for $f'(t)$ converges on I.

Proof: We apply Theorem 5.5 when the continuously differentiable functions $g_n(t)$ are the polynomials

$$g_n(t) = \sum_{k=0}^{n-1} a_k(t - \tau)^k.$$

By assumption the functions $g_n(t)$ converge to $g(t)$ pointwise. We need to verify that the sequence of derivatives $\{g_n'(t)\}$ converges uniformly on any interval $|t - \tau| \leq r < \rho$ to the convergent series

$$\sum_{k=1}^{\infty} ka_k(t - \tau)^{k-1}.$$

Let $r < R < \rho$. By assumption the series for f converges for $t - \tau = R$. Using the argument of Theorem 5.3 there is an $M > 0$ such that

$$|ka_k(t - \tau)^{k-1}| \leq kM(r/R)^{k-1}, \quad |t - \tau| \leq r, \quad k \geq 1.$$

By the ratio test the series

$$\sum_{k=1}^{\infty} kM(r/R)^{k-1}$$

converges, so

$$g_n'(t) \to \sum_{k=1}^{\infty} ka_k(t - \tau)^{k-1} = h$$

uniformly on any subinterval $|t - \tau| \leq r < \rho$.

By Theorem 5.5 the function f is differentiable, and f' is given by the desired series. Since this is true for all subintervals of the form $|t - \tau| \leq r < \rho$, the result holds for every point in I. $\square$

Having established Theorem 5.7 we can repeatedly differentiate $f(t)$, obtaining the general formula

$$f^{(l)}(t) = \sum_{k=l}^{\infty} k(k - 1) \cdots (k - l + 1)a_k(t - \tau)^{k-l}, \quad l = 1, 2, \ldots.$$

These differentiated series also converge for $|t - \tau| < \rho$. In particular

$$a_k = f^{(k)}(\tau)/k!, \quad k = 0, 1, 2, \ldots.$$

This formula for the coefficients immediately establishes the next theorem, that each analytic function has a unique power series representation with center τ.

Theorem 5.8: *If*

$$f(t) = \sum_{k=0}^{\infty} a_k(t - \tau)^k, \quad |t - \tau| < \rho, \quad a_k \in \mathcal{C},$$

and

$$f(t) = \sum_{k=0}^{\infty} b_k(t-\tau)^k, \quad |t-\tau| < \rho, \quad b_k \in \mathcal{C},$$

then $a_k = b_k$ for $k = 0, 1, 2, \ldots$.

Suppose

$$g(t) = \sum_{k=0}^{\infty} b_k(t-\tau)^k, \quad |t-\tau| < \rho, \quad b_k \in \mathcal{C},$$

where the series converges for $|t-\tau| < \rho$. Then $f+g$ and fg have series representations

$$f(t) + g(t) = \sum_{k=0}^{\infty} (a_k + b_k)(t-\tau)^k,$$

$$f(t)g(t) = \sum_{k=0}^{\infty} c_k(t-\tau)^k, \quad c_k = \sum_{j=0}^{k} a_j b_{k-j},$$

which are convergent for $|t-\tau| < \rho$.

We can also consider analytic functions whose values are $m \times n$ matrices with complex entries. If $\mathbf{F} : I \to M_{mn}(\mathcal{C})$, where I is an open interval, then $\mathbf{F}$ is *analytic* at $\tau \in I$ if for some $\rho > 0$

$$\mathbf{F}(t) = \sum_{k=0}^{\infty} \mathbf{A}_k(t-\tau)^k, \quad |t-\tau| < \rho, \quad \mathbf{A}_k \in M_{mn}(\mathcal{C}),$$

where the series converges for $|t-\tau| < \rho$. This means that if

$$\mathbf{F}_p(t) = \sum_{k=0}^{p} \mathbf{A}_k(t-\tau)^k,$$

then

$$|\mathbf{F}(t) - \mathbf{F}_p(t)| \to 0, \quad p \to \infty, \quad |t-\tau| < \rho. \tag{5.14}$$

Writing out the components of these matrices, $\mathbf{F} = (\mathbf{F}_{ij})$, $\mathbf{F}_p = ([\mathbf{F}_p]_{ij})$, $\mathbf{A}_k = ([\mathbf{A}_k]_{ij})$, it follows that

$$[\mathbf{F}_p]_{ij}(t) = \sum_{k=0}^{p} [\mathbf{A}_k]_{ij}(t-\tau)^k,$$

and (5.14) is true if and only if

$$|\mathbf{F}_{ij}(t) - [\mathbf{F}_p]_{ij}(t)| \to 0, \quad p \to \infty, \quad |t-\tau| < \rho,$$

for every $i = 1, \ldots, m$, and $j = 1, \ldots n$. From this fact it follows that the results outlined above are also valid for power series with matrix coefficients. Thus

$$\mathbf{F}^{(l)}(t) = \sum_{k=l}^{\infty} k(k-1)\cdots(k-l+1)\mathbf{A}_k(t-\tau)^{k-l}, \quad l = 1, 2, \ldots,$$

the differentiated series converge for $|t - \tau| < \rho$, and

$$\mathbf{A}_k = \mathbf{F}^{(k)}(\tau)/k!.$$

The sum and product formulas extend to the case of matrix-valued coefficients. If

$$\mathbf{G}(t) = \sum_{k=0}^{\infty} \mathbf{B}_k (t - \tau)^k, \quad \mathbf{B}_k \in M_{mn}(\mathcal{C}),$$

and

$$\mathbf{H}(t) = \sum_{k=0}^{\infty} \mathbf{C}_k (t - \tau)^k, \quad \mathbf{C}_k \in M_{np}(\mathcal{C}),$$

where the series converges for $|t - \tau| < \rho$, then

$$\mathbf{F}(t) + \mathbf{G}(t) = \sum_{k=0}^{\infty} (\mathbf{A}_k + \mathbf{B}_k)(t - \tau)^k$$

and

$$\mathbf{F}(t)\mathbf{H}(t) = \sum_{k=0}^{\infty} \mathbf{D}_k (t - \tau)^k, \quad \mathbf{D}_k = \sum_{j=0}^{k} \mathbf{A}_j \mathbf{C}_{k-j},$$

where these two series also converge for $|t - \tau| < \rho$.

5.4   First-order linear analytic systems

Let us now consider a first-order linear system

$$X' = \mathbf{A}(t)X + B(t), \tag{5.15}$$

where $\mathbf{A}$ and B are analytic at $\tau \in \mathcal{R}$. In this case τ is called an *ordinary point* for (5.15). Thus for some $\rho > 0$,

$$\mathbf{A}(t) = \sum_{k=0}^{\infty} \mathbf{A}_k (t - \tau)^k, \quad |t - \tau| < \rho, \quad \mathbf{A}_k \in M_n(\mathcal{C}), \tag{5.16}$$

$$B(t) = \sum_{k=0}^{\infty} B_k (t - \tau)^k, \quad |t - \tau| < \rho, \quad B_k \in \mathcal{C}^n,$$

where the two series are convergent for $|t-\tau| < \rho$. The main result about such systems is that all solutions are analytic at τ, with power series representations which also converge for $|t - \tau| < \rho$.

Theorem 5.9: *Let* $\mathbf{A}, B$ *in* (5.15) *be analytic at* $\tau \in \mathcal{R}$, *with power series expansions given by* (5.16) *which are convergent for* $|t - \tau| < \rho$, $\rho > 0$. *Given any* $\xi \in \mathcal{C}^n$, *there exists a solution* X *of* (5.15) *satisfying* $X(\tau) = \xi$ *with a power series representation*

$$X(t) = \sum_{k=0}^{\infty} C_k (t - \tau)^k, \quad C_k \in \mathcal{C}^n, \tag{5.17}$$

which is convergent for $|t - \tau| < \rho$. We have $C_0 = \xi$, and the C_k for $k \geq 1$ may be computed uniquely in terms of C_0 by substituting the series (5.17) into (5.15).

Proof: A simple change of variables $s = t - \tau$ will make zero the center of the power series. Thus there is no loss of generality in assuming $\tau = 0$ and considering (5.15) with

$$\mathbf{A}(t) = \sum_{k=0}^{\infty} \mathbf{A}_k t^k, \quad B(t) = \sum_{k=0}^{\infty} B_k t^k, \quad |t| < \rho, \tag{5.18}$$

$$\mathbf{A}_k \in M_n(\mathcal{C}), \quad B_k \in \mathcal{C}^n,$$

where the series converge for $|t| < \rho$.

Suppose X is a solution of (5.15), with $X(0) = \xi$ and

$$X(t) = \sum_{k=0}^{\infty} C_k t^k, \quad |t| < \rho, \quad C_k \in \mathcal{C}^n, \tag{5.19}$$

with the series being convergent for $|t| < \rho$. Then clearly $X(0) = C_0 = \xi$, and the C_k for $k \geq 1$ must satisfy a recursion relation. We have

$$X'(t) = C_1 + 2C_2 t + 3C_3 t^2 + \cdots = \sum_{k=0}^{\infty} (k+1)C_{k+1} t^k$$

and

$$\mathbf{A}(t)X(t) + B(t)$$

$$= [\mathbf{A}_0 + \mathbf{A}_1 t + \mathbf{A}_2 t^2 + \cdots][C_0 + C_1 t + C_2 t^2 + \cdots] + [B_0 + B_1 t + B_2 t^2 + \cdots]$$

$$= (\mathbf{A}_0 C_0 + B_0) + (\mathbf{A}_1 C_0 + \mathbf{A}_0 C_1 + B_1)t + (\mathbf{A}_2 C_0 + \mathbf{A}_1 C_1 + \mathbf{A}_0 C_2 + B_2)t^2 + \cdots$$

$$= \sum_{k=0}^{\infty} \left(\sum_{j=0}^{k} \mathbf{A}_{k-j} C_j + B_k \right) t^k.$$

Thus

$$X'(t) = \mathbf{A}(t)X(t) + B(t), \quad |t| < \rho,$$

if and only if

$$(k+1)C_{k+1} = \sum_{j=0}^{k} \mathbf{A}_{k-j} C_j + B_k, \quad k = 0, 1, 2, \ldots. \tag{5.20}$$

This shows that once $C_0 = \xi$ is determined, all other C_k are uniquely given by the recursion relation (5.20).

Now suppose X is given by (5.19) with $C_0 = \xi$, and the C_k for $k \geq 1$ are determined by (5.20). We next show that the series in (5.19) is convergent for $|t| < \rho$. Then the above computation shows that X satisfies (5.15) for $|t| < \rho$ and $X(0) = \xi$.

Let r be any number satisfying $0 < r < \rho$. Since the series in (5.18) are convergent for $|t| < \rho$, there is a constant $M > 0$ such that

$$|\mathbf{A}_j|r^j \leq M, \quad |B_j|r^j \leq M, \quad j = 0, 1, 2, \ldots.$$

Using this in (5.20) we find that

$$(k+1)|C_{k+1}| \leq \sum_{j=0}^{k} Mr^{j-k}|C_j| + Mr^{-k}, \quad k = 0, 1, 2, \ldots,$$

or

$$(k+1)|C_{k+1}| \leq Mr^{-k}\left[\sum_{j=0}^{k} |C_j|r^j + 1\right], \quad k = 0, 1, 2, \ldots. \qquad (5.21)$$

Let us define $d_k \geq 0$ by

$$d_0 = |C_0| = |\xi|, \qquad (5.22)$$

$$(k+1)d_{k+1} = Mr^{-k}\left[\sum_{j=0}^{k} d_j r^j + 1\right], \quad k = 0, 1, 2, \ldots.$$

From (5.21) it follows that

$$|C_k| \leq d_k, \quad k = 0, 1, 2, \ldots. \qquad (5.23)$$

We now investigate for which t the series

$$\sum_{k=0}^{\infty} d_k t^k \qquad (5.24)$$

converges. From (5.22) it follows that

$$d_1 = M(d_0 + 1),$$

$$(k+1)d_{k+1} = Md_k + r^{-1}kd_k, \quad k = 1, 2, \ldots.$$

An application of the ratio test gives

$$\left|\frac{d_{k+1}t^{k+1}}{d_k t^k}\right| = \left[\frac{M}{k+1} + \frac{k}{k+1}\frac{1}{r}\right]|t| \to \frac{|t|}{r}, \quad k \to \infty.$$

Thus the series (5.24) converges absolutely for $|t| < r$, and because of (5.23), this implies that the series (5.19) converges for $|t| < r$. Since r was any number satisfying $0 < r < \rho$, we have shown that (5.19) converges for $|t| < \rho$. $\square$

The uniqueness result for linear systems (Theorem 7.4) guarantees that there is only one solution X of (5.15) on $|t - \tau| < \rho$ satisfying $X(\tau) = \xi$ for each $\xi \in \mathcal{C}^n$. This yields the following result.

Corollary 5.10: *Every solution X of (5.15) on $|t - \tau| < \rho$ has a series representation of the form*

$$X(t) = \sum_{k=0}^{\infty} C_k(t - \tau)^k, \quad C_k \in \mathcal{C}^n,$$

which is convergent for $|t - \tau| < \rho$.

The simplest example of an equation with analytic coefficients occurs when $\mathbf{A}(t) = \mathbf{A}$, a constant matrix, and $B = 0$. In this case (5.15) becomes $X' = \mathbf{A}X$. Since $\mathbf{A}(t) = \mathbf{A}$ is a series trivially convergent for all $t \in \mathcal{R}$, every solution X of $X' = \mathbf{A}X$ has a power series representation, in powers of t, which converges for all $t \in \mathcal{R}$. The recursion relation (5.20) in this case is

$$(k + 1)C_{k+1} = \mathbf{A}C_k, \quad k = 0, 1, 2, \ldots,$$

and if $C_0 = \xi$ we see that

$$C_k = \frac{\mathbf{A}^k}{k!}\xi, \quad k = 0, 1, 2, \ldots.$$

Hence the solution X satisfying $X(0) = \xi$ has the well-known power series representation

$$X(t) = \left(\sum_{k=0}^{\infty} \frac{\mathbf{A}^k t^k}{k!} \right)\xi = e^{\mathbf{A}t}\xi,$$

convergent for all $t \in \mathcal{R}$.

The case

$$X' = (\mathbf{A}_0 + \mathbf{A}_1 t)X, \quad \mathbf{A}_j \in M_n(\mathcal{C})$$

provides a somewhat more complex example. Let $\mathbf{X}$ be the basis for the solutions of this equation satisfying $\mathbf{X}(0) = \mathbf{I}_n$. Again Theorem 5.9 guarantees that $\mathbf{X}$ has a power series representation

$$\mathbf{X}(t) = \sum_{k=0}^{\infty} \mathbf{C}_k t^k, \quad \mathbf{C}_k \in M_n(\mathcal{C}), \tag{5.25}$$

which is convergent for all $t \in \mathcal{R}$. This $\mathbf{X}$ satisfies

$$\mathbf{X}'(t) = (\mathbf{A}_0 + \mathbf{A}_1 t)\mathbf{X}(t), \quad t \in \mathcal{R},$$

and the recursion relation (5.20) shows that in this case

$$\mathbf{C}_0 = \mathbf{I}_n, \quad \mathbf{C}_1 = \mathbf{A}_0, \tag{5.26}$$

$$(k + 1)\mathbf{C}_{k+1} = \mathbf{A}_0\mathbf{C}_k + \mathbf{A}_1\mathbf{C}_{k-1}, \quad k = 1, 2, \ldots.$$

Although it is clear that all $\mathbf{C}_k$ are determined uniquely in terms of $\mathbf{C}_0, \mathbf{A}_0, \mathbf{A}_1$, a general formula for $\mathbf{C}_k$ is not so easy to determine. We have, for example,

$$\mathbf{C}_2 = \frac{1}{2}(\mathbf{A}_0^2 + \mathbf{A}_1),$$

$$\mathbf{C}_3 = \frac{1}{3!}(\mathbf{A}_0^3 + \mathbf{A}_0\mathbf{A}_1 + 2\mathbf{A}_1\mathbf{A}_0),$$

$$\mathbf{C}_4 = \frac{1}{4}(\mathbf{A}_0^4 + \mathbf{A}_0^2\mathbf{A}_1 + 2\mathbf{A}_0\mathbf{A}_1\mathbf{A}_0 + 3\mathbf{A}_1\mathbf{A}_0^2 + 3\mathbf{A}_1^2).$$

If $\mathbf{A}_0, \mathbf{A}_1$ commute, however, the series (5.25) with the $\mathbf{C}_k$ given by (5.26) is just the power series representation for

$$\mathbf{X}(t) = \exp(\mathbf{A}_0 t + \mathbf{A}_1 t^2/2).$$

5.5   Equations of order n

Theorem 5.9 has an immediate application to nth-order linear equations

$$x^{(n)} + a_{n-1}(t)x^{(n-1)} + \cdots + a_0(t)x = b(t), \tag{5.27}$$

where the a_j, b are analytic at $\tau \in \mathcal{R}$. We say τ is an *ordinary point* for (5.27).

Theorem 5.11: *Let $\tau \in \mathcal{R}$ and suppose the a_j and b in (5.27) have power series representations*

$$a_j(t) = \sum_{k=0}^{\infty} a_{jk}(t-\tau)^k, \quad j = 0, 1, 2, \ldots, n-1, \quad a_{jk} \in \mathcal{C},$$

$$b(t) = \sum_{k=0}^{\infty} b_k(t-\tau)^k, \quad b_k \in \mathcal{C},$$

which are convergent for $|t - \tau| < \rho$, $\rho > 0$. Given any $\xi \in \mathcal{C}^n$ there exists a solution x of (5.27) on $|t - \tau| < \rho$ which satisfies $\tilde{x}(\tau) = \xi$ and has a power series representation

$$x(t) = \sum_{k=0}^{\infty} c_k(t-\tau)^k, \quad c_k \in \mathcal{C}, \tag{5.28}$$

which is convergent on $|t - \tau| < \rho$. We have

$$k!c_k = \xi_{k+1}, \quad k = 0, 1, \ldots, n-1,$$

and the c_k for $k \geq n$ may be computed uniquely in terms of $c_0, c_1, \ldots, c_{n-1}$ by substituting the series (5.28) into (5.27).

Proof: Equation (5.27) has an associated system

$$Y' = \mathbf{A}(t)Y + B(t), \tag{5.29}$$

with

$$\mathbf{A} = \begin{pmatrix} 0 & 1 & 0 & \cdots & 0 \\ 0 & 0 & 1 & \cdots & 0 \\ \vdots & \vdots & \vdots & \cdots & \vdots \\ 0 & 0 & 0 & \cdots & 1 \\ -a_0 & -a_1 & -a_2 & \cdots & -a_{n-1} \end{pmatrix}, \quad B = \begin{pmatrix} 0 \\ \vdots \\ 0 \\ b \end{pmatrix}.$$

A, and B have convergent power series representations on $|t-\tau| < \rho$. Theorem 5.9 then guarantees that (5.29) has a solution Y which has a convergent power series representation, in powers of $t - \tau$, on $|t - \tau| < \rho$, and satisfies $Y(\tau) = \xi$. Now $Y = \tilde{x}$, where x is the first component of Y, and x will satisfy (5.27) with $\tilde{x}(\tau) = \xi$. $\square$

Again the uniqueness theorem provides an immediate corollary.

Corollary 5.12: *Every solution x of (5.27) on $|t - \tau| < \rho$ has a series representation*

$$x(t) = \sum_{k=0}^{\infty} c_k (t - \tau)^k, \quad c_k \in \mathcal{C},$$

which is convergent for $|t - \tau| < \rho$.

A simple example illustrating the result in Theorem 5.11 is the *Airy equation*

$$x'' - tx = 0. \tag{5.30}$$

Each solution x of this equation has a power series representation

$$x(t) = \sum_{k=0}^{\infty} c_k t^k$$

which is convergent for all $t \in \mathcal{R}$. Let us compute the coefficients c_k in terms of $c_0 = x(0)$, $c_1 = x'(0)$. We have

$$x''(t) = 2c_2 + 3 \cdot 2c_3 t + 4 \cdot 3c_4 t^2 + \cdots = \sum_{k=0}^{\infty} (k+2)(k+1)c_{k+2} t^k$$

and

$$tx(t) = c_0 t + c_1 t^2 + c_2 t^3 + \cdots = \sum_{k=1}^{\infty} c_{k-1} t^k.$$

Thus

$$x''(t) - tx(t) = 2c_2 + \sum_{k=1}^{\infty} [(k+2)(k+1)c_{k+2} - c_{k-1}]t^k = 0$$

for all $t \in \mathcal{R}$ if and only if

$$2c_2 = 0, \quad (k+2)(k+1)c_{k+2} = c_{k-1}, \quad k = 1, 2, \ldots.$$

For the first few terms we have

$$c_3 = \frac{c_0}{3 \cdot 2}, \quad c_4 = \frac{c_1}{4 \cdot 3}, \quad c_5 = \frac{c_2}{5 \cdot 4} = 0,$$

$$c_6 = \frac{c_3}{6 \cdot 5} = \frac{c_0}{6 \cdot 5 \cdot 3 \cdot 2}, \quad c_7 = \frac{c_4}{7 \cdot 6} = \frac{c_1}{7 \cdot 6 \cdot 4 \cdot 3}.$$

One may prove by induction that

$$c_{3m} = \frac{c_0}{2 \cdot 3 \cdot 5 \cdot 6 \cdots (3m-1) \cdot (3m)}, \quad m = 1, 2, \ldots,$$

$$c_{3m+1} = \frac{c_1}{3 \cdot 4 \cdot 6 \cdot 7 \cdots (3m) \cdot (3m+1)}, \quad m = 1, 2, \ldots,$$

$$c_{3m+2} = 0, \quad m = 0, 1, 2, \ldots.$$

Collecting together all terms with c_0 and c_1 as a factor we obtain $x = c_0 u + c_1 v$, where

$$u(t) = 1 + \sum_{m=1}^{\infty} \frac{t^{3m}}{2 \cdot 3 \cdot 5 \cdot 6 \cdots (3m-1) \cdot (3m)},$$

$$v(t) = t + \sum_{m=1}^{\infty} \frac{t^{3m+1}}{3 \cdot 4 \cdot 6 \cdot 7 \cdots (3m) \cdot (3m+1)}.$$

It is now clear that $\mathbf{X} = (u, v)$ is the basis for the solutions of (5.30) satisfying $\tilde{\mathbf{X}}(0) = (\tilde{u}(0), \tilde{v}(0)) = \mathbf{I}_2$.

5.6 The Legendre equation and its solutions

5.6.1 The Legendre equation

One of the important second-order equations with analytic coefficients is the *Legendre equation*

$$Lx = (1 - t^2)x'' - 2tx' + \alpha(\alpha + 1)x = 0, \tag{5.31}$$

where $\alpha \in \mathcal{C}$. This equation arises naturally in the study of the partial differential equations describing wave phenomena [24, pp. 257–263] or heat conduction [19, pp. 222–233] associated with spherical bodies.

Writing (5.31) as

$$x'' - \frac{2t}{1 - t^2}x' + \frac{\alpha(\alpha + 1)}{1 - t^2}x = 0,$$

the functions a_1, a_0 given by

$$a_1(t) = \frac{-2t}{1 - t^2}, \quad a_0(t) = \frac{\alpha(\alpha + 1)}{1 - t^2},$$

are analytic at $t = 0$. In fact,

$$\frac{1}{1 - t^2} = 1 + t^2 + t^4 + \cdots = \sum_{k=0}^{\infty} t^{2k},$$

this series converging for $|t| < \rho = 1$ and diverging for $|t| \geq 1$. Thus a_1, a_0 have the power representations

$$a_1(t) = \sum_{k=0}^{\infty} (-2)t^{2k+1}, \quad a_0(t) = \sum_{k-0}^{\infty} \alpha(\alpha + 1)t^{2k},$$

which converge for $|t| < 1$, and Theorem 5.11 implies that the solutions of (5.31) on $|t| < 1$ have convergent power series expansions there. Let us compute a basis for these solutions. If

$$x(t) = c_0 + c_1 t + c_2 t^2 + \cdots = \sum_{k=0}^{\infty} c_k t^k,$$

then

$$x'(t) = c_1 + 2c_2 t + 3c_3 t^2 + \cdots = \sum_{k=0}^{\infty} k c_k t^{k-1},$$

$$-2tx'(t) = -2c_1 t - (2)(2)c_2 t^2 - \cdots = \sum_{k=0}^{\infty} (-2k)c_k t^k,$$

$$x''(t) = 2c_2 + (3)(2)c_3 t + \cdots = \sum_{k=0}^{\infty} k(k-1)c_k t^{k-2} = \sum_{k=0}^{\infty} (k+2)(k+1)c_{k+2} t^k,$$

$$-t^2 x''(t) = -2c_2 t^2 - (3)(2)c_3 t^3 - \cdots = \sum_{k=0}^{\infty} -k(k-1)c_k t^k.$$

Thus

$$(Lx)(t) = (1 - t^2)x''(t) - 2tx'(t) + \alpha(\alpha+1)x(t)$$

$$= \sum_{k=0}^{\infty} [(k+2)(k+1)c_{k+2} - k(k-1)c_k - 2kc_k + \alpha(\alpha+1)c_k]t^k$$

$$= \sum_{k=0}^{\infty} [(k+2)(k+1)c_{k+2} + (\alpha+k+1)(\alpha-k)c_k]t^k,$$

and x satisfies $Lx = 0$ if and only if all the coefficients of the powers of t are zero. The coefficients must satisfy the recursion relation

$$(k+2)(k+1)c_{k+2} + (\alpha+k+1)(\alpha-k)c_k = 0, \quad k = 0, 1, 2, \ldots. \qquad (5.32)$$

The initial coefficients c_k are

$$c_2 = -\frac{(\alpha+1)\alpha}{2}c_0, \quad c_3 = -\frac{(\alpha+2)(\alpha-1)}{(3)(2)}c_1,$$

$$c_4 = -\frac{(\alpha+3)(\alpha-2)}{(4)(3)}c_2 = \frac{(\alpha+3)(\alpha+1)(\alpha)(\alpha-2)}{4!}c_0,$$

$$c_5 = -\frac{(\alpha+4)(\alpha-3)}{(5)(4)}c_3 = \frac{(\alpha+4)(\alpha+2)(\alpha-1)(\alpha-3)}{5!}c_1.$$

An induction argument can now be used to show that for $m = 1, 2, \ldots$,

$$c_{2m} = (-1)^m \frac{(\alpha+2m-1)\cdots(\alpha+1)(\alpha)(\alpha-2)\cdots(\alpha-2m+2)}{(2m)!}c_0,$$

$$c_{2m+1} = (-1)^m \frac{(\alpha + 2m) \cdots (\alpha + 2)(\alpha - 1)(\alpha - 3) \cdots (\alpha - 2m + 1)}{(2m + 1)!} c_1.$$

Therefore all c_k are determined uniquely by c_0 and c_1, and $x = c_0 u + c_1 v$, where

$$u(t) = 1 + \sum_{m=1}^{\infty} d_{2m} t^{2m}, \tag{5.33}$$

$$d_{2m} = (-1)^m \frac{(\alpha + 2m - 1) \cdots (\alpha + 1)(\alpha)(\alpha - 2) \cdots (\alpha - 2m + 2)}{(2m)!},$$

$$v(t) = t + \sum_{m=1}^{\infty} d_{2m+1} t^{2m+1}, \tag{5.34}$$

$$d_{2m+1} = (-1)^m \frac{(\alpha + 2m) \cdots (\alpha + 2)(\alpha - 1)(\alpha - 3) \cdots (\alpha - 2m + 1)}{(2m + 1)!}.$$

Note that $x(0) = c_0$, $x'(0) = c_1$, so that $\mathbf{X} = (u, v)$ is the basis for the solutions of $Lx = 0$ for $|t| < 1$ such that $\tilde{\mathbf{X}}(0) = (\tilde{u}(0), \tilde{v}(0)) = \mathbf{I}_2$.

If α is a nonnegative even integer, $\alpha = 2m$, $m = 0, 1, 2, \ldots$, then $u(t)$ has only a finite number of terms which are not zero. In this case we see that u is a polynomial of degree $2m$ which contains only even powers of t. Thus

$$u(t) = 1, \quad \alpha = 0,$$

$$u(t) = 1 - 3t^2, \quad \alpha = 2,$$

$$u(t) = 1 - 10t^2 + \frac{35}{3}t^4, \quad \alpha = 4.$$

The solution v is not a polynomial in this case since none of the coefficients in the series (5.34) vanish.

Similarly, if α is a nonnegative odd integer, $\alpha = 2m + 1$, $m = 0, 1, 2, \ldots$, then v reduces to a polynomial of degree $2m + 1$ containing only odd powers of t. The first few of these are

$$v(t) = t, \quad \alpha = 1,$$

$$v(t) = t - \frac{5}{3}t^3, \quad \alpha = 3,$$

$$v(t) = t - \frac{14}{3}t^3 + \frac{21}{5}t^5, \quad \alpha = 5.$$

The solution u is not a polynomial in this case.

We know from Theorem 5.11 that the two series (5.33), (5.34) are convergent for $|t| < 1$. If $\alpha = 2m$ is an even integer, the series for u reduces to a polynomial which is convergent for all $t \in \mathcal{R}$, and this u gives a solution of the Legendre equation for all $t \in \mathcal{R}$. In this case when $\alpha = 2m$, the ratio test shows that the series for v diverges for $|t| > 1$, and thus we can only guarantee that v is a solution of $Lx = 0$ for $|t| < 1$. A similar situation prevails in case $\alpha = 2m + 1$, with the roles of u and v reversed.

The polynomial solutions of the Legendre equation when $\alpha = n$ is a nonnegative integer are of great importance, and so we develop some of their properties in the next section.

5.6.2   The Legendre polynomials

Let $\alpha = n$, a nonnegative integer, and consider the Legendre equation

$$L_n x = (1 - t^2) x'' - 2t x' + n(n+1) x = 0. \qquad (5.35)$$

For each $n = 0, 1, 2, \ldots$ there is a polynomial solution of (5.35) of degree n. The polynomial solution P_n of degree n satisfying $P_n(1) = 1$ is called the *nth Legendre polynomial*. After obtaining a formula for a solution having these properties, we will show that there is only one such solution for each $n = 0, 1, 2, \ldots$.

The polynomial

$$p(t) = D^n (t^2 - 1)^n, \quad D = \frac{d}{dt}$$

of degree n satisfies (5.35). To see this, let

$$v(t) = (t^2 - 1)^n,$$

and verify that

$$(t^2 - 1) v'(t) - 2nt v(t) = 0.$$

Now differentiating this expression $n + 1$ times leads to

$$(t^2 - 1) v^{(n+2)} + 2t(n+1) v^{(n+1)}$$

$$+ n(n+1) v^{(n)} - 2nt v^{(n+1)} - 2n(n+1) v^{(n)} = 0.$$

Since $p = v^{(n)}$ we have

$$(1 - t^2) p'' - 2t p' + n(n+1) p = 0,$$

so that p does satisfy (5.35).

This polynomial p satisfies

$$p(t) = D^n (t^2 - 1)^n = D^n [(t+1)^n (t-1)^n]$$

$$= (t+1)^n D^n (t-1)^n + \text{terms with } (t-1) \text{ as a factor}$$

$$= n!(t+1)^n + \text{terms with } (t-1) \text{ as a factor},$$

from which it is clear that $p(1) = 2^n n!$. This shows that the polynomial P_n of degree n given by

$$P_n(t) = \frac{1}{2^n n!} D^n (t^2 - 1)^n \qquad (5.36)$$

is a solution of (5.35) satisfying $P_n(1) = 1$. This expression (5.36) for P_n is called the *Rodrigues formula*.

We will show next that there is no other such polynomial solution of (5.35), and hence (5.36) gives the nth Legendre polynomial P_n. Suppose q is any polynomial solution of (5.35). We give the argument for the case when n is even, $n = 2m$. For $|t| < 1$ we have $q = c_0 u + c_1 v$ for some $c_0, c_1 \in \mathcal{C}$, where (u, v) is the basis for (5.35) given by (5.33), (5.34) with $\alpha = n$. Now u is a polynomial of degree n, and hence $q - c_0 u$ is a polynomial. But $c_1 v$ is not a polynomial unless $c_1 = 0$. Hence $c_1 = 0$ and $q = c_0 u$. In particular, the P_n given by (5.36) satisfies $P_n = c_0 u$ for some $c_0 \in \mathcal{C}$. In case n is odd, the roles of u and v are reversed.

Now $1 = P_n(1) = c_0 u(1)$ shows that $u(1) \neq 0$. Similarly, $v(1) \neq 0$ if n is odd. Thus no nontrivial polynomial solution of (5.35) can be zero at $t = 1$. From this we can deduce that there is only one polynomial P_n satisfying (5.35) and $P_n(1) = 1$. If p_n is another such polynomial, then $p_n - P_n = q_n$ is a polynomial solution of (5.35) such that $q_n(1) = 0$. Hence, $q_n(t) = 0$ for all $t \in \mathcal{R}$, or $p_n = P_n$.

The first five Legendre polynomials are

$$P_0(t) = 1, \quad P_1(t) = t, \quad P_2(t) = \frac{1}{2}(3t^2 - 1),$$

$$P_3(t) = \frac{1}{2}(5t^3 - 3t), \quad P_4(t) = \frac{1}{8}(35t^4 - 30t^2 + 3).$$

The Legendre polynomials satisfy a number of interesting recurrence formulas, one of which will be used in the next section. It states that

$$P_n' - P_{n-2}' = (2n - 1)P_{n-1}, \quad n = 2, 3, \ldots. \tag{5.37}$$

This can be obtained directly from the Rodrigues formula (5.36). Note that

$$D^2(t^2 - 1)^n = D[n(t^2 - 1)^{n-1}(2t)] \tag{5.38}$$

$$= 2n[(t^2-1)^{n-1}+2(n-1)t^2(t^2-1)^{n-2}] = 2n(t^2-1)^{n-2}[2(n-1)(t^2-1)+2(n-1)].$$

Then (5.36) gives

$$P_n'(t) - P_{n-2}'(t) = \frac{1}{2^n n!} D^{n+1}(t^2 - 1)^n - \frac{1}{2^{n-2}(n - 2)!} D^{n-1}(t^2 - 1)^{n-2}$$

$$= \frac{1}{2^n n!} D^{n-1}[D^2(t^2 - 1)^n - 4n(n - 1)(t^2 - 1)^{n-2}]$$

$$= \frac{1}{2^n n!} D^{n-1}[2n(2n - 1)(t^2 - 1)^{n-1}] = \frac{2n - 1}{2^{n-1}(n - 1)!} D^{n-1}(t^2 - 1)^{n-1}$$

$$= (2n - 1)P_{n-1}(t),$$

where we have used (5.38).

The recurrence formula (5.37) implies that, if $n = 2m$ is even,

$$P'_{2m} - P'_{2m-2} = (4m - 1)P_{2m-1},$$

$$P'_{2m-2} - P'_{2m-4} = (4m - 5)P_{2m-3},$$

$$\vdots$$

$$P'_4 - P'_2 = 7P_3,$$

$$P'_2 - P'_0 = 3P_1.$$

Using $P'_0 = 0$ and adding gives

$$P'_{2m} = (4m - 1)P_{2m-1} + (4m - 5)P_{2m-3} + \cdots + 7P_3 + 3P_1.$$

Similarly, if $n = 2m + 1$,

$$P'_{2m+1} = (4m + 1)P_{2m} + (4m - 3)P_{2m-2} + \cdots + 5P_2 + P_0.$$

A formula covering both the even and the odd cases is

$$P'_n = \sum_{k=0}^{\lfloor (n-1)/2 \rfloor} (2n - 4k - 1)P_{n-2k-1}, \quad n = 1, 2, \ldots, \tag{5.39}$$

where $\lfloor (n-1)/2 \rfloor$ is the greatest integer which is less than or equal to $(n-1)/2$,

$$\lfloor (n-1)/2 \rfloor = \begin{matrix} m - 1, & n = 2m, \\ m, & n = 2m + 1. \end{matrix}$$

The Legendre polynomials are sometimes referred to as *Legendre functions of the first kind.*

5.6.3   Legendre functions of the second kind

We continue our investigation of the Legendre equation (5.35) when $\alpha = n$, a nonnegative integer. If $\alpha = 0$, the series (5.33) gives $u = P_0$, whereas (5.34) becomes

$$v(t) = t + t^3/3 + t^5/5 + \cdots = \sum_{m=0}^{\infty} t^{2m+1}/(2m + 1), \quad |t| < 1.$$

This series represents a familiar function. Since

$$\log(1 + t) = t - t^2/2 + t^3/3 - \cdots, \quad |t| < 1,$$

and

$$\log(1 - t) = -t - t^2/2 - t^3/3 - \cdots, \quad |t| < 1,$$

it follows that

$$\frac{1}{2}\log\left(\frac{1+t}{1-t}\right) = \frac{1}{2}\log(1+t) - \frac{1}{2}\log(1-t)$$

$$= t + t^3/3 + t^5/5 + \cdots = v(t), \quad |t| < 1.$$

This solution of the equation $L_0 x = 0$ will be denoted by Q_0,

$$Q_0(t) = \frac{1}{2}\log\left(\frac{1+t}{1-t}\right), \quad |t| < 1. \tag{5.40}$$

When $\alpha = 1$, the series (5.34) reduces to a polynomial $v = P_1$, whereas the solution u becomes

$$u(t) = 1 - t^2 - t^4/4t^6/5 - \cdots = 1 - t(t + t^3/3 + t^5/5 + \cdots) = 1 - tQ_0(t).$$

Denote the negative of this solution by Q_1:

$$Q_1(t) = tQ_0(t) - 1, \quad |t| < 1.$$

The simple formulas

$$Q_0 = P_0 Q_0, \quad Q_1 = P_1 Q_0 - 1$$

generalize. It turns out that for $n = 1, 2, \ldots$, there is a unique solution Q_n of $L_n x = 0$ of the form

$$Q_n = P_n Q_0 - p, \quad |t| < 1,$$

where p is a polynomial of degree $n - 1$. This solution Q_n is called the nth *Legendre function of the second kind*. After finding an explicit formula for such a Q_n, we will show that such a solution is unique.

Observe that if

$$x = P_n Q_0 - p,$$

then

$$x' = P_n' Q_0 + P_n Q_0' - p',$$

$$x'' = P_n'' Q_0 + 2P_n' Q_0' + P_n Q_0'' - p'',$$

and

$$L_n x(t) = Q_0(t) L_n P_n(t) + P_n(t) L_0 Q_0(t) + 2(1 - t^2)Q_0'(t)P_n'(t) - L_n p(t).$$

Now

$$L_n P_n(t) = 0, \quad L_0 Q_0(t) = 0, \quad (1 - t^2)Q_0'(t) = 1,$$

and these imply that $L_n x = 0$ if and only if $L_n p = 2P_n'$. We define Q_n by

$$Q_n = P_n Q_0 - \sum_{k=0}^{\lfloor (n-1)/2 \rfloor} \frac{2n - 4k - 1}{(2k+1)(n-k)} P_{n-2k-1}, \quad n = 1, 2, \ldots, \tag{5.41}$$

which has the form $Q_n = P_n Q_0 - p$, where p is a polynomial of degree $n - 1$. To show that $L_n Q_n = 0$, that is, $L_n p = 2 P'_n$, note that

$$L_n P_j = L_n P_j - L_j P_j$$

$$= [n(n+1) - j(j+1)]P_j = (n-j)(n+j+1)P_j, \quad j = 0, 1, 2, \ldots,$$

and therefore

$$L_n P_{n-2k-1} = 2(2k+1)(n-k)P_{n-2k-1}.$$

Using (5.39), it follows that

$$L_n p = 2 \sum_{k=0}^{\lfloor (n-1)/2 \rfloor} (2n - 4k - 1) P_{n-2k-1} = 2 P'_n.$$

To establish the uniqueness of Q_n, suppose that $P_n Q_0 - q$ is another solution of $L_n x = 0$, where q is a polynomial of degree $n - 1$. Then $w = q - p$ is a polynomial of degree $\leq n - 1$ which satisfies

$$L_n w = L_n q - L_n p = 2 P'_n - 2 P'_n = 0.$$

Thus $w = c_0 u + c_1 v$ for $c_j \in \mathcal{C}$, where u, v are given by (5.33), (5.34) with $\alpha = n$. If $n = 2m$, u and w are polynomials but v is not, and hence $c_1 = 0$, which means $w = c_0 u$. But $deg(w) \leq n-1$ and $deg(u) = n$ implies that $c_0 = 0$. Thus $w = 0$, or $q = p$. A similar argument is valid in case $n = 2m + 1$ with the roles of u and v reversed. Thus the polynomial p is unique, and the Q_n given by (5.41), with Q_0 given by (5.40), is the unique solution of $L_n x = 0$ having the form $P_n Q_0 - p$, where p is a polynomial having $deg(p) = n - 1$.

Observe next that the function

$$q_0(t) = \frac{1}{2} \log \left(\frac{t+1}{t-1} \right), \quad |t| > 1,$$

obtained by reversing the sign of the argument of Q_0, satisfies the equation

$$L_0 x = (1 - t^2)x'' - 2tx' = 0,$$

as a direct computation shows. This function may also be written as

$$q_0(t) = \frac{1}{2} \log \left| \frac{t+1}{t-1} \right|, \quad |t| > 1.$$

If the definition of Q_0 is extended by putting

$$Q_0(t) = \frac{1}{2} \log \left| \frac{t+1}{t-1} \right|, \quad |t| \neq 1, \tag{5.42}$$

then $L_0 Q_0 = 0$ for $|t| \neq 1$, and if Q_n is defined by the formula (5.41) for $n = 1, 2, \ldots$, we obtain two solutions P_n, Q_n of $L_n x = 0$ for $|t| > 1$ as well as for $|t| < 1$. Now (P_n, Q_n) will be a basis for the solutions of $L_n x = 0$ for $|t| \neq 1$ if they are linearly independent. Suppose there are constants c_1, c_2 such that

$$c_1 P_n(t) + c_2 Q_n(t) = 0$$

for $|t| < 1$ or for $|t| > 1$. Since $|Q_0(t)| \to \infty$ as $t \to 1$ and $P_n(1) = 1$, we have $|Q_n(t)| \to \infty$ as $t \to 1$, and

$$c_2 = -c_1 \frac{P_n(t)}{Q_n(t)}$$

for t near 1 shows that as $t \to 1$, $|P_n(t)|/|Q_n(t)| \to 0$, or $c_2 = 0$. Then $c_1 P_n(t) = 0$ implies $c_1 P_n(1) = c_1 = 0$.

Note that if x is any solution of $L_n x = 0$ which exists at $t = 1$, then the differential equation implies that x is restricted by the condition

$$-2x'(1) + n(n+1)x(1) = 0.$$

Similarly, if x exists at $t = -1$, then

$$2x'(-1) + n(n+1)x(-1) = 0.$$

Thus not all initial value problems for $L_n x = 0$ can be solved using the initial points $\tau = \pm 1$.

5.7 Notes

The reader may find it helpful to review the material in analysis, which is covered in [17], or in the classic [22].

The use of power series methods and the related analytic function theory for attacking differential equations flourished in the nineteenth century. An interesting historical overview can be found in Chapter 29 of [12]. The classic work [27] contains a vast amount of material on this subject, and on Legendre functions in particular. A more modern treatment of the interplay between differential equations and the theory of analytic functions is [8]. This work also has many historical notes. The elementary text [19, pp. 222–231] shows how Legendre functions may arise from consideration of partial differential equations. In addition to material on Legendre functions, the book [24, pp. 239–240] discusses the connection between the Hermite polynomials, discussed in the exercises, and partial differential equations.

The term *ordinary point*, for a point where the coefficients of an equation (5.15) or (5.37) are analytic, is not very suggestive, but it is well entrenched in the literature. The term *analytic point* is used in [3, p. 111].

5.8   Exercises

1. Let $X, Y \in \mathcal{C}^3$ be given by $X = (i, -i, 1)$, $Y = (1+i, -2, i)$. Compute

(a) $X + Y$,   (b) $X - Y$,   (c) $|X| = \|X\|_1$,   (d) (X, Y).

Verify that
$$\big\| |X| - |Y| \big\| \leq |X \pm Y| \leq |X| + |Y|.$$

2. Use the inequality
$$2|a||b| \leq |a|^2 + |b|^2$$

to show that
$$\|X\|_2 \leq |X| \leq \sqrt{n}\|X\|_2$$

for all $X \in \mathcal{C}^n$.

3. A complex number was defined as an ordered pair of real numbers, $z = (a, b) \in \mathcal{R}^2$. Show that in general the absolute value $|z|$, $z \in \mathcal{C}$, is different from the norm $|(a, b)| = \|(a, b)\|_1$, $(a, b) \in \mathcal{R}^2$.

4. (a) If $\| \ \|$ is a norm on $\mathcal{C}^n$, show that the function

$$f : X \in \mathcal{C}^n \to \|X\| \in \mathcal{R}$$

is continuous at each $x \in \mathcal{C}^n$; that is,

$$|f(Y) - f(X)| \to 0 \quad \text{as } |Y - X| \to 0.$$

(b) If $\| \ \|$ is any norm on $\mathcal{C}^n$, show that there is an $M > 0$ such that

$$\|X\| \leq M|X|, \quad X = (x_1, \ldots, x_n) \in \mathcal{C}^n.$$

(c) Show that $\| \ \|$ and $| \ |$ are equivalent norms on $\mathcal{C}^n$.
(d) Show that any two norms on $\mathcal{C}^n$ are equivalent.
5. Prove Theorem 5.2.
6. Let $f, g \in C(I, \mathcal{C})$, where $I = [0, 1]$, and

$$f(t) = t - it^2, \quad g(t) = t^3 + 2it, \quad t \in I.$$

Compute

(a) $f + g$,   (b) fg,   (c) $\int_0^1 f(t)\, dt$,   (d) $|f(t)|$,

(e) $\|f\|_\infty$,   (f) $\|f\|_1$,   (g) $\|f\|_2$,   (h) $\|f - g\|_\infty$.

7. (a) Show that $\|f\|_1 \leq \|f\|_\infty$ for all $f \in C(I, \mathcal{C})$, where $I = [0, 1]$.
(b) Show that there is no $M > 0$ such that $\|f\|_\infty \leq M\|f\|_1$ for all $f \in C(I, \mathcal{C})$, $I = [0, 1]$. (Hint: Consider the functions $f_n(t) = t^n$.)
8. Are $\| \ \|_1$ and $\| \ \|_2$ equivalent norms on $C(I, \mathcal{C})$?

9. (a) For $n = 1, 2, \ldots$, let

$$f_n(t) = \begin{array}{ll} 1, & 0 \le t \le 1/2, \\ 1 - (n+1)(t - 1/2), & 1/2 < t < 1/2 + 1/[n+1], \\ 0, & 1/2 + 1/[n+1] \le t \le 1. \end{array}$$

Thus f_n is a piecewise linear function on $I = [0, 1]$ and $f_n \in C(I, \mathcal{R})$. Draw the graph of f_n. Show that the sequence $\{f_n\}$ is Cauchy in the metric $d_1(f, g) = \|f - g\|_1$.

(b) Let

$$f(t) = \begin{array}{ll} 1, & 0 \le t \le 1/2, \\ 0, & 1/2 < t \le 1. \end{array}$$

Show that $\|f_n - f\|_1 \to 0$ as $n \to \infty$. Note that $f \notin C(I, \mathcal{R})$.

(c) Show that $C(I, \mathcal{R})$ with the metric d_1 is not complete.

10. Let $\mathbf{A}, \mathbf{B} \in C(I, M_2(\mathcal{C}))$, where $I = [0, 1]$, and

$$\mathbf{A}(t) = \begin{pmatrix} t & t^2 \\ -t^2 & t \end{pmatrix}, \quad \mathbf{B}(t) = \begin{pmatrix} 1 & t^2 \\ it & -t \end{pmatrix}, \quad t \in I.$$

Compute

(a) $\mathbf{A} + \mathbf{B}$,   (b) $\mathbf{AB}$,   (c) $\mathbf{A}'$,   (d) $\displaystyle\int_0^1 \mathbf{A}(t) \, dt$,

(e) $\displaystyle\int_0^1 |\mathbf{A}(t)| \, dt$,   (f) $\|\mathbf{A}\|_\infty$,   (g) $\|\mathbf{A} - \mathbf{B}\|_\infty$.

Verify that

$$\left| \int_0^1 \mathbf{A}(t) \, dt \right| \le \int_0^1 |\mathbf{A}(t)| \, dt.$$

11. Let $\mathbf{A} \in M_n(\mathcal{C})$ satisfy $|\mathbf{A}| < 1$.

(a) Show that $(\mathbf{I}_n - \mathbf{A})^{-1}$ exists.

(b) Show that the series

$$S = \sum_{k=0}^\infty \mathbf{A}^k$$

converges and has the sum

$$S = (\mathbf{I}_n - \mathbf{A})^{-1},$$

by noting that the partial sums

$$S_p = \mathbf{I}_n + \mathbf{A} + \mathbf{A}^2 + \cdots + \mathbf{A}^p$$

satisfy

$$(\mathbf{I}_n - \mathbf{A})S_p = S_p(\mathbf{I}_n - \mathbf{A}) = \mathbf{I}_n - \mathbf{A}^{p+1}.$$

(c) Show that

$$|S| \le (n - 1) + (1 - |\mathbf{A}|)^{-1}.$$

12. Let $\mathbf{A} \in M_n(\mathcal{C})$.

(a) Show that the series

$$S(t) = \sum_{k=0}^{\infty} \mathbf{A}^k t^k$$

is convergent for $t \in I$, where

$$I = \{t \in \mathcal{R} \mid |t| < 1/|\mathbf{A}|\}.$$

What is the sum $S(t)$ of this series for $t \in I$?

(b) Let $\mathbf{B}(t) = \mathbf{A}(\mathbf{I}_n - \mathbf{A}t)^{-1}$ on the interval I. Show that $\mathbf{B}$ is analytic on I by computing its power series expansion

$$\mathbf{B}(t) = \sum_{k=0}^{\infty} \mathbf{B}^k t^k,$$

and showing that this series converges on I.

(c) Compute the basis $\mathbf{X}$,

$$\mathbf{X}(t) = \sum_{k=0}^{\infty} \mathbf{C}^k t^k,$$

for the solutions of

$$X' = \mathbf{B}(t)X, \quad \mathbf{B}(t) = \mathbf{A}(\mathbf{I}_n - \mathbf{A}t)^{-1}$$

on I such that $\mathbf{X}(0) = \mathbf{I}_n$.

13. Find the basis $\mathbf{X}$ for the solutions of

$$X' = \mathbf{A}(t)X, \quad \mathbf{A}(t) = \begin{pmatrix} 2 & t \\ 0 & 2 \end{pmatrix}, \quad t \in \mathcal{R},$$

such that $\mathbf{X}(0) = \mathbf{I}_2$.

14. (a) Show that the basis $\mathbf{X}$ of

$$X' = \mathbf{A}_1 t X, \quad t \in \mathcal{R}, \quad \mathbf{A}_1 \in M_n(\mathcal{C}),$$

satisfying $\mathbf{X}(0) = \mathbf{I}_n$, is given by

$$\mathbf{X}(t) = \exp(\mathbf{A}_1 t^2/2).$$

(b) Let $\mathbf{A}_0, \mathbf{A}_1 \in M_n(\mathcal{C})$ and $\mathbf{A}_0\mathbf{A}_1 = \mathbf{A}_1\mathbf{A}_0$. Show that

$$X' = (\mathbf{A}_0 + \mathbf{A}_1 t)X, \quad t \in \mathcal{R},$$

has a basis

$$\mathbf{X}(t) = \exp(\mathbf{A}_0 t + \mathbf{A}_1 t^2/2) = \exp(\mathbf{A}_0 t)\exp(\mathbf{A}_1 t^2/2)$$

and that this satisfies $\mathbf{X}(0) = \mathbf{I}_n$.

15. Let $\mathbf{A}_0, \mathbf{A}_1, \ldots, \mathbf{A}_m \in M_n(\mathcal{C})$ be pairwise commutative,

$$\mathbf{A}_j\mathbf{A}_k = \mathbf{A}_k\mathbf{A}_j, \quad j, k = 0, 1, 2, \ldots m.$$

Show that the basis $\mathbf{X}$ of

$$X' = (\mathbf{A}_0 + \mathbf{A}_1 t + \cdots + \mathbf{A}_m t^m)X, \quad t \in \mathcal{R},$$

satisfying $\mathbf{X}(0) = \mathbf{I}_n$, is given by

$$\mathbf{X}(t) = \exp[\mathbf{A}_0 + \mathbf{A}_1 t^2/2 + \cdots + \mathbf{A}_m t^{m+1}/(m+1)].$$

16. Find the basis $\mathbf{X}$ for the solutions of the system

$$X' = \mathbf{A}(t)X, \quad \mathbf{A}(t) = \begin{pmatrix} 0 & 1 \\ t & 0 \end{pmatrix}, \quad t \in \mathcal{R},$$

such that $\mathbf{X}(0) = \mathbf{I}_2$ as follows. $\mathbf{X} = (U, V)$, where $U(0) = E_1$, $V(0) = E_2$. Compute U and then V. Thus find that

$$U(t) = \sum_{k=0}^{\infty} C_k t^k,$$

where $C_0 = E_1$ and

$$C_{3m} = \frac{1}{2 \cdot 3 \cdot 5 \cdot 6 \cdots (3m-1) \cdot (3m)} E_1, \quad m = 1, 2, \ldots,$$

$$C_{3m+1} = 0, \quad m = 0, 1, 2, \ldots,$$

$$C_{3m+2} = \frac{1}{2 \cdot 3 \cdot 5 \cdot 6 \cdots (3m-1) \cdot (3m) \cdot (3m+2)} E_2, \quad m = 0, 1, 2, \ldots,$$

etc.

17. Find the basis $\mathbf{X}$ for the solutions of

$$x'' + tx = 0, \quad t \in \mathcal{R},$$

satisfying $\tilde{\mathbf{X}}(0) = \mathbf{I}_2$. (Hint: Let $t = -s$, $y(s) = x(-s)$, and show that y satisfies $y'' - sy = 0$, which is the Airy equation.)

18. Find a basis $\mathbf{X} = (u, v)$ for each of the following equations by computing solutions x in the form of power series

$$x(t) = \sum_{k=0}^{\infty} c_k t^k, \quad c_k \in \mathcal{C}.$$

(a) $x'' + x = 0$, (b) $x'' - t^2 x = 0$,

(c) $x'' - tx' + x = 0$, (d) $x'' + 3t^2 x' - tx = 0$.

19. Find the solution u of

$$x'' + (t-1)^2 x' - (t-1)x = 0$$

in the form

$$u(t) = \sum_{k=0}^{\infty} c_k(t-1)^k,$$

which satisfies $\tilde{u}(1) = E_1$, i.e., $u(1) = 1$, $u'(1) = 0$. (Hint: Let $s = t - 1$.)

20. Compute the solution u of

$$x''' - tx = 0,$$

which satisfies $\tilde{u}(0) = E_1$, i.e., $u(0) = 1$, $u'(0) = 0$, $u''(0) = 0$.

21. The equation

$$x'' + e^t x = 0$$

has a solution v satisfying $\tilde{v}(0) = E_2$ which can be represented as a series

$$v(t) = \sum_{k=0}^{\infty} c_k t^k$$

convergent for all $t \in \mathcal{R}$. Compute $c_0, \dots, c_5$. (Hint: $c_k = v^{(k)}(0)/k!$ and $v''(t) = -e^t v(t)$.)

22. The equation

$$(1 - t^2)x'' - tx' + \alpha^2 x = 0, \quad \alpha \in \mathcal{C}, \tag{5.43}$$

is called the *Chebyshev equation* (also spelled *Tschebycheff*). According to Theorem 5.11 there is a solution x of the form

$$x(t) = \sum_{k=0}^{\infty} c_k t^k,$$

where the series converges for $|t| < 1$.

(a) Show that the c_k satisfy the recursion relation

$$(k+1)(k+2)c_{k+2} = (k^2 - \alpha^2)c_k, \quad k = 0, 1, 2, \dots.$$

(b) From (a) it follows that the basis $\mathbf{X} = (u, v)$ for (5.43) such that $\tilde{\mathbf{X}}(0) = \mathbf{I}_2$ has the form

$$u(t) = 1 + \sum_{m=1}^{\infty} d_{2m} t^{2m}, \quad v(t) = t + \sum_{m=1}^{\infty} d_{2m+1} t^{2m+1}.$$

Compute d_{2m} and d_{2m+1} for $m = 1, 2, \dots$.

(c) Show directly that the series converge for $|t| < 1$. (Hint: Use the ratio test.)

(d) If $\alpha = n$, a nonnegative integer, show that u is a polynomial of degree n if n is even and v is a polynomial of degree n if n is odd. When appropriately normalized these are called the *Chebyshev polynomials* .

23. For $|t| < 1$ consider the Chebyshev equation of exercise 22 with $\alpha = n$, a nonnegative integer.

$$(1 - t^2)x'' - tx' + n^2x = 0. \tag{5.44}$$

(a) For each $n = 0, 1, 2, \ldots$, let

$$T_n(t) = \cos(n \, \cos^{-1}(t)).$$

Show that T_n is a polynomial of degree n satisfying $T_n(1) = 1$. This T_n is called the *nth Chebyshev polynomial*. (Hint: Let $t = \cos(\theta)$ and note that

$$\cos(n\theta) = \text{Re}(e^{in\theta}) = \text{Re}([\cos(\theta) + i \, \sin(\theta)]^n)$$

$$= \text{Re}\left(\sum_{k=0}^{n} \binom{n}{k} i^k \cos(\theta)^{n-k} \sin(\theta)^k\right),$$

where $\binom{n}{k}$ is the binomial coefficient $\binom{n}{k} = \frac{n!}{k!(n-k)!}$.)
(b) Verify that T_n satisfies (5.44).
(c) Verify that the first five Chebyshev polynomials are given by

$$T_0(t) = 1, \quad T_1(t) = t, \quad T_2(t) = 2t^2 - 1,$$

$$T_3(t) = 4t^3 - 3t, \quad T_4(t) = 8t^4 - 8t^2 + 1.$$

24. The equation

$$x'' - 2tx' + 2\alpha x = 0, \quad \alpha \in \mathcal{C}, \tag{5.45}$$

is called the *Hermite equation*. It has a solution x of the form

$$x(t) = \sum_{k=0}^{\infty} c_k t^k,$$

where the series converges for all $t \in \mathcal{R}$.
(a) Show that the c_k satisfy the recursion relation

$$(k+1)(k+2)c_{k+2} = 2(k - \alpha)c_k, \quad k = 0, 1, 2, \ldots.$$

(b) From (a) it follows that the basis $\mathbf{X} = (u, v)$ for (5.45) satisfying $\tilde{\mathbf{X}}(0) = \mathbf{I}_2$ has the form

$$u(t) = 1 + \sum_{k=0}^{\infty} d_{2m}t^{2m}, \quad v(t) = t + \sum_{k=0}^{\infty} d_{2m+1}t^{2m+1}.$$

Compute d_{2m} and d_{2m+1}.

(c) If $\alpha = n$ is a nonnegative integer, show that u is a polynomial of degree n if n is even, and v is a polynomial of degree n if n is odd.

25. Consider that Hermite equation (5.45) of exercise 24 with $\alpha = n$, a nonnegative integer

$$x'' - 2tx' + 2nx = 0. \tag{5.46}$$

(a) For $n = 0, 1, 2, \ldots$, let

$$H_n(t) = (-1)^n e^{t^2} D^n e^{-t^2}, \quad D = \frac{d}{dt}.$$

Show that H_n is a polynomial of degree n. This H_n is called the *nth Hermite polynomial*.

(b) Show that H_n satisfies (5.46) as follows. If $w(t) = \exp(-t^2)$, show that $w'(t) + 2tw(t) = 0$. Differentiate this equation n times to obtain

$$H_{n+1}(t) - 2tH_n(t) + 2nHn - 1(t) = 0, \quad n \geq 1.$$

Differentiate H_n to obtain

$$H'_n(t) = 2tH_n(t) - H_{n+1}(t), \quad n \geq 0.$$

Use these two equations to show H_n satisfies (5.46).

(c) Verify that the first five Hermite polynomials are given by

$$H_0(t) = 1, \quad H_1(t) = 2t, \quad H_2(t) = 4t^2 - 2,$$

$$H_3(t) = 8t^3 - 12t, \quad H_4(t) = 16t^4 - 48t^2 + 12.$$

26. (a) Show that the series (5.33), (5.34) for u and v converge for $|t| < 1$.

(b) If the series for u or v does not reduce to a polynomial, show that it diverges for $|t| > 1$. (Hint: Use the ratio test.)

27. Find two linearly independent solutions u, v of

$$(1 - t^2)x'' - 2tx' = 0$$

for $|t| < 1$. (Hint: Let $y = x'$.)

28. (a) Verify that Q_1 given by

$$Q_1(t) = \frac{t}{2} \log\left(\frac{1+t}{1-t}\right) - 1, \quad |t| < 1,$$

is a solution of the Legendre equation when $\alpha = 1$.

(b) Express Q_1 as a linear combination of the basis u, v given by (5.33), (5.34) with $\alpha = 1$. (Hint: Compute $Q_1(0), Q'_1(0)$.)

29. (a) Show that $P_n(-t) = (-1)^n P_n(t)$ and hence that $P_n(-1) = (-1)^n$.

(b) Show that $P'_n(1) = n(n+1)/2$ and $P'_n(-1) = (-1)^{n+1} n(n+1)/2$.

30. Show that the coefficient c_n of t^n in $P_n(t)$,

$$P_n(t) = c_n t^n + \cdots + c_0,$$

is given by

$$c_n = \frac{(2n)!}{2^n (n!)^2}.$$

(Hint: The Rodrigues formula implies

$$P_n(t) = \frac{1}{2^n n!} D^n \left[t^{2n} - \frac{n(t^2)^{n-1}}{1!} + \cdots \right].)$$

31. Show that P_n is given by the explicit formula

$$P_n(t) = \frac{1}{2^n} \sum_{k=0}^{\lfloor n/2 \rfloor} \frac{(-1)^k (2n - 2k)!}{k!(n-k)!(n-2k)!} t^{n-2k}.$$

(Hint: Use the Rodrigues formula and note that

$$(t^2 - 1)^n = \sum_{k=0}^{n} (-1)^k \binom{n}{k} t^{2n-2k},$$

and

$$D^n t^{2n-2k} = \frac{(2n - 2k)!}{(n - 2k)!} t^{n-2k}.$$

Here $\lfloor n/2 \rfloor$ is the largest integer less than or equal to $n/2$.)

32. Show that

$$\int_{-1}^{1} P_n(t) P_m(t) \, dt = 0, \quad n \neq m$$

as follows. Note that

$$[(1 - t^2) P_m']' = -m(m+1) P_m,$$

$$[(1 - t^2) P_n']' = -n(n+1) P_n.$$

Hence

$$P_n[(1 - t^2) P_m']' - P_m[(1 - t^2) P_n']'$$

$$= [(1 - t^2)(P_n P_m' - P_m P_n')]' = [n(n+1) - m(m+1)] P_n P_m.$$

Integrate from -1 to 1.

33. Show that

$$\int_{-1}^{1} P_n^2(t) \, dt = \frac{2}{2n+1}$$

as follows. Let $u(t) = (t^2 - 1)^n$ in the Rodrigues formula

$$P_n(t) = \frac{1}{2^n n!} u^{(n)}(t).$$

Show that $u^{(k)}(1) = u^{(k)}(-1) = 0$ if $0 \leq k < n$. Integrate by parts to obtain

$$\int_{-1}^{1} u^{(n)}(t) u^{(n)}(t) \, dt = u^{(n)}(t) u^{(n-1)}(t) \big|_{-1}^{1} - \int_{-1}^{1} u^{(n+1)}(t) u^{(n-1)}(t) \, dt$$

$$= - \int_{-1}^{1} u^{(n+1)}(t)u^{(n-1)}(t) \ dt = \cdots$$

$$= (-1)^n \int_{-1}^{1} u^{(2n)}(t)u(t) \ dt = (2n)! \int_{-1}^{1} (1 - t^2)^n \ dt.$$

Let $t = \sin(\theta)$ in this last integral, and obtain

$$\int_{-1}^{1} (1 - t^2)^n \ dt = 2 \int_{0}^{\pi/2} \cos(\theta)^{2n+1} \ d\theta = \frac{2(2^n n!)^2}{(2n + 1)!}.$$

34. Show that for $n = 0, 1, 2, \ldots$, there are constants $a_0, a_1, \ldots, a_n \in \mathcal{R}$ such that

$$t^n = a_0 P_0(t) + a_1 P_1(t) + \cdots + a_n P_n(t).$$

35. Let P be any polynomial of degree n,

$$P(t) = b_n t^n + \cdots + b_0, \quad b_n \neq 0, \quad b_j \in \mathcal{C}.$$

Show that P is a linear combination of $P_0, P_1, \ldots, P_n$,

$$P = \alpha_0 P_0 + \alpha_1 P_1 + \cdots + \alpha_n P_n, \quad \alpha_j \in \mathcal{C}.$$

(Hint: see exercise 34.)

36. If P is a polynomial of degree n, with

$$P = \alpha_0 P_0 + \alpha_1 P_1 + \cdots + \alpha_n P_n$$

as in exercise 35, show that

$$\alpha_k = \frac{2k + 1}{2} \int_{-1}^{1} P(t) P_k(t) \ dt, \quad k = 0, 1, \ldots, n.$$

(Hint: Use the results of exercise 32.)

37. Show that

$$\int_{-1}^{1} t^m P_n(t) \ dt = 0$$

if m is an integer, $0 \le m < n$. (Hint: Use exercise 32.)

38. Show that

$$(n + 1)P_{n+1}(t) - (2n + 1)t P_n(t) + n P_{n-1}(t) = 0$$

for $n = 1, 2, \ldots$, as follows. Comparing the coefficients of t^{n+1} in P_{n+1} and $t P_n$ (see exercise 30), show that

$$(n + 1)P_{n+1}(t) - (2n + 1)t P_n(t) = P(t),$$

where P is a polynomial of degree at most n. By exercise 35,

$$P = \alpha_0 P_0 + \alpha_1 P_1 + \cdots + \alpha_n P_n$$

for some α_j. Using exercise 32 show that $\alpha_k = 0$, $k = 0, 1, 2, \ldots, n - 2$. Hence

$$(n + 1)P_{n+1}(t) - (2n + 1)tP_n(t) = \alpha_{n-1}P_{n-1}(t) + \alpha_n P_n(t).$$

Evaluate α_{n-1}, α_n by putting $t = \pm 1$.

39. Show that

$$P_n'(t) - tP_{n-1}'(t) - nP_{n-1}(t) = 0, \quad n = 1, 2, \ldots.$$

(Hint: Use (5.47) and exercise 38.)

40. (a) Suppose $P_n(\tau) = 0$ for some τ, $-1 < \tau < 1$. Show that τ is a simple zero of P_n, that is, its multiplicity m is 1. (Hint: Apply the uniqueness theorem to $L_n x = 0$ with initial values at τ.)

(b) Show that P_n for $n \geq 1$ has n distinct zeros on $(-1, 1)$ as follows. Suppose P_n has k zeros $\tau_1, \ldots, \tau_k$ in $(-1, 1)$, where $0 \leq k < n$. If

$$P(t) = (t - \tau_1) \cdots (t - \tau_k),$$

show that the polynomial PP_n has a constant sign in $(-1, 1)$, and hence

$$\int_{-1}^{1} P(t)P_n(t) \, dt \neq 0.$$

But (by exercise 37) $deg(P) = k < n$ implies

$$\int_{-1}^{1} P(t)P_n(t) \, dt = 0,$$

a contradiction. Thus P_n has at least n zeros on $(-1, 1)$, and since $deg(P_n) = n$, it has all of its zeros on $(-1, 1)$, and they are distinct by (a).

41. (a) Show that Q_0 given by (5.42) satisfies

$$Q_0(-t) = -Q_0(t), \quad |t| \neq 1.$$

(b) From (5.41) we have $Q_n = P_n Q_0 - p_{n-1}$, where p_{n-1} is a polynomial, $deg(p_{n-1}) = n - 1$. Show that

$$p_{n-1}(-t) = (-1)^{n+1} p_{n-1}(t).$$

(c) Show that

$$Q_n(-t) = (-1)^{n+1} Q_n(t), \quad |t| \neq 1.$$

42. Verify that

$$Q_1(t) = tQ_0(t) - 1, \quad Q_2(t) = \frac{1}{2}(3t^2 - 1)Q_0(t) - \frac{3}{2}t,$$

$$Q_3(t) = \frac{1}{2}(5t^3 - 3t)Q_0(t) - \frac{1}{6}(15t^2 - 4).$$

43. If m, n are nonnegative integers, the *associated Legendre equation* is

$$L_{m,n}x = (1 - t^2)x'' - 2tx' + \left[n(n+1) - \frac{m^2}{1 - t^2}\right]x = 0, \quad |t| \neq 1.$$

(a) If x is any solution of the Legendre equation $L_n x = 0$, show that $y = (1-t^2)^{m/2} D^m x$ is a solution of the associated Legendre equation $L_{m,n} y = 0$ as follows. Compute y' and y'' to obtain

$$L_{m,n}y = (1 - t^2)^{m/2} z,$$

where

$$z = (1 - t^2)x^{(m+2)} - 2(m + 1)tx^{(m+1)} + [n(n+1) - m(m+1)]x^{(m)}.$$

Differentiate the equation $L_n x = 0$ m times and show $z = 0$.

(b) Define

$$P_n^m(t) = (1 - t^2)^{m/2} D^m P_n(t), \quad Q_n^m(t) = (1 - t^2)^{m/2} D^m Q_n(t), \quad |t| \neq 1.$$

These are the *associated Legendre functions* of the first and second kind. Note that P_n^m is the zero function if $m > n$. For any $m = 0, 1, \ldots, n$ and nonnegative integer n, show that $X = (P_n^m, Q_n^m)$ is a basis for $L_{m,n}x = 0$ for $|t| < 1$ and for $|t| > 1$.

44. (a) Show that

$$\int_{-1}^{1} P_n^m(t) P_p^m(t) \, dt = 0, \quad n \neq p.$$

(b) Show that

$$\int_{-1}^{1} [P_n^m(t)]^2 \, dt = \frac{(n + m)!}{(n - m)!} \frac{2}{2n + 1}, \quad m = 0, 1, \ldots, n.$$

(Hint: The case $m = 0$ is considered in exercise 33.)

Chapter 6

Singular Points

6.1 Introduction

While the power series method of the previous chapter applies to a number of important problems, the theory requires modification to handle many other interesting examples. In such examples the matrix-valued function $\mathbf{A}(t)$ or the vector-valued function $B(t)$ in the system

$$X' = \mathbf{A}(t)X + B(t)$$

may be analytic at every point $t \in I$ except at some particular point τ. In such cases the behavior in a neighborhood of τ is often most significant.

A physically motivated example, the vibrations of a circular drumhead, illustrates how such equations arise. Choose coordinates so that the drumhead rests in the $x - y$ plane (Figure 6.1). Small vibrations in the vertical direction are modeled by the partial differential equation

$$\frac{\partial^2 u}{\partial t^2} = c^2 \left(\frac{\partial^2 u}{\partial x^2} + \frac{\partial^2 u}{\partial y^2} \right), \qquad (W_E)$$

where $u(x, y, t)$ represents the vertical displacement of the drumhead at position (x, y) and time t. If the membrane is fixed at the boundary, the condition $u(x, y, t) = 0$ must be satisfied for (x, y) on the boundary. The constant c depends on the physical parameters of the membrane.

Since the value of c will not play an important role in this discussion, assume that $c = 1$. The equation (W_E) applies regardless of the shape of the drumhead. If the shape is a disk of radius R, the problem is simplified by using polar coordinates (r, θ). By using the chain rule for differentiation, the form of the equation for $u(r, \theta, t)$ in polar coordinates becomes

$$\frac{\partial^2 u}{\partial t^2} = \frac{\partial^2 u}{\partial r^2} + \frac{1}{r} \frac{\partial u}{\partial r} + \frac{1}{r^2} \frac{\partial^2 u}{\partial \theta^2}. \qquad (W_P)$$

The condition $u(R, \theta, t) = 0$ expresses the fixing of the membrane at the boundary $r = R$.

In some cases special solutions of partial differential equations can be written as products of functions of a single variable. For the vibrating circular drum this approach, called *separation of variables*, leads us to look for solutions of the form

$$u(r, \theta, t) = f(r)g(\theta)h(t).$$

If this form for u is inserted into equation (W_P), and both sides are divided by $u = f(r)g(\theta)h(t)$, we find

$$\frac{h''(t)}{h(t)} = \frac{f''(r)}{f(r)} + \frac{1}{r}\frac{f'(r)}{f(r)} + \frac{1}{r^2}\frac{g''(\theta)}{g(\theta)}.$$

This equation says that the value of the function $h''(t)/h(t)$ is actually independent of t, or there is a constant λ such that

$$h''(t) + \lambda h(t) = 0.$$

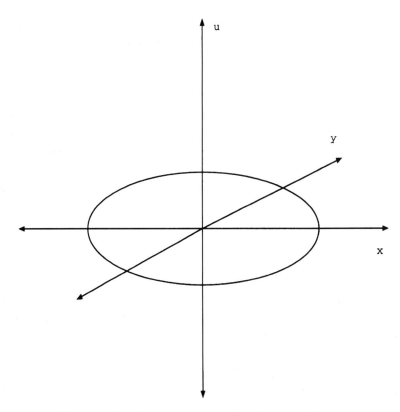

Figure 6.1: Geometry for a circular drumhead.

A similar argument shows that there is a constant α such that

$$g''(\theta) + \alpha g(\theta) = 0,$$

and finally that $f(r)$ satisfies

$$f''(r) + \frac{1}{r}f'(r) + \left[\lambda - \frac{\alpha}{r^2}\right]f(r) = 0.$$

In addition, the condition $u(R, \theta, t) = 0$ takes the form $f(R) = 0$.

The equations for g and h have constant coefficients, so these can be solved explicitly by elementary methods. On the other hand the equation for f has coefficients which are analytic when $r > 0$ but have a singularity when $r = 0$. This equation falls into the class of problems considered in this chapter. In fact, we will be able to express f in terms of solutions to Bessel's equation, the main example taken up later in the chapter (see exercise 43).

While we have identified the equations which must be satisfied by $f(r)$, $g(\theta)$, and $h(t)$, there are additional conditions which must be satisfied if $u(r, \theta, t) = f(r)g(\theta)h(t)$ is to be a meaningful solution of the physical problem. The boundary condition $f(R) = 0$ has been noted. In addition, since the coordinates (r, θ) and $(r, \theta + 2\pi)$ describe the same points, the function $g(\theta)$ must satisfy $g(\theta) = g(\theta + 2\pi)$. For a more complete discussion of these additional constraints, including conditions on the behavior of f as $r \to 0$, the reader may consult [19, pp. 195–200] or [24, pp. 251–257].

Turning to a more general discussion, consider a first-order linear system

$$\mathbf{A}_1(t)X' + \mathbf{A}_0(t)X = B(t), \tag{6.1}$$

where $\mathbf{A}_0, \mathbf{A}_1 \in C(I, M_n(\mathcal{C}))$, and $B \in C(I, \mathcal{C}^n)$ for some real interval I. If $\mathbf{A}_1$ is invertible on I, then X is a solution of (6.1) on I if and only if

$$X' = -\mathbf{A}_1^{-1}(t)\mathbf{A}_0(t)X + \mathbf{A}_1^{-1}(t)B. \tag{6.2}$$

Now (6.2) is an equation for which our existence and uniqueness results hold. Thus, given any $\tau \in I$, $\xi \in \mathcal{C}^n$ there exists a unique solution X of (6.1) on I satisfying $X(\tau) = \xi$. A point $\tau \in I$ where $\mathbf{A}_1$ is invertible is called a *regular point* for (6.1), and a point $\tau \in I$ where $\mathbf{A}_1$ is not invertible is called a *singular point* for (6.1).

If τ is a singular point for (6.1), existence or uniqueness of solutions may fail for the initial value problem at τ. The following two simple examples illustrate this in the case $n = 1$:

$$tx' + x = 0, \tag{6.3}$$

$$tx' - x = 0. \tag{6.4}$$

In both cases $\tau = 0$ is the only singular point. The solutions of (6.3) have the form

$$x(t) = c/t, \quad t \neq 0, \quad c \in \mathcal{C},$$

and the only solution which exists at $\tau = 0$ is the trivial one, where $c = 0$. No initial condition at $t = 0$ other than $x(0) = 0$ is possible. On the other hand, the solutions x of (6.4) have the form

$$x(t) = ct, \quad c \in \mathcal{C}.$$

These all exist at $\tau = 0$, with $x(\tau) = 0$. Thus not only is $\xi = 0$ the only initial value possible at $\tau = 0$, but there are infinitely many solutions satisfying $x(\tau) = 0$.

In general it is rather difficult to determine the nature of the solutions in the vicinity of a singular point. However, there is a large class of equations for which the singularity is not too bad, in that slight modifications of the power series method used in Chapter 5 for solving equations with analytic coefficients can be employed to provide solutions near the singularity. After a thorough analysis is provided for systems with these mild singularities, the general techniques are brought to bear on two important examples, the Euler equation, which has explicit elementary solutions, and Bessel's equation, one of the most famous of differential equations.

We will concentrate on the homogeneous equation

$$\mathbf{A}_1(t)X' + \mathbf{A}_0(t)X = 0, \tag{6.5}$$

since once a basis for such an equation is known, the nonhomogeneous equation (6.1) can be solved using the variation of parameters formula. A point $\tau \in \mathcal{R}$ is a *singular point of the first kind* for (6.5) if the equation can be written in the form

$$(t - \tau)X' = \mathbf{A}(t)X \tag{6.6}$$

on some open interval containing τ, where $\mathbf{A}$ is analytic at τ and $\mathbf{A}(\tau) \neq 0$. Thus $\mathbf{A}$ has a power series representation

$$\mathbf{A}(t) = \sum_{k=0}^{\infty} \mathbf{A}_k(t - \tau)^k, \quad \mathbf{A}_k \in M_n(\mathcal{C}),$$

where the series is convergent for $|t - \tau| < \rho$ for some $\rho > 0$ and $\mathbf{A}_0 \neq 0$. If $\mathbf{A}_0 = 0$, division of (6.6) by $t - \tau$ results in an equation with analytic coefficients.

If $\mathbf{A}_0 \neq 0$ then new phenomenon appear, but these turn out to be manageable. The simplest example occurs when $\mathbf{A}(t) = \mathbf{A}$ is a constant matrix, and we take $\tau = 0$,

$$tX' = \mathbf{A}X, \quad \mathbf{A} \in M_n(\mathcal{C}). \tag{6.7}$$

If $n = 1$ it is easy to see that

$$X(t) = e^{\mathbf{A} \log(t)} = t^{\mathbf{A}}$$

gives a solution of (6.7) in case $t > 0$, and, similarly, a basis $\mathbf{X}$ for (6.7) in the general case is given by

$$\mathbf{X}(t) = e^{\mathbf{A} \log(t)}$$

if $t > 0$.

We define $t^{\mathbf{A}}$ for any $\mathbf{A} \in M_n(\mathcal{C})$ by

$$t^{\mathbf{A}} = e^{\mathbf{A} \log(t)}, \quad \mathbf{A} \in M_n(\mathcal{C}), \quad t > 0.$$

Suppose $\mathbf{J} = \operatorname{diag}(\mathbf{J}_1, \ldots, \mathbf{J}_q)$ is a Jordan canonical form for $\mathbf{A}$, with $\mathbf{A} = \mathbf{Q}\mathbf{J}\mathbf{Q}^{-1}$. Then

$$\mathbf{X}(t) = \mathbf{Q}e^{\mathbf{J}\log(t)}\mathbf{Q}^{-1} = \mathbf{Q}t^{\mathbf{J}}\mathbf{Q}^{-1}$$

and

$$\mathbf{Y}(t) = \mathbf{Q}t^{\mathbf{J}}$$

is another basis for (6.7) for $t > 0$. The explicit structure of $t^{\mathbf{J}} = \exp(\mathbf{J}\log(t))$ shows that the columns $Y_1, \ldots, Y_n$ of $\mathbf{Y}$ have the form

$$Y_j(t) = e^{\lambda \log(t)} P(\log(t)) = t^\lambda P(\log(t)),$$

where λ is an eigenvalue of $\mathbf{A}$, and P is a vector polynomial of degree less than or equal to $m - 1$, where m is the multiplicity of λ. For example, if $\mathbf{J}_1$ is the $r_1 \times r_1$ matrix

$$\mathbf{J}_1 = \begin{pmatrix} \lambda_1 & 1 & 0 & \cdots & 0 & 0 \\ 0 & \lambda_1 & 1 & \cdots & 0 & 0 \\ \vdots & \vdots & \vdots & \cdots & \vdots & \vdots \\ 0 & 0 & 0 & \cdots & \lambda_1 & 1 \\ 0 & 0 & 0 & \cdots & 0 & \lambda_1 \end{pmatrix}, \quad r_1 > 1,$$

and if the columns of $\mathbf{Q}$ are $Q_1, \ldots, Q_n$, then the first r_1 columns of $\mathbf{Y}$ are

$$Y_1(t) = t^{\lambda_1} Q_1, \quad Y_2(t) = t^{\lambda_1}(Q_1 \log(t) + Q_2),$$

$$\ldots, \quad Y_{r_1}(t) = t^{\lambda_1}\left(\frac{Q_1 [\log(t)]^{r_1-1}}{(r_1-1)!} + \cdots + Q_{r_1} \right).$$

Corresponding to each eigenvalue λ of $\mathbf{A}$ there is at least one solution Y of the form

$$Y(t) = t^\lambda Q, \quad t > 0, \quad Q \in \mathcal{C}^n.$$

If $\lambda_1, \ldots, \lambda_k$ are the distinct eigenvalues of $\mathbf{A}$ with multiplicities $m_1, \ldots, m_k$, respectively, every solution $X(t)$ of (6.7) can be written as

$$X(t) = t^{\lambda_1} P_1(\log(t)) + \cdots + t^{\lambda_k} P_k(\log(t)), \quad t > 0,$$

where P_j is a vector polynomial of degree at most $m_j - 1$.

For $t < 0$,

$$\mathbf{X}(t) = |t|^{\mathbf{A}} = e^{\mathbf{A}\,\log(|t|)} \tag{6.8}$$

is a basis for (6.7), since $|t| = -t$ for $t < 0$ and

$$\frac{d}{dt}\log(|t|) = \frac{1}{t}, \quad t \neq 0.$$

Thus (6.8) gives a basis for (6.7) on any interval not containing the singular point $\tau = 0$. For each eigenvalue λ of $\mathbf{A}$ with eigenvector Q, there is a solution Y of the form

$$Y(t) = |t|^\lambda Q, \quad t \neq 0, \quad Q \in \mathcal{C}^n,$$

and every solution X of (6.7) can be written as

$$X(t) = |t|^{\lambda_1} P_1(\log|t|) + \cdots + |t|^{\lambda_k} P_k(\log|t|), \quad t \neq 0,$$

where the λ_j are the eigenvalues of $\mathbf{A}$ and P_j is a vector polynomial of degree at most $m_j - 1$. If the eigenvalues λ_j are all distinct, with nontrivial eigenvectors Q_j, then we have a basis

$$X = (|t|^{\lambda_1} Q_1, \ldots, |t|^{\lambda_n} Q_n) = |t|^{\mathbf{A}} \mathbf{Q}, \quad \mathbf{Q} = (Q_1, \ldots, Q_n).$$

The only difference between the example (6.7) and the more general case

$$tX' = \left(\sum_{k=0}^{\infty} \mathbf{A}_k t^k \right) X, \quad \mathbf{A}_k \in M_n(\mathcal{C}), \tag{6.9}$$

where the series converges for $|t| < \rho$ for some $\rho > 0$, is that a basis $\mathbf{X}$ for (6.9) has the form

$$\mathbf{X}(t) = \mathbf{P}(t)|t|^{\mathbf{S}}, \quad \mathbf{S} \in M_n(\mathcal{C}),$$

where

$$\mathbf{P}(t) = \sum_{k=0}^{\infty} \mathbf{P}_k t^k, \quad \mathbf{P}_k \in M_n(\mathcal{C}),$$

and the series for $\mathbf{P}$ converges for $|t| < \rho$. In fact, in many cases we can choose $\mathbf{S}$ to be the constant term $\mathbf{A}_0$ in the power series expansion of $\mathbf{A}$. This is the central result concerning systems with a singular point of the first kind.

6.2 Systems of equations with singular points

6.2.1 Singular points of the first kind: A special case

Consider a first-order system with a singular point of the first kind at τ. Such a system may be written as

$$X' = \left[\frac{\mathbf{R}}{t - \tau} + \mathbf{A}(t) \right] X, \quad t \neq \tau, \tag{6.10}$$

where $\mathbf{R} \in M_n(\mathcal{C})$ and $\mathbf{A}$ is analytic at τ, so that

$$\mathbf{A}(t) = \sum_{k=0}^{\infty} \mathbf{A}_k (t - \tau)^k, \quad \mathbf{A}_k \in M_n(\mathcal{C}),$$

where the series converges for $|t - \tau| < \rho$ for some $\rho > 0$. The next result shows that there are always solutions X of (6.10) of the form $X(t) = |t - \tau|^{\lambda} P(t)$, where λ is an eigenvalue of $\mathbf{R}$ and P has a power series expansion in powers of $t - \tau$ which is convergent for $|t - \tau| < \rho$. It is not necessarily the case that such solutions exist for each eigenvalue of $\mathbf{R}$.

Theorem 6.1: *Let λ be an eigenvalue of $\mathbf{R}$ with the property that $\lambda + k$ is not an eigenvalue of $\mathbf{R}$ for any positive integer k. Then (6.10) has a solution X of the form*

$$X(t) = |t - \tau|^\lambda P(t), \quad 0 < |t - \tau| < \rho, \tag{6.11}$$

where P has a convergent power series representation

$$P(t) = \sum_{k=0}^{\infty} P_k (t - \tau)^k, \quad P_0 \neq 0, \quad P_k \in \mathcal{C}^n, \quad |t - \tau| < \rho. \tag{6.12}$$

The coefficients P_k can be computed uniquely in terms of P_0 by substituting (6.11), (6.12) into (6.10).

Proof: There is no loss of generality in assuming $\tau = 0$, which reduces the problem to

$$X' = \left[\frac{\mathbf{R}}{t} + \mathbf{A}(t) \right] X, \tag{6.13}$$

where

$$\mathbf{A}(t) = \sum_{k=0}^{\infty} \mathbf{A}_k t^k, \quad |t| < \rho. \tag{6.14}$$

Suppose X is a solution of (6.13) of the form

$$X(t) = |t|^\lambda P(t), \quad 0 < |t| < \rho,$$

where

$$P(t) = \sum_{k=0}^{\infty} P_k t^k, \quad P_0 \neq 0, \quad P_k \in \mathcal{C}^n, \quad |t| < \rho. \tag{6.15}$$

Then

$$tX'(t) = \lambda |t|^\lambda P(t) + t|t|^\lambda P'(t) = \mathbf{R}|t|^\lambda P(t) + t\mathbf{A}(t)|t|^\lambda P(t),$$

and P must satisfy

$$\lambda P(t) + tP'(t) - \mathbf{R}P(t) = t\mathbf{A}(t)P(t), \quad |t| < \rho. \tag{6.16}$$

The terms in (6.16) have the form

$$\lambda P(t) = \sum_{k=0}^{\infty} \lambda P_k t^k = \lambda P_0 + \sum_{k=1}^{\infty} \lambda P_k t^k,$$

$$tP'(t) = \sum_{k=0}^{\infty} k P_k t^k = \sum_{k=1}^{\infty} k P_k t^k, \quad \mathbf{R}P(t) = \mathbf{R}P_0 + \sum_{k=1}^{\infty} \mathbf{R}P_k t^k,$$

$$t\mathbf{A}(t)P(t) = t[\mathbf{A}_0 P_0 + (\mathbf{A}_1 P_0 + \mathbf{A}_0 P_1)t + \cdots] = t\sum_{k=0}^{\infty} C_k t^k = \sum_{k=1}^{\infty} C_{k-1} t^k,$$

where

$$C_k = \sum_{j=0}^{\infty} \mathbf{A}_{k-j} P_j, \quad k = 0, 1, 2, \dots.$$

Thus P satisfies (6.16) if and only if

$$(\lambda \mathbf{I}_n - \mathbf{R}) P_0 + \sum_{k=1}^{\infty} [(\lambda + k)\mathbf{I}_n - \mathbf{R}] P_k t^k = \sum_{k=1}^{\infty} C_{k-1} t^k,$$

and this is true if and only if

$$(\lambda \mathbf{I}_n - \mathbf{R}) P_0 = 0, \tag{6.17}$$

$$[(\lambda + k)\mathbf{I}_n - \mathbf{R}] P_k = C_{k-1}, \quad k = 1, 2, \dots. \tag{6.18}$$

The equation (6.17) just says that P_0 is an eigenvector of $\mathbf{R}$ for the eigenvalue λ. The assumption about λ comes into play with equation (6.18). Since $\lambda + k$ is not an eigenvalue of $\mathbf{R}$ for any positive integer k, the matrix $(\lambda + k)\mathbf{I}_n - \mathbf{R}$ is invertible for all $k = 1, 2, \dots$. Thus (6.18) may be rewritten as

$$P_k = [(\lambda + k)\mathbf{I}_n - \mathbf{R}]^{-1} C_{k-1}, \quad k = 1, 2, \dots,$$

which is a recursion relation for the P_k, since C_{k-1} only uses $P_0, P_1, \dots, P_{k-1}$. Once P_0 is selected, all other P_k are uniquely given by (6.18).

The remaining problem is to show that if the $P_k \in \mathcal{C}^n$ satisfy (6.17), (6.18), with $P_0 \neq 0$, then the series (6.15) is convergent for $|t| < \rho$. From the above computation it then follows that X satisfies (6.13) for $0 < |t| < \rho$.

By writing (6.18) in the form

$$k P_k = (\mathbf{R} - \lambda \mathbf{I}_n) P_k + C_{k-1}, \quad k = 1, 2, \dots,$$

we obtain the inequality

$$|k P_k| \leq |\mathbf{R} - \lambda \mathbf{I}_n| |P_k| + |C_{k-1}|, \quad k = 1, 2, \dots. \tag{6.19}$$

If l is a positive integer satisfying $|\mathbf{R} - \lambda \mathbf{I}_n| \leq l/2$, then $|\mathbf{R} - \lambda \mathbf{I}_n| \leq k/2$ for $k \geq l$, and (6.19) yields

$$k|P_k| \leq 2|C_{k-1}| = 2 \left| \sum_{j=0}^{k-1} \mathbf{A}_{k-1-j} P_j \right|, \quad k \geq l. \tag{6.20}$$

The series (6.14) converges for $|t| < \rho$. Thus if r is any number such that $0 < r < \rho$, then

$$|\mathbf{A}_j| r^j \leq M, \quad j = 0, 1, 2, \dots,$$

for some constant $M > 0$. Using this in (6.20) leads to

$$k|P_k| \leq 2 \sum_{j=0}^{k-1} |\mathbf{A}_{k-1-j}| \, |P_j| \leq 2 M r^{1-k} \sum_{j=0}^{k-1} |P_j| r^j, \quad k \geq l. \tag{6.21}$$

Let $d_k \geq 0$ be defined by

$$d_k = |P_k|, \quad k = 0, 1, \ldots, l - 1,$$

$$k d_k = 2M r^{1-k} \sum_{j=0}^{k-1} d_j r^j, \quad k \geq l. \tag{6.22}$$

It then follows from (6.21) that

$$|P_k| \leq d_k, \quad k = 0, 1, \ldots. \tag{6.23}$$

The proof that the series (6.15) converges for $|t| < \rho$ now proceeds just as in the proof of Theorem 5.1. We repeat the brief argument. The equation (6.22) implies

$$(k+1) d_{k+1} = (2M + r^{-1}k) d_k, \quad k \geq l.$$

Applying the ratio test for $k \geq l$ gives

$$\left| \frac{d_{k+1} t^{k+1}}{d_k t^k} \right| = \frac{2M}{k+1} + \frac{k}{k+1} \frac{1}{r} \to \frac{|t|}{r}, \quad k \to \infty.$$

Thus the series

$$\sum_{k=0}^{\infty} d_k t^k$$

converges absolutely for $|t| < r$, and (6.23) then implies that the series for $P(t)$ given in (6.15) is convergent for $|t| < r$. Since r was any number such that $0 < r < \rho$, it follows that (6.15) converges for $|t| < \rho$. $\square$

Corollary 6.2: *If the distinct eigenvalues of* $\mathbf{R}$ *are* $\lambda_1, \ldots, \lambda_k$ *and* λ *is any eigenvalue such that*

$$\mathrm{Re}(\lambda) = \max[\mathrm{Re}(\lambda_j)], \quad j = 1, \ldots, k,$$

there is a solution X *of* (6.10) *of the form*

$$X(t) = |t - \tau|^\lambda P(t), \quad 0 < |t - \tau| < \rho,$$

where P *has a convergent power series representation* (6.12) *for* $|t - \tau| < \rho$ *and* $P(\tau) \neq 0$.

Corollary 6.3: *If* $\mathbf{R}$ *has* n *distinct eigenvalues* $\lambda_1, \ldots, \lambda_n$ *with the property that no two differ by a positive integer, then* (6.10) *has a basis* $\mathbf{X} = (X_1, \ldots, X_n)$, *where the* X_j *have the form*

$$X_j(t) = |t - \tau|^{\lambda_j} P_j(t), \quad 0 < |t - \tau| < \rho.$$

The P_j *have convergent power series expansions in powers of* $t - \tau$ *for* $|t-\tau| < \rho$ *and* $P_j(\tau) \neq 0$.

Proof: Choosing $P_j(\tau)$ to be an eigenvector of $\mathbf{R}$ for the eigenvalue λ_j, we obtain the existence of the solutions by Theorem 6.1. The matrix $\mathbf{X}$ has the form

$$\mathbf{X}(t) = \mathbf{P}(t)|t - \tau|^{\mathbf{S}},$$

where

$$\mathbf{P} = (P_1, \dots, P_n), \quad \mathbf{S} = \mathrm{diag}(\lambda_1, \dots, \lambda_n).$$

The n eigenvectors $P_1(\tau), \dots, P_n(\tau)$, belonging to distinct eigenvalues form a basis for $\mathcal{C}^n$, so that $\det(\mathbf{P}(\tau)) \neq 0$. The continuity of $\mathbf{P}$ implies that $\det(\mathbf{P}(t)) \neq 0$ for t near τ, and thus $\mathbf{X}(t)$ is invertible for t near τ. Hence $\mathbf{X}$ is a basis for (6.10) for $0 < |t - \tau| < \rho$. $\square$

A simple example is given by

$$X' = \left(\frac{\mathbf{R}}{t} + \mathbf{A}\right) X, \quad \mathbf{R} = \frac{1}{6}\begin{pmatrix} 5 & -6 \\ 4 & -6 \end{pmatrix}, \quad \mathbf{A} = 3\begin{pmatrix} 1 & 0 \\ 0 & 1 \end{pmatrix}, \quad t \neq 0.$$

Here the eigenvalues of $\mathbf{R}$ are $\lambda_1 = -1/2$, $\lambda_2 = 1/3$, and, since $\lambda_2 - \lambda_1 = 5/6$ is not a positive integer, there is a basis $\mathbf{X} = (U, V)$ of the form

$$U(t) = |t|^{-1/2} P(t), \quad V(t) = |t|^{1/3} Q(t),$$

where P, Q have power series expansions

$$P(t) = \sum_{k=0}^{\infty} P_k t^k, \quad Q(t) = \sum_{k=0}^{\infty} Q_k t^k$$

convergent for all $t \in \mathcal{R}$. The constant terms P_0, Q_0 are eigenvectors for $\mathbf{R}$ for λ_1, λ_2, respectively. Thus we may take

$$P_0 = \begin{pmatrix} 3 \\ 4 \end{pmatrix}, \quad Q_0 = \begin{pmatrix} 2 \\ 1 \end{pmatrix}.$$

Let us compute $P(t)$. Its coefficients satisfy the recursion relation (6.18), which in this case reduces to

$$\mathbf{R}(\lambda, k)P_k = C_{k-1} = AP_{k-1} = 3P_{k-1}, \quad k = 1, 2, \dots,$$

where

$$\mathbf{R}(\lambda, k) = (\lambda + k)\mathbf{I}_2 - \mathbf{R}, \quad \lambda = \lambda_1 = -1/2.$$

Thus

$$\mathbf{R}(\lambda, 1)P_1 = 3P_0, \quad P_1 = 3[\mathbf{R}(\lambda, 1)]^{-1} P_0.$$

When λ is an eigenvalue of $\mathbf{R}$ with eigenvector P_0,

$$\mathbf{R}(\lambda, k)P_0 = (\lambda \mathbf{I}_2 - R)P_0 + kP_0 = kP_0,$$

so that

$$[\mathbf{R}(\lambda, k)]^{-1} P_0 = P_0/k, \quad k = 1, 2, \dots.$$

Consequently,

$$P_1 = 3P_0, \quad \mathbf{R}(\lambda, 2)P_2 = 3P_1 = 9P_0, \quad P_2 = 9P_0/2,$$

and, in general,

$$P_k = \frac{3^k}{k!}P_0.$$

Therefore

$$P(t) = \sum_{k=0}^{\infty} \left(\frac{3^k t^k}{k!} \right) P_0 = e^{3t}P_0,$$

and similarly $Q(t) = e^{3t}Q_0$. A basis $\mathbf{X} = (U, V)$ is given by

$$U(t) = |t|^{-1/2}e^{3t} \begin{pmatrix} 3 \\ 4 \end{pmatrix}, \quad V(t) = |t|^{1/3}e^{3t} \begin{pmatrix} 2 \\ 1 \end{pmatrix}, \quad |t| > 0,$$

or

$$\mathbf{X}(t) = e^{3t} \begin{pmatrix} 3 & 2 \\ 4 & 1 \end{pmatrix} \begin{pmatrix} |t|^{-1/2} & 0 \\ 0 & |t|^{1/3} \end{pmatrix}, \quad |t| > 0.$$

The equations (6.17), (6.18) in the proof of Theorem 6.1 may have a nontrivial solution if $\lambda + m$ are eigenvalues of $\mathbf{R}$ for some positive integers m. That is, the equations $[(\lambda + m)\mathbf{I}_n - \mathbf{R}]P_m = C_{m-1}$ may have solutions for P_m even though $\det[(\lambda + m)\mathbf{I}_n - \mathbf{R}] = 0$. This would lead to solutions X of the form (6.11), for we have proved that if the $P_k \in \mathcal{C}^n$ are any vectors satisfying (6.17), (6.18), then the series

$$\sum_{k=0}^{\infty} P_k t^k$$

is convergent for $|t - \tau| < \rho$. This is illustrated by the simple example

$$tX' = \mathbf{R}X, \quad \mathbf{R} = \begin{pmatrix} 1 & 0 \\ 0 & 0 \end{pmatrix}.$$

Here $\lambda_1 = 1$, $\lambda_2 = 0$, and we have the basis

$$\mathbf{X} = (U, V), \quad U(t) = E_1 t, \quad V(t) = E_2.$$

A contrasting example is given by

$$tX' = [\mathbf{R} + \mathbf{A}_0 t]X, \quad \mathbf{R} = \begin{pmatrix} 1 & 0 \\ 0 & 0 \end{pmatrix}, \quad \mathbf{A}_0 = \begin{pmatrix} 0 & 1 \\ 0 & 0 \end{pmatrix}.$$

Again the eigenvalues of $\mathbf{R}$ are $\lambda_1 = 1$, $\lambda_2 = 0$. For $\lambda_2 = 0$ the equation (6.17) has a solution $P_0 = E_2$. However, the equation (6.18) for $k = 1$ becomes

$$(\mathbf{I}_2 - \mathbf{R})P_1 = \begin{pmatrix} 0 & 0 \\ 0 & 1 \end{pmatrix} \begin{pmatrix} \alpha \\ \beta \end{pmatrix} = \mathbf{A}_0 P_0 = E_1 = \begin{pmatrix} 1 \\ 0 \end{pmatrix}$$

for

$$P_1 = \begin{pmatrix} \alpha \\ \beta \end{pmatrix},$$

which has no solution. It is not difficult to see that a basis for the solutions is
$\mathbf{X} = (U, V)$, where

$$U(t) = \begin{pmatrix} t \\ 0 \end{pmatrix}, \quad V(t) = \begin{pmatrix} t \ \log(|t|) \\ 1 \end{pmatrix}.$$

We are thus faced with the problem of finding a basis $\mathbf{X}$ for (6.10) in case
the eigenvalues of $\mathbf{R}$ do not satisfy the conditions in Corollary 6.3. Further
progress will require two results about matrices, given in the next section.

6.2.2   Two results about matrices

If $\mathbf{A} \in M_n(\mathcal{C})$ and p is any polynomial,

$$p(\lambda) = a_k \lambda^k + \cdots + a_1 \lambda + a_0, \quad a_j \in C,$$

then the matrix $p(\mathbf{A})$ is defined by

$$p(\mathbf{A}) = a_k \mathbf{A}^k + \cdots + a_1 \mathbf{A} + a_0 \mathbf{I}_n.$$

Cayley–Hamilton theorem: *If $\mathbf{A} \in M_n(\mathcal{C})$ has the characteristic polynomial $p_{\mathbf{A}}$,*

$$p_{\mathbf{A}}(\lambda) = \det(\lambda \mathbf{I}_n - \mathbf{A}) = \lambda^n + a_{n-1}\lambda^{n-1} + \cdots + a_1 \lambda + a_0, \quad a_j \in \mathcal{C},$$

then
$$p_{\mathbf{A}}(\mathbf{A}) = \mathbf{A}^n + a_{n-1}\mathbf{A}^{n-1} + \cdots + a_1 \mathbf{A} + a_0 \mathbf{I}_n = 0.$$

Proof: A proof can be based on the primary decomposition theorem (Chapter 3). Let $\lambda_1, \ldots, \lambda_k$ be the distinct eigenvalues of $\mathbf{A}$, with multiplicities $m_1, \ldots, m_k$, respectively. Then

$$p_{\mathbf{A}}(\lambda) = (\lambda - \lambda_1)^{m_1} \cdots (\lambda - \lambda_k)^{m_k},$$

and every $\xi \in \mathcal{C}^n$ can be written uniquely as

$$\xi = \xi_1 + \cdots + \xi_k, \tag{6.24}$$

where
$$(\mathbf{A} - \lambda_j \mathbf{I}_n)^{m_j} \xi_j = 0, \quad j = 1, \ldots, k. \tag{6.25}$$

Thus
$$p_{\mathbf{A}}(\mathbf{A}) = (\mathbf{A} - \lambda_1 \mathbf{I}_n)^{m_1} \cdots (\mathbf{A} - \lambda_k \mathbf{I}_n)^{m_k},$$

and all the factors in this expression commute. If $\xi \in \mathcal{C}^n$ is written as in (6.24),

$$p_{\mathbf{A}}(\mathbf{A})\xi = (\mathbf{A} - \lambda_2 \mathbf{I}_n)^{m_2} \cdots (\mathbf{A} - \lambda_k \mathbf{I}_n)^{m_k} (\mathbf{A} - \lambda_1)^{m_1} \xi_1$$

$$+ \cdots + (\mathbf{A} - \lambda_1 \mathbf{I}_n)^{m_1} \cdots (\mathbf{A} - \lambda_{k-1} \mathbf{I}_n)^{m_{k-1}} (\mathbf{A} - \lambda_k)^{m_k} \xi_k = 0,$$

since (6.25) is valid. It follows that the columns of $p_{\mathbf{A}}(\mathbf{A})$, which are just $p_{\mathbf{A}}(\mathbf{A})E_1, \ldots, p_{\mathbf{A}}(\mathbf{A})E_n$, are all zero, and hence $p_{\mathbf{A}}(\mathbf{A}) = 0$. $\square$

The second result concerns solutions $\mathbf{U} \in M_n(\mathcal{C})$ of a matrix equation

$$\mathbf{UA} - \mathbf{BU} = \mathbf{C}, \tag{6.26}$$

where $\mathbf{A}, \mathbf{B}, \mathbf{C} \in M_n(\mathcal{C})$ are given matrices. Note that (6.26) is a linear system of equations for the n^2 components $\mathbf{U}_{ij}$ of the matrix $\mathbf{U}$, for the map

$$f : \mathbf{U} \in M_n(\mathcal{C}) \to f(\mathbf{U}) = \mathbf{UA} - \mathbf{BU} \in M_n(\mathcal{C})$$

is linear:

$$f(\alpha_1\mathbf{U}_1 + \alpha_2\mathbf{U}_2) = \alpha_1 f(\mathbf{U}_1) + \alpha_2 f(\mathbf{U}_2), \quad \alpha_j \in \mathcal{C}, \quad \mathbf{U}_j \in M_n(\mathcal{C}).$$

Thus (6.26) has a unique solution for all $\mathbf{C} \in M_n(\mathcal{C})$ if and only if the homogeneous equation

$$\mathbf{UA} - \mathbf{BU} = 0 \tag{6.27}$$

has only the trivial solution $\mathbf{U} = 0$.

Theorem 6.4: *Let $\mathbf{A}, \mathbf{B} \in M_n(\mathcal{C})$ and suppose that $\mathbf{A}$ and $\mathbf{B}$ have no eigenvalue in common. Then (6.27) has only the trivial solution $\mathbf{U} = 0$, and (6.26) has a unique solution for every $\mathbf{C} \in M_n(\mathcal{C})$.*

Proof: Let $\lambda_1, \ldots, \lambda_k$ be the distinct eigenvalues of $\mathbf{A}$, and let $\mu_1, \ldots, \mu_p$ be the distinct eigenvalues of $\mathbf{B}$. Let μ_j have multiplicity m_j. Then the characteristic polynomial $p_{\mathbf{B}}$ of $\mathbf{B}$ is

$$p_{\mathbf{B}}(\lambda) = (\lambda - \mu_1)^{m_1} \cdots (\lambda - \mu_p)^{m_p} = \lambda^n + b_{n-1}\lambda^{n-1} + \cdots + b_0.$$

Since $\lambda_l \neq \mu_j$ for $l = 1, \ldots, k$ and $j = 1, \ldots, p$, the matrices

$$(\mathbf{A} - \mu_j\mathbf{I}_n), \quad j = 1, \ldots, p,$$

are all invertible, and so are the matrices

$$(\mathbf{A} - \mu_j\mathbf{I}_n)^{m_j}, \quad j = 1, \ldots, p.$$

Thus

$$p_{\mathbf{B}}(\mathbf{A}) = (\mathbf{A} - \mu_1)^{m_1} \cdots (\mathbf{A} - \mu_p)^{m_p}$$

is invertible, whereas the Cayley–Hamilton theorem implies that $p_{\mathbf{B}}(\mathbf{B}) = 0$.

Now suppose $\mathbf{U}$ satisfies (6.27). Then

$$\mathbf{UA} = \mathbf{BU}, \quad \mathbf{UA}^2 = \mathbf{BUA} = \mathbf{B}^2\mathbf{U}, \ldots, \quad \mathbf{UA}^n = \mathbf{B}^n\mathbf{U}$$

implies that

$$\mathbf{U}p_{\mathbf{B}}(\mathbf{A}) = \mathbf{U}(\mathbf{A}^n + b_{n-1}\mathbf{A}^{n-1} + \cdots + b_0\mathbf{I}_n) = \mathbf{UA}^n + b_{n-1}\mathbf{UA}^{n-1} + \cdots + \mathbf{U}b_0\mathbf{I}_n$$

$$= \mathbf{B}^n\mathbf{U} + b_{n-1}\mathbf{B}^{n-1}\mathbf{U} + \cdots + b_0\mathbf{U} = (\mathbf{B}^n + b_{n-1}\mathbf{B}^{n-1} + \cdots + b_0)\mathbf{U} = p_{\mathbf{B}}(\mathbf{B})\mathbf{U} = 0,$$

and since $(p_{\mathbf{B}}(\mathbf{A}))^{-1}$ exists, $\mathbf{U} = 0$. $\square$

6.2.3   The special case, continued

Returning to the system

$$X' = \left[\frac{\mathbf{R}}{t-\tau} + \mathbf{A}(t)\right] X, \quad t \neq \tau, \tag{6.10}$$

suppose $\mathbf{R} \in M_n(\mathcal{C})$ and $\mathbf{A}$ has the convergent series representation

$$\mathbf{A}(t) = \sum_{k=0}^{\infty} \mathbf{A}_k (t-\tau)^k, \quad |t-\tau| < \rho, \quad \mathbf{A}_k \in M_n(\mathcal{C}).$$

We now consider the case where the eigenvalues of $\mathbf{R}$ may not be distinct but no two differ by a positive integer.

Theorem 6.5: *Suppose that $\mathbf{R}$ has eigenvalues with the property that no two differ by a positive integer. Then (6.10) has a basis $\mathbf{X}$ of the form*

$$\mathbf{X}(t) = \mathbf{P}(t)|t-\tau|^{\mathbf{R}}, \quad 0 < |t-\tau| < \rho, \tag{6.28}$$

where $\mathbf{P}$ has a convergent power series representation

$$\mathbf{P}(t) = \sum_{k=0}^{\infty} \mathbf{P}_k (t-\tau)^k, \quad |t-\tau| < \rho, \quad \mathbf{P}_0 = \mathbf{I}_n, \quad \mathbf{P}_k \in M_n(\mathcal{C}). \tag{6.29}$$

The coefficients $\mathbf{P}_k$ can be computed uniquely by substituting (6.28), (6.29) into (6.10).

Proof: Assuming that $\tau = 0$, the system is

$$X' = [\mathbf{R}t^{-1} + \mathbf{A}(t)]X, \tag{6.30}$$

where

$$\mathbf{A}(t) = \sum_{k=0}^{\infty} \mathbf{A}_k t^k, \quad |t| < \rho. \tag{6.31}$$

Suppose $\mathbf{X}$ is a basis for (6.30) of the form

$$\mathbf{X}(t) = \mathbf{P}(t)|t|^{\mathbf{R}}, \quad 0 < |t| < \rho,$$

where

$$\mathbf{P}(t) = \sum_{k=0}^{\infty} \mathbf{P}_k t^k, \quad \mathbf{P}_0 = \mathbf{I}_n, \quad \mathbf{P}_k \in M_n(\mathcal{C}), \tag{6.32}$$

and this series converges for $|t| < \rho$. Then

$$t\mathbf{X}'(t) = [\mathbf{R} + t\mathbf{A}(t)]\mathbf{X}(t)$$

or

$$t\mathbf{P}'(t)|t|^{\mathbf{R}} + \mathbf{P}(t)\mathbf{R}|t|^{\mathbf{R}} = \mathbf{R}\mathbf{P}(t)|t|^{\mathbf{R}} + t\mathbf{A}(t)\mathbf{P}(t)|t|^{\mathbf{R}}.$$

Since $|t|^{\mathbf{R}}$ is invertible for $|t| > 0$,

$$t\mathbf{P}'(t) + \mathbf{P}(t)\mathbf{R} = \mathbf{R}\mathbf{P}(t) + t\mathbf{A}(t)\mathbf{P}(t), \quad |t| < \rho. \tag{6.33}$$

Now

$$t\mathbf{P}'(t) = \mathbf{P}_1 t + 2\mathbf{P}_2 t^2 + \cdots = \sum_{k=1}^{\infty} k\mathbf{P}_k t^k,$$

$$\mathbf{P}(t)\mathbf{R} = \mathbf{P}_0\mathbf{R} + \sum_{k=1}^{\infty} \mathbf{P}_k\mathbf{R}t^k, \quad \mathbf{R}\mathbf{P}(t) = \mathbf{R}\mathbf{P}_0 + \sum_{k=1}^{\infty} \mathbf{R}\mathbf{P}_k t^k,$$

$$t\mathbf{A}(t)\mathbf{P}(t) = t[\mathbf{A}_0\mathbf{P}_0 + (\mathbf{A}_1\mathbf{P}_0 + \mathbf{A}_0\mathbf{P}_1)t + \cdots] = t\sum_{k=0}^{\infty} \mathbf{C}_k t^k = \sum_{k=1}^{\infty} \mathbf{C}_{k-1} t^k,$$

where

$$\mathbf{C}_k = \sum_{j=0}^{k} \mathbf{A}_{k-j}\mathbf{P}_j, \quad k = 0, 1, 2, \ldots.$$

Thus $\mathbf{P}$ satisfies (6.33) if and only if

$$\mathbf{P}_0\mathbf{R} = \mathbf{R}\mathbf{P}_0, \tag{6.34}$$

$$k\mathbf{P}_k + \mathbf{P}_k\mathbf{R} = \mathbf{R}\mathbf{P}_k + \mathbf{C}_{k-1}, \quad k = 1, 2, \ldots. \tag{6.35}$$

Clearly $\mathbf{P}_0 = \mathbf{I}_n$ satisfies (6.34). The equation (6.35), which can be written as

$$\mathbf{P}_k(k\mathbf{I}_n + \mathbf{R}) - \mathbf{R}\mathbf{P}_k = \mathbf{C}_{k-1}, \quad k = 1, 2, \ldots, \tag{6.36}$$

represents a recursion relation for the $\mathbf{P}_k$, since

$$\mathbf{C}_{k-1} = \mathbf{A}_{k-1}\mathbf{P}_0 + \mathbf{A}_{k-2}\mathbf{P}_1 + \cdots + \mathbf{A}_0\mathbf{P}_{k-1}.$$

For a given k this equation (6.36) has the form $\mathbf{U}\mathbf{A} - \mathbf{B}\mathbf{U} = \mathbf{C}$ if we put

$$\mathbf{U} = \mathbf{P}_k, \quad \mathbf{A} = k\mathbf{I}_n + \mathbf{R}, \quad \mathbf{B} = \mathbf{R}, \quad \mathbf{C} = \mathbf{C}_{k-1}.$$

Now μ is an eigenvalue of $\mathbf{A} = k\mathbf{I}_n + \mathbf{R}$ if and only if $\mu = k + \lambda$, where λ is an eigenvalue of $\mathbf{B} = \mathbf{R}$, since

$$p_{\mathbf{A}}(\mu) = \det(\mu\mathbf{I}_n - \mathbf{A}) = \det([\mu - k]\mathbf{I}_n - \mathbf{R}) = p_{\mathbf{B}}(\mu - k).$$

The assumption that no two eigenvalues of $\mathbf{R}$ differ by a positive integer implies that $\mathbf{A} = k\mathbf{I}_n + \mathbf{R}$ and $\mathbf{B} = \mathbf{R}$ have no eigenvalues in common. Thus Theorem 6.5 is applicable, and (6.36) can be solved uniquely for $\mathbf{P}_k$ if we know $\mathbf{C}_{k-1}$. An induction then guarantees that all the $\mathbf{P}_k$, $k = 1, 2, \ldots$, are uniquely determined by (6.36).

Now suppose that the $\mathbf{P}_k$ satisfy (6.36), where $\mathbf{P}_0 = \mathbf{I}_n$. If the series in (6.32) is convergent for $|t| < \rho$, then our computation above shows that $\mathbf{X}$ satisfies

$$\mathbf{X}'(t) = [\mathbf{R}t^{-1} + \mathbf{A}(t)]\mathbf{X}(t), \quad 0 < |t| < \rho.$$

The continuity of $\mathbf{P}$ and the fact that $\mathbf{P}(0) = \mathbf{I}_n$ imply that $\mathbf{P}(t)$ is invertible for t near 0. Since $|t|^{\mathbf{R}}$ is invertible for $t \neq 0$, the matrix $\mathbf{X}(t)$ is invertible for t near 0, $t \neq 0$. This implies that $\mathbf{X}$ is a basis for (6.30) for $0 < |t| < \rho$.

The proof of the convergence of the series in (6.32) is a simple adaptation of the corresponding convergence proof in Theorem 6.1, and so this is just briefly sketched. If (6.36) is written as

$$k\mathbf{P}_k = \mathbf{R}\mathbf{P}_k - \mathbf{P}_k\mathbf{R} + \mathbf{C}_{k-1},$$

then

$$k|\mathbf{P}_k| \leq |\mathbf{R}||\mathbf{P}_k| + |\mathbf{P}_k||\mathbf{R}| + |\mathbf{C}_{k-1}| = 2|\mathbf{R}||\mathbf{P}_k| + |\mathbf{C}_{k-1}|. \tag{6.37}$$

Let l be a positive integer such that $2|\mathbf{R}| \leq l/2$. Then $2|\mathbf{R}| \leq k/2$ for all $k \geq l$, and (6.37) gives

$$k|\mathbf{P}_k| \leq 2|\mathbf{C}_{k-1}| \leq 2\sum_{j=0}^{k-1} |\mathbf{A}_{k-1-j}||\mathbf{P}_j|, \quad k \geq l. \tag{6.38}$$

Let r satisfy $0 < r < \rho$. Since the series (6.31) converges for $|t| < \rho$, there is an $M > 0$ such that $|\mathbf{A}_j|r^j \leq M$ for $j = 0, 1, 2, \ldots$. Using this in (6.38) we find that

$$k|\mathbf{P}_k| \leq 2Mr^{1-k}\sum_{j=0}^{k-1} |\mathbf{P}_j|r^j, \quad k \geq l. \tag{6.39}$$

Let $d_k \geq 0$ be defined by

$$d_k = |\mathbf{P}_k|, \quad k = 0, 1, \ldots, l-1,$$

$$kd_k = 2Mr^{1-k}\sum_{j=0}^{k-1} d_j r^j, \quad k \geq l. \tag{6.40}$$

Comparing this definition of d_k with (6.39) we see that

$$|\mathbf{P}_k| \leq d_k, \quad k = 0, 1, \ldots. \tag{6.41}$$

Now (6.40) implies that

$$(k+1)d_{k+1} = (2M + r^{-1}k)d_k, \quad k \geq l,$$

and the ratio test for $k \geq l$ gives

$$\left| \frac{d_{k+1}t^{k+1}}{d_k t^k} \right| = \left[\frac{2M}{k+1} + \frac{k}{k+1}\frac{1}{r} \right] |t| \to \frac{|t|}{r}, \quad k \to \infty.$$

This shows that the series

$$\sum_{k=0}^{\infty} d_k t^k$$

converges for $|t| < r$, and (6.41) then implies that the series (6.32) for $\mathbf{P}(t)$ converges for $|t| < r$. Since r was any number satisfying $0 < r < \rho$, the series (6.32) converges for $|t| < \rho$. $\square$

Theorem 6.5 applies to the example

$$X' = [\mathbf{R}t^{-1} + \mathbf{A}(t)]X, \quad t \neq 0,$$

where

$$\mathbf{R} = \begin{pmatrix} 1 & 2 \\ -2 & -3 \end{pmatrix}, \quad \mathbf{A}(t) = t\mathbf{I}_2.$$

The matrix $\mathbf{R}$ has the single eigenvalue -1 with multiplicity 2, and, since there is only one linearly independent eigenvector for -1, a Jordan canonical form for $\mathbf{R}$ is

$$\mathbf{J} = \begin{pmatrix} -1 & 1 \\ 0 & -1 \end{pmatrix}.$$

Thus we have a basis $\mathbf{X}$ of the form

$$\mathbf{X}(t) = \mathbf{P}(t)|t|^{\mathbf{R}}, \quad |t| > 0,$$

where

$$\mathbf{P}(t) = \sum_{k=0}^{\infty} \mathbf{P}_k t^k, \quad \mathbf{P}_0 = \mathbf{I}_2,$$

and the series is convergent for all $t \in \mathcal{R}$.

The coefficients $\mathbf{P}_k$ may be computed by substituting into the equation

$$X' = [\mathbf{R}t^{-1} + t\mathbf{I}_2]X.$$

However, a small trick can be used to simplify matters here. If we put $X(t) = U(t)\exp(t^2/2)$, then X satisfies our system if and only if U satisfies $U' = \mathbf{R}t^{-1}U$. A basis for the latter system is $\mathbf{X}(t) = \exp(t^2/2)|t|^{\mathbf{R}}$, and thus $\mathbf{P}(t) = \exp(t^2/2)\mathbf{I}_2$. A matrix $\mathbf{Q}$ such that $\mathbf{R} = \mathbf{Q}\mathbf{J}\mathbf{Q}^{-1}$ is given by

$$\mathbf{Q} = \begin{pmatrix} 2 & 1 \\ -2 & 0 \end{pmatrix},$$

and so

$$|t|^{\mathbf{R}} = \mathbf{Q}|t|^{\mathbf{J}}\mathbf{Q}^{-1}, \quad |t|^{\mathbf{J}} = |t|^{-1}\begin{pmatrix} 1 & \log(|t|) \\ 0 & 1 \end{pmatrix}.$$

Another basis is given by $\mathbf{Y}$, where

$$\mathbf{Y}(t) = \mathbf{X}(t)\mathbf{Q} = \exp(t^2/2)\mathbf{Q}|t|^{\mathbf{J}}$$

$$= |t|^{-1}\exp(t^2/2)\begin{pmatrix} 2 & 1 + 2\log(|t|) \\ -2 & -2\log(|t|) \end{pmatrix}, \quad |t| > 0.$$

6.2.4   Singular points of the first kind: The general case

Now consider a system

$$X' = \left[\frac{\mathbf{R}}{t - \tau} + \mathbf{A}(t)\right] X, \quad t \neq \tau,$$

with no restrictions on the eigenvalues of $\mathbf{R}$. This general case can be reduced to the special case treated in Theorem 6.5 by an appropriate change in the variables. Again putting $\tau = 0$, suppose that $\lambda_1, \ldots, \lambda_k$ are the distinct eigenvalues of $\mathbf{R}$ in (6.30), with algebraic multiplicities $m_1, \ldots, m_k$, respectively. If $\mathbf{J} = \mathbf{Q}^{-1}\mathbf{R}\mathbf{Q} = \text{diag}(\mathbf{J}_1, \ldots, \mathbf{J}_q)$ is a Jordan canonical form for $\mathbf{R}$, we can write $\mathbf{J}$ as $\mathbf{J} = \text{diag}(\mathbf{R}_1, \mathbf{R}_2)$, where $\mathbf{R}_1$ is the $m_1 \times m_1$ matrix containing all the $\mathbf{J}_i$ having the eigenvalue λ_1. It has the $m_1 \times m_1$ upper triangular form

$$\mathbf{R}_1 = \begin{pmatrix} \lambda_1 & * & \cdots & * \\ 0 & \lambda_1 & \cdots & * \\ 0 & \vdots & \cdots & * \\ 0 & 0 & \cdots & \lambda_1 \end{pmatrix}.$$

If $X = \mathbf{Q}Y$ in (6.30), the resulting equation for Y is

$$Y' = [\mathbf{J}t^{-1} + \mathbf{B}(t)]Y, \quad \mathbf{J} = \mathbf{Q}^{-1}\mathbf{R}\mathbf{Q}, \quad \mathbf{B} = \mathbf{Q}^{-1}\mathbf{A}\mathbf{Q}. \tag{6.42}$$

Now make a further change by putting $Y = \mathbf{U}Z$, where $\mathbf{U}$ is given by

$$\mathbf{U}(t) = \text{diag}(t\mathbf{I}_{m_1}, \mathbf{I}_{n-m_1}).$$

This $\mathbf{U}$ is linear in t, and thus analytic for all t, and $\mathbf{U}(t)$ is invertible for $t \neq 0$, with

$$\mathbf{U}^{-1}(t) = \text{diag}(t^{-1}\mathbf{I}_{m_1}, \mathbf{I}_{n-m_1}).$$

The function $Y(t)$ satisfies (6.42) if and only if

$$Y'(t) = \mathbf{U}(t)Z'(t) + \mathbf{U}'(t)Z(t) = [\mathbf{J}t^{-1} + \mathbf{B}(t)]\mathbf{U}(t)Z(t)$$

or if and only if $Z(t)$ is a solution of the equation

$$Z' = \mathbf{U}^{-1}(t)[\mathbf{J}\mathbf{U}(t)t^{-1} - \mathbf{U}'(t) + \mathbf{B}(t)\mathbf{U}(t)]Z. \tag{6.43}$$

To compute the coefficient of Z in this equation, note that

$$\mathbf{U}^{-1}(t)\mathbf{J}\mathbf{U}(t) = \mathbf{J} = \text{diag}(\mathbf{R}_1, \mathbf{R}_2),$$

$$\mathbf{U}^{-1}(t)\mathbf{U}'(t) = \text{diag}(t^{-1}\mathbf{I}_{m_1}, 0).$$

Now

$$\mathbf{B}(t) = \sum_{k=0}^{\infty} \mathbf{B}_k t^k, \quad \mathbf{B}_k = \mathbf{Q}^{-1}\mathbf{A}_k\mathbf{Q},$$

where this series converges for $|t| < \rho$. Partitioning $\mathbf{B}_k$ as

$$\mathbf{B}_k = \begin{pmatrix} (\mathbf{B}_k)_{11} & (\mathbf{B}_k)_{12} \\ (\mathbf{B}_k)_{21} & (\mathbf{B}_k)_{22} \end{pmatrix},$$

where $(\mathbf{B}_k)_{11}$ is $m_1 \times m_1$ and $(\mathbf{B}_k)_{22}$ is $(n - m_1) \times (n - m_1)$, a short calculation shows that

$$\mathbf{U}^{-1}\mathbf{B}(t)\mathbf{U}(t) = \mathbf{C}_{-1}t^{-1} + \sum_{k=0}^{\infty} \mathbf{C}_k t^k,$$

where

$$\mathbf{C}_{-1} = \begin{pmatrix} 0 & (\mathbf{B}_0)_{12} \\ 0 & 0 \end{pmatrix}, \quad \mathbf{C}_0 = \begin{pmatrix} (\mathbf{B}_0)_{11} & (\mathbf{B}_1)_{12} \\ 0 & (\mathbf{B}_0)_{22} \end{pmatrix},$$

$$\mathbf{C}_k = \begin{pmatrix} (\mathbf{B}_k)_{11} & (\mathbf{B}_{k+1})_{12} \\ (\mathbf{B}_{k-1})_{21} & (\mathbf{B}_k)_{22} \end{pmatrix}, \quad k = 1, 2, \ldots.$$

Thus (6.43) can be written as

$$Z' = [\mathbf{K}t^{-1} + \mathbf{C}(t)]Z, \tag{6.44}$$

where

$$\mathbf{K} = \begin{pmatrix} \mathbf{R}_1 - \mathbf{I}_{m_1} & (\mathbf{B}_0)_{12} \\ 0 & \mathbf{R}_2 \end{pmatrix}, \quad \mathbf{C}(t) = \sum_{k=0}^{\infty} \mathbf{C}_k t^k,$$

and the series $\mathbf{C}(t)$ converges for $|t| < \rho$.

The total effect of the change from X to Z given by $X = \mathbf{Q}\mathbf{U}Z$ is to change the original equation $X' = [\mathbf{R}t^{-1} + \mathbf{A}(t)]X$ into the equation (6.44), having a singular point of the first kind at zero, but with the matrix $\mathbf{R}$ with eigenvalues $\lambda_1, \ldots, \lambda_k$ replaced by $\mathbf{K}$, which has eigenvalues $\lambda_1 - 1, \lambda_2, \ldots, \lambda_k$. By a finite number of these changes $X = \mathbf{Q}_1\mathbf{U}_1\mathbf{Q}_2\mathbf{U}_2 \cdots \mathbf{Q}_s\mathbf{U}_s W$ we can arrive at an equation

$$W' = [\mathbf{S}t^{-1} + \mathbf{D}(t)]W, \tag{6.45}$$

where $\mathbf{S}$ has the property that no two of its eigenvalues differ by a positive integer. $\mathbf{D}(t)$ will have a power series representation converging for $|t| < \rho$. Theorem 6.5 may be applied to (6.45) to yield a basis of the form $\mathbf{W}(t) = \mathbf{V}(t)|t|^{\mathbf{S}}$, where $\mathbf{V}(t)$ has a power series representation in powers of t which is convergent for $|t| < \rho$. Then $\mathbf{X}$ given by

$$\mathbf{X}(t) = \mathbf{P}(t)|t|^{\mathbf{S}}, \quad \mathbf{P}(t) = \mathbf{Q}_1\mathbf{U}_1(t) \cdots \mathbf{Q}_s\mathbf{U}_s(t)\mathbf{V}(t)$$

is a basis for (6.30) for $0 < |t| < \rho$, and $\mathbf{P}$ has a power series expansion in powers of t which is convergent for $|t| < \rho$.

Thus for a singular point τ of the first kind the following result holds in the general case.

Theorem 6.6: *Consider the system*

$$X' = \left[\frac{\mathbf{R}}{t - \tau} + \mathbf{A}(t) \right] X, \quad t \neq \tau,$$

where $\mathbf{R} \in M_n(\mathcal{C})$ *and* $\mathbf{A}(t)$ *has a convergent power series representation,*

$$\mathbf{A}(t) = \sum_{k=0}^{\infty} \mathbf{A}_k(t-\tau)^k, \quad |t-\tau| < \rho, \quad \mathbf{A}_k \in M_n(\mathcal{C}).$$

There exists a basis $\mathbf{X}$ *for* (6.10) *of the form*

$$\mathbf{X}(t) = \mathbf{P}(t)|t-\tau|^{\mathbf{S}}, \quad 0 < |t-\tau| < \rho, \tag{6.46}$$

where $\mathbf{P}$ *has a convergent power series representation*

$$\mathbf{P}(t) = \sum_{k=0}^{\infty} \mathbf{P}_k(t-\tau)^k, \quad |t-\tau| < \rho, \quad \mathbf{P}_k \in M_n(\mathcal{C}).$$

If a Jordan canonical form for $\mathbf{S}$ *in* (6.46) *is* $\mathbf{H}$ *and* $\mathbf{S} = \mathbf{THT}^{-1}$, *then*

$$\mathbf{X}(t) = \mathbf{P}(t)\mathbf{T}|t-\tau|^{\mathbf{H}}\mathbf{T}^{-1},$$

and another basis for (6.10) *is* $\mathbf{Y}$, *where*

$$\mathbf{Y}(t) = \mathbf{X}(t)\mathbf{T} = \mathbf{P}(t)\mathbf{T}|t-\tau|^{\mathbf{H}}.$$

Let $\lambda_1, \ldots, \lambda_k$ be the distinct eigenvalues of $\mathbf{S}$, with multiplicities $m_1, \ldots, m_k$, respectively. Every solution X of (6.10) can be written as $X = \mathbf{Y}C$ for some $C \in \mathcal{C}^n$, and the explicit nature of $\mathbf{H}$ shows that such an X has the form

$$X(t) = |t-\tau|^{\lambda_1} P_1(t) + \cdots + |t-\tau|^{\lambda_k} P_k(t),$$

where each P_j is a vector polynomial in $\log(|t-\tau|)$ of degree at most $m_j - 1$ with coefficient functions of t having power series representations in powers of $t - \tau$ which are convergent for $|t-\tau| < \rho$. Thus for $j = 1, \ldots, k$ we have

$$P_j(t) = P_{j,1}(t)[\log(|t-\tau|)]^{m_j-1} + \cdots + P_{j,m_j-1}(t)[\log(|t-\tau|)] + P_{j,m_j}(t),$$

where

$$P_{j,l} = \sum_{q=0}^{\infty} P_{j,l,q}(t-\tau)^q, \quad P_{j,l,q} \in \mathcal{C}^n,$$

and the series are convergent for $|t-\tau| < \rho$.

To illustrate Theorem 6.5, consider the example

$$X' = [\mathbf{R}t^{-1} + \mathbf{A}]X, \quad \mathbf{R} = \begin{pmatrix} -1 & 1 \\ 0 & -2 \end{pmatrix}, \quad \mathbf{A} = \begin{pmatrix} 3 & -1 \\ 0 & 3 \end{pmatrix}$$

for $t > 0$. The eigenvalues of $\mathbf{R}$ are $-1, -2$, so that a Jordan canonical form $\mathbf{J}$ for $\mathbf{R}$ is

$$\mathbf{J} = \begin{pmatrix} -1 & 0 \\ 0 & -2 \end{pmatrix}.$$

An invertible matrix $\mathbf{Q}$ such that $\mathbf{R} = \mathbf{QJQ}^{-1}$ is given by

$$\mathbf{Q} = \begin{pmatrix} 1 & 1 \\ 0 & -1 \end{pmatrix} = \mathbf{Q}^{-1}.$$

Letting $X = \mathbf{Q}Y$ our system is transformed into

$$Y' = (\mathbf{J}t^{-1} + \mathbf{B})Y, \quad \mathbf{B} = \mathbf{Q}^{-1}\mathbf{AQ} = \begin{pmatrix} 3 & 1 \\ 0 & 3 \end{pmatrix},$$

and putting $Y = \mathbf{U}(t)Z$, where

$$\mathbf{U}(t) = \begin{pmatrix} t & 0 \\ 0 & 1 \end{pmatrix},$$

we obtain the system

$$Z' = (\mathbf{K}t^{-1} + \mathbf{C})Z, \quad \mathbf{K} = \begin{pmatrix} -2 & 1 \\ 0 & -2 \end{pmatrix}, \quad \mathbf{C} = 3\mathbf{I}_2.$$

Since $\mathbf{K}$ has the eigenvalue -2 with multiplicity 2, there is a basis $\mathbf{Z}$ for this system of the form $\mathbf{Z}(t) = \mathbf{V}(t)t^{\mathbf{K}}$, $t > 0$, where $\mathbf{V}$ has a power series expansion

$$\mathbf{V}(t) = \sum_{k=0}^{\infty} \mathbf{V}_k t^{\mathbf{K}}, \quad \mathbf{V}_0 = \mathbf{I}_2,$$

which is convergent for all $t \in \mathcal{R}$. The coefficients $\mathbf{V}_k$ may be computed by substituting this $\mathbf{Z}$ into the equation $\mathbf{Z}' = (\mathbf{K}t^{-1} + 3\mathbf{I}_2)\mathbf{Z}$. However, the substitution $\mathbf{Z}(t) = e^{3t}\mathbf{W}(t)$ leads to the equation

$$\mathbf{W}' = \mathbf{K}t^{-1}\mathbf{W}$$

for $\mathbf{W}$, and so $\mathbf{Z}(t) = e^{3t}t^{\mathbf{K}}$ is a basis for the system for Z, and

$$\mathbf{X}(t) = e^{3t}\mathbf{QU}(t)t^{\mathbf{K}}$$

gives a basis $\mathbf{X}$ for the original system. Since

$$t^{\mathbf{K}} = t^{-2} \begin{pmatrix} 1 & \log(t) \\ 0 & 1 \end{pmatrix}, \quad t > 0,$$

it follows that

$$\mathbf{X}(t) = t^{-2}e^{3t} \begin{pmatrix} t & 1 + \log(t) \\ 0 & -1 \end{pmatrix}, \quad t > 0.$$

6.2.5 Singular points of the second kind

A point $\tau \in \mathcal{R}$ is a *singular point of the second kind* for a first-order linear system

$$\mathbf{A}_1(t)X' + \mathbf{A}_0(t)X = 0$$

if it can be written in the form

$$(t - \tau)^{r+1}X' = \mathbf{A}(t)X \qquad (6.47)$$

on some open interval containing τ, where r is a positive integer and $\mathbf{A}$ is analytic at τ with $\mathbf{A}(\tau) \neq 0$.

The simplest example occurs when $\mathbf{A}(t) = \mathbf{A}$ is a constant matrix and $\tau = 0$,

$$t^{r+1}X' = \mathbf{A}X, \quad \mathbf{A} \in M_n(\mathcal{C}). \qquad (6.48)$$

It is easy to check directly that $\mathbf{X}$ given by

$$\mathbf{X}(t) = \exp(-(\mathbf{A}/r)t^{-r}), \quad |t| > 0,$$

is a basis for (6.48) for $|t| > 0$.

The results valid for systems with a singular point of the second kind are considerably more involved than those for systems having a singular point of the first kind. For example, it can be proved that (6.47) has certain formal solutions involving series. However, it may happen that these formal series converge only at one point. This is illustrated by the example

$$t^2 X' = \left(\begin{pmatrix} 0 & 0 \\ 0 & 1 \end{pmatrix} + \begin{pmatrix} 0 & 1 \\ -1 & -2 \end{pmatrix} t \right) X. \qquad (6.49)$$

Thus, if

$$X = \begin{pmatrix} x_1 \\ x_2 \end{pmatrix},$$

we have

$$t^2 x_1' = t x_2, \quad t x_2' = x_2 - (x_1 + 2x_2)t.$$

A straightforward computation shows that the series

$$x_1(t) = \sum_{k=0}^{\infty} k! t^k, \quad x_2(t) = \sum_{k=1}^{\infty} (k)(k!) t^k$$

satisfy these equations in a formal sense, but neither of these series converge for any $t \neq 0$. The series are not useless, however, since they contain valuable information about the actual solutions of (6.49) near the singular point $\tau = 0$. For a discussion about the solutions of a system with a singular point of the second kind, consult [26, in particular starting on p. 49].

6.3   Single equations with singular points

6.3.1   Equations of order n

Consider a homogeneous linear equation of order n,

$$c_n(t)x^{(n)} + c_{n-1}(t)x^{(n-1)} + \cdots + c_0(t)x = 0, \qquad (6.50)$$

where the $c_j \in C(I, \mathcal{C})$ for some interval I. Say that $\tau \in I$ is a *regular point* for (6.50) if $c_n(\tau) \neq 0$, while τ is a *singular point* if $c_n(\tau) = 0$. Of particular interest are equations which can be written in the form

$$(t-\tau)^n x^{(n)} + a_{n-1}(t)(t-\tau)^{n-1} x^{(n-1)} + \cdots + a_1(t)(t-\tau)x' + a_0(t)x = 0, \quad (6.51)$$

where the functions $a_0, \ldots, a_{n-1}$ are analytic at τ and at least one of $a_0(\tau)$, $\ldots, a_{n-1}(\tau)$ is not zero. In this case τ is a *regular singular point* .

The first-order system associated with (6.51) can be written as

$$(t - \tau)^n Y' = \mathcal{A}(t)Y,$$

where $Y = \tilde{x}$ and

$$\mathcal{A}(t) = \begin{pmatrix} 0 & 1 & 0 & \cdots & 0 \\ 0 & 0 & 1 & \cdots & 0 \\ \vdots & \vdots & \vdots & \cdots & \vdots \\ 0 & 0 & \cdots & 0 & 1 \\ -a_0(t) & -a_1(t)(t-\tau) & \cdots & \cdots & -a_{n-1}(t)(t-\tau)^{n-1} \end{pmatrix}.$$

If $n \geq 2$ this is a system having τ as a singular point of the second kind unless each a_j can be written as

$$a_j(t) = (t - \tau)^{n-j-1} b_j(t), \quad j = 0, 1, \ldots, n-1,$$

where the b_j are analytic at τ, in which case the system has τ as a singular point of the first kind. However there is a first-order system generated by (6.51) with the property that if τ is a regular singular point for (6.51) then τ is a singular point of the first kind for the system.

If x is a solution of (6.51), define the vector $Z = (z_1, \ldots, z_n)$:

$$z_j(t) = (t - \tau)^{j-1} x^{(j-1)}(t), \quad j = 1, \ldots n. \qquad (6.52)$$

Then

$$(t - \tau)z_j' = (j - 1)z_j(t) + z_{j+1}(t), \quad j = 1, \ldots, n-1,$$

$$(t - \tau)z_n' = -a_0(t)z_1(t) - a_1(t)z_2(t) - \cdots + [(n - 1) - a_{n-1}(t)]z_n(t),$$

so that Z is a solution of the system

$$(t - \tau)Z' = \mathbf{B}(t)Z, \qquad (6.53)$$

where

$$\mathbf{B}(t) = \begin{pmatrix} 0 & 1 & 0 & 0 & \cdots & & 0 \\ 0 & 1 & 1 & 0 & \cdots & & 0 \\ 0 & 0 & 2 & 1 & \cdots & & 0 \\ \vdots & \vdots & \vdots & \vdots & & \ddots & \vdots \\ 0 & 0 & \cdots & 0 & n-2 & & 1 \\ -a_0(t) & -a_1(t) & \cdots & \cdots & -a_{n-2}(t) & (n-1) - a_{n-1}(t) \end{pmatrix}.$$

Conversely, if Z is a solution of (6.53) and $x = z_1$ is its first component, then the other components of Z satisfy (6.52), where x satisfies (6.51).

For any x having $n-1$ derivatives, define $\hat{x}$ by

$$\hat{x}(t) = \begin{pmatrix} x(t) \\ (t-\tau)x'(t) \\ \vdots \\ (t-\tau)^{n-1}x^{(n-1)}(t) \end{pmatrix}.$$

Using this notation, x is a solution of (6.51) if and only if $\hat{x}$ is a solution of (6.53). Also $\mathbf{X} = (x_1, \ldots, x_n)$ is a basis for (6.51) on some interval if and only if $\hat{\mathbf{X}} = (\hat{x}_1, \ldots, \hat{x}_n)$ is a basis for (6.53) on this interval. We call (6.53) the *first-order system generated by* (6.51).

If τ is a regular singular point for (6.51), then τ is a singular point of the first kind for (6.53). Theorems 6.1, 6.5, and 6.6 are thus applicable to (6.53), giving information about the solutions of (6.51). Writing (6.53) as a system

$$Z' = \left[\frac{\mathbf{R}}{t - \tau} + \mathbf{A}(t) \right] Z, \quad t \neq \tau, \tag{6.54}$$

the matrix $\mathbf{R}$ has the form

$$\mathbf{R} = \begin{pmatrix} 0 & 1 & 0 & 0 & \cdots & & 0 \\ 0 & 1 & 1 & 0 & \cdots & & 0 \\ 0 & 0 & 2 & 1 & \cdots & & 0 \\ \vdots & \vdots & \vdots & \vdots & & \ddots & \vdots \\ 0 & 0 & \cdots & 0 & n-2 & & 1 \\ -a_0(\tau) & -a_1(\tau) & \cdots & \cdots & -a_{n-2}(\tau) & (n-1) - a_{n-1}(\tau) \end{pmatrix}. \tag{6.55}$$

The complexity of the solutions near τ depends on the eigenvalues of $\mathbf{R}$.

Theorem 6.7: *The characteristic polynomial $p_{\mathbf{R}}$ of the matrix $\mathbf{R}$ in (6.55) is given by*

$$p_{\mathbf{R}}(\lambda) = \lambda(\lambda - 1) \cdots (\lambda - n + 1) + a_{n-1}(\tau)\lambda(\lambda - 1) \cdots (\lambda - n + 2)$$

$$+ \cdots + a_1(\tau)\lambda + a_0(\tau). \tag{6.56}$$

If $\lambda_1, \ldots, \lambda_k$ are the distinct eigenvalues of $\mathbf{R}$, then

$$\dim(\mathcal{E}(\mathbf{R}, \lambda_j)) = 1, \quad j = 1, \ldots, k. \tag{6.57}$$

The proof of (6.56) proceeds by induction on n. The result is clearly true for 1×1 matrices, where $\mathbf{R} = (-a_0(\tau))$ and $p_{\mathbf{R}}(\lambda) = \lambda + a_0(\tau)$. Assume it is true for all $(n-1) \times (n-1)$ matrices of the form (6.55). The characteristic polynomial is

$$p_{\mathbf{R}}(\lambda) = \det(\lambda \mathbf{I}_n - \mathbf{R})$$

$$= \det \begin{pmatrix} \lambda & -1 & 0 & 0 & \cdots & & 0 \\ 0 & \lambda-1 & -1 & 0 & \cdots & & 0 \\ 0 & 0 & \lambda-2 & -1 & \cdots & & 0 \\ \vdots & \vdots & \vdots & \vdots & \cdots & & \vdots \\ 0 & \cdots & \cdots & 0 & \lambda-(n-2) & & -1 \\ -a_0(\tau) & -a_1(\tau) & \cdots & \cdots & -a_{n-2}(\tau) & & \lambda-(n-1)+a_{n-1}(\tau) \end{pmatrix}.$$

Expanding by the first column we obtain

$$p_{\mathbf{R}}(\lambda) = \lambda \, \det(\mu \mathbf{I}_{n-1} - \mathbf{R}_1) + a_0(\tau), \qquad (6.58)$$

where $\mu = \lambda - 1$ and $\mathbf{R}_1$ is an $(n-1) \times (n-1)$ matrix

$$\mathbf{R}_1(t) = \begin{pmatrix} 0 & 1 & 0 & 0 & \cdots & & 0 \\ 0 & 1 & 1 & 0 & \cdots & & 0 \\ 0 & 0 & 2 & 1 & \cdots & & 0 \\ \vdots & \vdots & \vdots & \vdots & \cdots & & \vdots \\ 0 & 0 & \cdots & 0 & n-3 & & 1 \\ -a_1(\tau) & -a_2(\tau) & \cdots & \cdots & -a_{n-2}(\tau) & & (n-2)-a_{n-1}(\tau) \end{pmatrix}$$

which has the same structure as $\mathbf{R}$. The induction assumption implies that

$$\det(\mu \mathbf{I}_n - \mathbf{R}_1) = \mu(\mu-1)\cdots(\mu-n+2)+a_{n-1}(\tau)\mu(\mu-1)\cdots(\mu-n+3)+\cdots+a_1(\tau)$$

$$= (\lambda-1)(\lambda-2)\cdots(\lambda-n+1)+a_{n-1}(\tau)(\lambda-1)(\lambda-2)\cdots(\lambda-n+2)+\cdots+a_1(\tau),$$

and putting this into (6.58) yields (6.56).

As for (6.57), suppose α, with components $\alpha_1, \ldots, \alpha_n$, is an eigenvector of $\mathbf{R}$ for the eigenvalue λ. Then $\mathbf{R}\alpha = \lambda\alpha$ if and only if

$$\alpha_2 = \lambda\alpha_1, \quad \alpha_2 + \alpha_3 = \lambda\alpha_2, \quad 2\alpha_3 + \alpha_4 = \lambda\alpha_3,$$

$$\ldots, (n-2)\alpha_{n-1} + \alpha_n = \lambda\alpha_{n-1}$$

and

$$-a_0(\tau)\alpha_1 - a_1(\tau)\alpha_2 - \cdots + (n-1)\alpha_n - a_{n-1}(\tau)\alpha_n = \lambda\alpha_n$$

or

$$\alpha_j = \lambda(\lambda-1)\cdots(\lambda-j+2)\alpha_1, \quad j = 2, \ldots n,$$

and $p_{\mathbf{R}}(\lambda)\alpha_1 = 0$. Since $\mathcal{E}(\mathbf{R}, \lambda)$ has a basis consisting of the single vector

$$\alpha = \begin{pmatrix} 1 \\ \lambda \\ \lambda(\lambda-1) \\ \vdots \\ \lambda(\lambda-1)\ldots(\lambda-n+2) \end{pmatrix},$$

it follows that $\dim(\mathcal{E}(\mathbf{R}, \lambda)) = 1$. $\square$

The polynomial $p_{\mathbf{R}}$ is called the *indicial polynomial* of (6.51) relative to the point τ. A direct consequence of (6.57) is that a Jordan canonical form $\mathbf{J}$ for $\mathbf{R}$ has the form $\mathbf{J} = \operatorname{diag}(\mathbf{J}_1, \ldots, \mathbf{J}_k)$, where there is only one block $\mathbf{J}_j$ containing the eigenvalue λ_j and this block is $m_j \times m_j$, where m_j is the multiplicity of λ_j.

Suppose the coefficients $a_0, \ldots, a_{n-1}$ in (6.51) all have convergent power series representations

$$a_j(t) = \sum_{k=0}^{\infty} a_{jk}(t - \tau)^k, \quad |t - \tau| < \rho, \quad a_{jk} \in \mathcal{C}.$$

Then Corollary 6.2 implies that if λ is a root of the indicial polynomial $p_{\mathbf{R}}$ such that

$$\operatorname{Re}(\lambda) = \max(\operatorname{Re}(\lambda_j)), \quad j = 1, \ldots, k,$$

there is a solution of (6.51) of the form

$$x(t) = |t - \tau|^\lambda p(t), \quad 0 < |t - \tau| < \rho, \tag{6.59}$$

where p has a power series representation

$$p(t) = \sum_{k=0}^{\infty} p_k(t - \tau)^k, \quad p_k \in \mathcal{C}, \quad p_0 \neq 0,$$

which is convergent for $|t - \tau| < \rho$. Corollary 6.3 implies that if $p_{\mathbf{R}}$ has n distinct roots $\lambda_1, \ldots, \lambda_n$, no two of which differ by a positive integer, then (6.51) has a basis $\mathbf{X} = (x_1, \ldots, x_n)$, where

$$x_j(t) = |t - \tau|^{\lambda_j} p_j(t), \quad 0 < |t - \tau| < \rho, \tag{6.60}$$

and the p_j have power series expansions in powers of $t - \tau$ which are convergent for $|t - \tau| < \rho$, $p_j(\tau) \neq 0$. All the coefficients in these series for p and the p_j can be computed by substituting (6.59) or (6.60) into (6.51).

From Theorem 6.5 it follows that if $p_{\mathbf{R}}$ has roots with the property that no two differ by a positive integer, then corresponding to each root λ with multiplicity m there are m solutions $y_1, \ldots, y_m$ of (6.51) of the form

$$y_1(t) = |t - \tau|^\lambda p_1(t),$$

$$y_2(t) = |t - \tau|^\lambda \left[p_1(t) \log(|t - \tau|) + p_2(t) \right],$$

$$\vdots$$

$$y_m(t) = |t - \tau|^\lambda \left[\frac{p_1(t)}{(m-1)!} [\log(|t - \tau|)]^{m-1} + \cdots + p_m(t) \right],$$

where the p_j have power series expansions in powers of $t - \tau$ which are convergent for $|t - \tau| < \rho$. The collection of all such solutions corresponding to the distinct roots $\lambda_1, \ldots, \lambda_k$ of $p_{\mathbf{R}}$ forms a basis for the solutions of (6.51).

6.3.2   The Euler equation

The simplest example of an nth-order equation with a regular singular point at $\tau = 0$ is the *Euler equation*

$$Lx = t^n x^{(n)} + a_{n-1} t^{n-1} x^{(n-1)} + \cdots + a_1 t x' + a_0 x = 0, \qquad (6.61)$$

where the $a_j \in \mathcal{C}$ are constants, not all of which are 0. The system (6.54) generated by (6.61) is then just

$$Z' = \frac{\mathbf{R}}{t} Z, \quad t \neq 0, \qquad (6.62)$$

where $\mathbf{R}$ is given by (6.55) with $a_j(\tau) = a_j$. This system has a basis $\mathbf{Z}$ given by $\mathbf{Z}(t) = |t|^{\mathbf{R}}$ for $|t| > 0$. Using a Jordan canonical form for $\mathbf{R}$ and Theorem 6.6 we can obtain a basis for (6.61) by taking the first components of the columns of a basis for (6.62). Alternatively, a basis can be obtained in a direct manner, which also shows how the indicial polynomial of (6.61) arises in a natural way.

For any $\lambda \in \mathcal{C}$ and $t > 0$ we have

$$tDt^\lambda = \lambda t^\lambda, \quad t^2 D^2 t^\lambda = \lambda(\lambda - 1) t^\lambda,$$

and in general

$$t^j D^j t^\lambda = \lambda(\lambda - 1) \cdots (\lambda - j + 1) t^\lambda,$$

so that

$$L(t^\lambda) = q(\lambda) t^\lambda, \quad t > 0, \quad \lambda \in \mathcal{C}, \qquad (6.63)$$

where q is the indicial polynomial of (6.61). Differentiation of (6.63) l times with respect to λ leads to

$$\frac{\partial^l}{\partial \lambda^l} L(t^\lambda) = L \left(\frac{\partial^l}{\partial \lambda^l} e^{\lambda \log(t)} \right) = L(t^\lambda [\log(t)]^l) \qquad (6.64)$$

$$= [q^{(l)}(\lambda) + l q^{(l-1)}(\lambda) \log(t) + \cdots + q(\lambda)[\log(t)]^l] t^\lambda.$$

If λ_1 is a root of q of multiplicity m_1, then

$$q(\lambda_1) = q'(\lambda_1) = \cdots = q^{(m_1-1)}(\lambda_1) = 0, \quad q^{(m_1)}(\lambda_1) \neq 0,$$

and (6.64) implies that

$$t^{\lambda_1}, t^{\lambda_1} \log(t), \ldots, t^{\lambda_1} [\log(t)]^{m_1-1}$$

give solutions of (6.61) for $t > 0$. If $t < 0$ then t can be replaced by $-t = |t|$, yielding

$$L(|t|^\lambda) = q(\lambda) |t|^\lambda, \quad t \neq 0, \quad \lambda \in \mathcal{C}.$$

Then (6.64) is valid with t replaced by $|t|$, and the following result is obtained.

Theorem 6.8: *Let* $\lambda_1, \ldots, \lambda_k$ *be the distinct roots of the indicial polynomial* q *of* (6.61),

$$q(\lambda) = \lambda(\lambda - 1) \cdots (\lambda - n + 1) + a_{n-1}\lambda(\lambda - 1) \cdots (\lambda - n + 2) + \cdots + a_0,$$

and suppose λ_j *has multiplicity* m_j. *Then the* n *functions* x_{ij}, *where*

$$x_{ij}(t) = |t|^{\lambda_j}[\log|t|]^{i-1}, \quad i = 1, \ldots, m_j, \quad j = 1, \ldots k, \qquad (6.65)$$

form a basis for the solutions of the Euler equation (6.61) *on any interval not containing* $t = 0$.

Proof: The x_{ij} satisfy $Lx_{ij} = 0$ for $|t| > 0$. To see that they form a linearly independent set, let $s = \log|t|$, $|t| = e^s$ in (6.65). Then

$$x_{ij}(t) = s^{i-1}e^{\lambda_j s} = y_{ij}(s).$$

The independence of these n functions y_{ij} on any interval was established in Theorem 3.13, and hence the x_{ij} are linearly independent on any interval not containing $t = 0$. $\square$

As an example consider the equation

$$t^2 x'' + tx' + x = 0, \quad t \neq 0.$$

The indicial polynomial q is given by

$$q(\lambda) = \lambda(\lambda - 1) + \lambda + 1 = \lambda^2 + 1,$$

and its roots are $\lambda_1 = i$, $\lambda_2 = -i$. Thus a basis is given by $\mathbf{X} = (x_1, x_2)$, where

$$x_1(t) = |t|^i, \quad x_2(t) = |t|^{-i}, \quad t \neq 0.$$

Since
$$|t|^i = e^{i\log(|t|)} = \cos(\log(|t|)) + i\sin(\log(|t|)),$$

a real basis is given by $\mathbf{Y} = (y_1, y_2)$, where

$$y_1(t) = \cos(\log(|t|)), \quad y_2(t) = \sin(\log(|t|)).$$

6.3.3 The second-order equation

Since many important equations arising in applications turn out to be second-order equations with a regular singular point, it is worth summarizing here what we know about their solutions. Consider an equation of order 2 with a regular singular point at $\tau = 0$,

$$t^2 x'' + a(t)tx' + b(t)x = 0, \qquad (6.66)$$

where $a(t)$ and $b(t)$ have convergent power series representations,

$$a(t) = \sum_{k=0}^{\infty} a_k t^k, \quad b(t) = \sum_{k=0}^{\infty} b_k t^k, \quad |t| < \rho.$$

The indicial polynomial of (6.66) relative to $\tau = 0$ is given by

$$q(\lambda) = \lambda(\lambda - 1) + a_0 \lambda + b_0.$$

Let λ_1, λ_2 be the two roots of q, with $\mathrm{Re}(\lambda_1) \geq \mathrm{Re}(\lambda_2)$. Corollary 6.3 and Theorem 6.5 imply the following result.

Theorem 6.9: (i) *If $\lambda_1 - \lambda_2$ is not zero or a positive integer, there is a basis $\mathbf{X} = (x_1, x_2)$ for (6.66) of the form*

$$x_1(t) = |t|^{\lambda_1} p_1(t), \quad x_2(t) = |t|^{\lambda_2} p_2(t), \quad 0 < |t| < \rho,$$

where

$$p_j(t) = 1 + \sum_{k=1}^{\infty} p_{j,k} t^k, \quad j = 1, 2, \quad p_{j,k} \in \mathcal{C},$$

and these series converge for $|t| < \rho$.

(ii) *If $\lambda_1 = \lambda_2$, there is a basis $\mathbf{X} = (x_1, x_2)$ for (6.66) of the form*

$$x_1(t) = |t|^{\lambda_1} p_1(t), \quad x_2(t) = x_1(t) \log(|t|) + |t|^{\lambda_1} p_2(t), \quad 0 < |t| < \rho,$$

where p_1 is as in (i), and

$$p_2(t) = \sum_{k=0}^{\infty} p_{2,k} t^k, \quad p_{j,k} \in \mathcal{C},$$

the series converging for $|t| < \rho$.

In case (ii) the coefficient $p_{2,0}$ in the series for p_2 may be zero.

An example illustrating Theorem 6.9 (i) is

$$t^2 x'' + \frac{3}{2} t x' + t x = 0, \quad t \neq 0,$$

which has the origin as a regular singular point. The indicial polynomial q is given by

$$q(\lambda) = \lambda(\lambda - 1) + \frac{3}{2}\lambda = \lambda(\lambda + 1/2),$$

with the two roots $\lambda_1 = 0$, $\lambda_2 = -1/2$. A calculation shows that

$$x(t) = |t|^{\lambda} \sum_{k=0}^{\infty} c_k t^k, \quad c_0 = 1,$$

will satisfy this equation for $|t| > 0$ if and only if

$$q(\lambda)|t|^{\lambda} + |t|^{\lambda} \sum_{k=1}^{\infty} [q(\lambda + k)c_k + c_{k-1}] t^k = 0, \quad |t| > 0.$$

This is true if and only if

$$q(\lambda) = 0, \quad q(\lambda + k)c_k + c_{k-1} = 0, \quad k = 1, 2, \ldots.$$

Thus $\lambda = 0$ or $\lambda = -1/2$, and

$$c_k = \frac{(-1)^k}{q(\lambda + k)q(\lambda + k - 1) \cdots q(\lambda + 1)}, \quad k = 1, 2, \ldots.$$

Letting $\lambda = 0$, we obtain a solution

$$x_1(t) = 1 + \sum_{k=1}^{\infty} \frac{(-1)^k t^k}{q(k)q(k-1) \cdots q(1)},$$

and letting $\lambda = -1/2$ we obtain the solution

$$x_2(t) = |t|^{-1/2} \left[1 + \sum_{k=1}^{\infty} \frac{(-1)^k t^k}{q(k - 1/2)q(k - 3/2) \cdots q(1/2)} \right].$$

The linear independence of $\mathbf{X} = (x_1, x_2)$ can be verified directly. Suppose there are constants $a_1, a_2 \in \mathcal{C}$ such that for $t \neq 0$,

$$a_1 x_1(t) + a_2 x_2(t) = 0.$$

Then

$$|t|^{1/2} a_1 x_1(t) + |t|^{1/2} a_2 x_2(t) = 0$$

for $t \neq 0$, and letting $t \to 0$ we get $a_2 = 0$, and then $a_1 x_1(t) = 0$ implies, on letting $t \to 0$, that $a_1 = 0$. Thus $\mathbf{X} = (x_1, x_2)$ is a basis for $|t| > 0$. The ratio test shows that the series involved in x_1 and x_2 converge for all $t \in \mathcal{R}$.

Case (ii) of Theorem 6.9 is illustrated by the Bessel equation of order 0:

$$t^2 x'' + t x' + t^2 x = 0.$$

The indicial polynomial is

$$q(\lambda) = \lambda(\lambda - 1) + \lambda = \lambda^2,$$

with the root $\lambda_1 = 0$ having multiplicity 2. This equation is considered in detail in the next section.

The remaining case occurs when $\lambda_1 - \lambda_2 = m$, a positive integer. In this case, the steps used to prove Theorem 6.5, when applied to (6.66), show that every solution x of (6.66) has the form

$$x(t) = |t|^{\lambda_2}[q_1(t) \log(|t|) + q_2(t)], \tag{6.67}$$

where the q_j have power series expansions which converge for $|t| < \rho$. Corollary 6.2 implies that there is always a solution x_1 of the form given in Theorem

6.9(i). Thus there is a basis $\mathbf{X} = (x_1, x_2)$ for (6.66), with x_2 having the form given in (6.67). In fact, as we now show, such an x_2 must have the form

$$x_2(t) = cx_1(t) \log(|t|) + |t|^{\lambda_2} p_2(t),$$

where p_2 is as in Theorem 6.9 (ii), and c is a constant, which could be zero.

Recall the formula for the Wronskian $W_{\mathbf{X}}$ (Theorem 2.10):

$$W_{\mathbf{X}}(t) = x_1(t)x_2'(t) - x_1'(t)x_2(t) = W_{\mathbf{X}}(t_0) \exp\left(-\int_{t_0}^t \frac{a(s)}{s}\, ds\right), \qquad (6.68)$$

where $0 < t_0 < \rho$, $0 < t < \rho$. Now

$$\frac{a(s)}{s} = \frac{a_0}{s} + a_1 + a_2 s + a_3 s^2 + \cdots = \frac{a_0}{s} + \alpha(s),$$

where the function α is defined on $|s| < \rho$ by its convergent series expansion

$$\alpha(s) = \sum_{k=0}^{\infty} a_{k+1} s^k, \qquad |s| < \rho.$$

The right side of (6.68) has the form

$$W_{\mathbf{X}}(t_0) \exp\left(-\int_{t_0}^t \frac{a(s)}{s}\, ds\right) = Kt^{-a_0} r(t), \qquad (6.69)$$

where

$$K = W_{\mathbf{X}}(t_0) t_0^{a_0} \neq 0, \quad r(t) = \exp\left(-\int_{t_0}^t \alpha(s)\, ds\right).$$

The function r is defined for $|t| < \rho$ and has all derivatives there. Now consider the left side of (6.68). Since x_1, x_2 have the form

$$x_1(t) = t^{\lambda_1} p_1(t), \quad x_2(t) = t^{\lambda_2}[q_1(t) \log(t) + q_2(t)], \quad 0 < t < \rho,$$

$W_{\mathbf{X}}$ is given by

$$W_{\mathbf{X}}(t) = t^{\lambda_1 + \lambda_2 - 1}[r_1(t) \log(t) + r_2(t)], \qquad (6.70)$$

where

$$r_1(t) = tp_1(t)q_1'(t) - [mp_1(t) + tp_1'(t)]q_1(t), \qquad (6.71)$$

$$r_2(t) = tp_1(t)q_2'(t) - [mp_1(t) + tp_1'(t)]q_2(t) + p_1(t)q_1(t).$$

In calculating r_1, r_2 we have used $\lambda_1 - \lambda_2 = m$.

Since λ_1, λ_2 are the roots of the indicial polynomial q,

$$q(\lambda) = \lambda^2 + (a_0 - 1)\lambda + b_0 = (\lambda - \lambda_1)(\lambda - \lambda_2) = \lambda^2 - (\lambda_1 + \lambda_2)\lambda + \lambda_1\lambda_2,$$

it follows that $-a_0 = \lambda_1 + \lambda_2 - 1$. Since the expression in (6.69) equals that in (6.70) we must have

$$r_1(t) \log(t) + r_2(t) = Kr(t), \quad 0 < t < \rho. \qquad (6.72)$$

In fact, $r_1(t) = 0$ for $|t| < \rho$. The form of r_1 given in (6.71) shows that it is analytic at $t = 0$ with a power series expansion

$$r_1(t) = \sum_{k=0}^{\infty} c_k t^k$$

which is convergent for $|t| < \rho$. If r_1 is not the zero function, there is a smallest k such that $c_k \neq 0$, so that $r_1(t) = t^k u_1(t)$, where u_1 is analytic at $t = 0$, $u_1(0) = c_k \neq 0$. Differentiating the relation (6.72) k times, and observing that

$$t^l D^l \log(t) = (-1)^{l-1}(l-1)!, \quad l = 1, 2, \ldots,$$

it follows that

$$k! u_1(t) \log(t) + \frac{k}{t} r_1^{(k-1)}(t)$$

$$+ \cdots + \frac{(-1)^{k-1}(k-1)!}{t^k} r_1(t) + r_2^{(k)}(t) = K r^{(k)}(t), \quad 0 < t < \rho.$$

Letting $t \to +0$ we see that all the terms tend to finite limits except the first one, whose magnitude tends to $+\infty$. This gives a contradiction. Therefore $r_1(t) = 0$ for $|t| < \rho$, which implies that q_1 satisfies the first-order linear equation

$$t p_1(t) q_1' - [m p_1(t) + t p_1'(t)] q_1 = 0, \quad |t| < \rho.$$

This is easily solved. Multiplying by t^{m-1} and dividing by $t^{2m} p_1^2(t)$, the result is

$$\left(\frac{q_1}{t^m p_1} \right)' = 0, \quad 0 < |t| < \rho.$$

Hence $q_1(t) = c t^m p_1(t)$ for $|t| < \rho$, where c is a constant. Putting this into the expression for x_2 gives

$$x_2(t) = |t|^{\lambda_2} [q_1(t) \log(|t|) + q_2(t)]$$

$$= |t|^{\lambda_2} [c t^m p_1(t) \log(|t|) + q_2(t)] = c x_1(t) \log(|t|) + |t|^{\lambda_2} q_2(t),$$

since $\lambda_2 + m = \lambda_1$. The following result is thus demonstrated.

Theorem 6.10: *If $\lambda_1 - \lambda_2 = m$, a positive integer, there is a basis $\mathbf{X} = (x_1, x_2)$ for (6.66) of the form*

$$x_1(t) = |t|^{\lambda_1} p_1(t), \quad x_2(t) = c x_1(t) \log(|t|) + |t|^{\lambda_2} q_2(t)$$

for $0 < |t| < \rho$. Here

$$p_1(t) = 1 + \sum_{k=1}^{\infty} p_{1,k} t^k, \quad p_{1,k} \in \mathcal{C}, \quad p_2(t) = \sum_{k=0}^{\infty} p_{2,k} t^k, \quad p_{2,k} \in \mathcal{C},$$

and the series converge for $|t| < \rho$. The constant c may be zero.

Theorems 6.9 and 6.10 give rather precise information about the form that a basis $\mathbf{X} = (x_1, x_2)$ for (6.66) must take. Given any particular second-order equation, substitution of x_1 into (6.66) allows us to compute the series for p_1. Then assuming the relevant form for x_2 we can substitute it into (6.66) and determine the series for p_2. This will be illustrated with the Bessel equation of order α in subsequent sections.

6.3.4   The Bessel equation

One of the most celebrated and thoroughly studied equations is the *Bessel equation of order* α,

$$t^2 x'' + t x' + (t^2 - \alpha^2) x = 0, \tag{6.73}$$

where α is a complex constant. This equation arises in problems connected with heat flow and wave motion where there is circular or spherical symmetry (see [19, pp. 195–207] and [24, pp. 251–261]). Bessel's equation has the form (6.66) with

$$a(t) = 1, \quad b(t) = t^2 - \alpha^2.$$

Both a and b are analytic at 0, $a(0) \neq 0$, and the power series describing a and b are polynomials, which are thus convergent for $|t| < \infty$. Hence $\tau = 0$ is a regular singular point for (6.73). The indicial polynomial q is given by

$$q(\lambda) = \lambda(\lambda - 1) + \lambda - \alpha^2 = \lambda^2 - \alpha^2,$$

whose two roots are $\lambda_1 = \alpha$, $\lambda_2 = -\alpha$.

Theorem 6.9 (ii) implies that there is a basis $\mathbf{X} = (x_1, x_2)$ of (6.73) of the form

$$x_1(t) = p_1(t), \quad x_2(t) = x_1(t) \log(|t|) + p_2(t), \quad t \neq 0,$$

where p_1, p_2 have power series expansions in powers of t which converge for all $t \in \mathcal{R}$. Let us consider the case $\alpha = 0$ with $t > 0$ and compute p_1, p_2. When $\alpha = 0$ the Bessel equation is equivalent to

$$Lx = tx'' + x' + tx = 0.$$

If

$$x_1(t) = c_0 + c_1 t + c_2 t^2 + \cdots = \sum_{k=0}^{\infty} c_k t^k,$$

then

$$x_1'(t) = c_1 + 2 c_2 t + 3 c_3 t^2 + \cdots,$$
$$x_1''(t) = c_2 + (3)(2) c_3 t + (4)(3) c_4 t^2 + \cdots,$$

so that

$$t x_1''(t) = c_2 t + (3)(2) c_3 t^2 + (4)(3) c_4 t^3 + \cdots = \sum_{k=1}^{\infty} k(k+1) c_{k+1} t^k,$$

$$x_1'(t) = c_1 + \sum_{k=1}^{\infty} (k+1) c_{k+1} t^k,$$

$$t x_1(t) = c_0 t + c_1 t^2 + \cdots = \sum_{k=1}^{\infty} c_{k-1} t^k.$$

Thus

$$Lx_1(t) = c_1 + \sum_{k=1}^{\infty}([k(k+1) + (k+1)]c_{k+1} + c_{k-1})t^k = 0$$

for $t > 0$ if and only if $c_1 = 0$ and

$$(k+1)^2 c_{k+1} + c_{k-1} = 0, \quad k = 1, 2, \ldots.$$

This is a recursion relation for the c_k. The choice $c_0 = 1$ leads to

$$c_2 = -\frac{1}{2^2}, \quad c_4 = -\frac{c_2}{4^2} = \frac{1}{2^2 4^2}, \ldots,$$

and in general

$$c_{2m} = \frac{(-1)^m}{2^2 4^2 \ldots (2m)^2} = \frac{(-1)^m}{2^{2m}(m!)^2}, \quad m = 1, 2, \ldots.$$

Since $c_1 = 0$ it follows that $c_{2m+1} = 0$ for $m = 1, 2, \ldots$; therefore, x_1 contains only even powers of t. This function x_1, usually denoted J_0, is called the *Bessel function of the first kind of order zero*,

$$J_0(t) = \sum_{m=0}^{\infty} \frac{(-1)^m}{(m!)^2} \left(\frac{t}{2}\right)^{2m}.$$

For $t > 0$ the solution x_2 has the form

$$x_2(t) = J_0(t)\log(t) + p_2(t),$$

where

$$p_2(t) = \sum_{k=0}^{\infty} d_k t^k.$$

Differentiation of the series gives

$$x_2'(t) = \frac{J_0(t)}{t} + J_0'(t)\log(t) + d_1 + \sum_{k=1}^{\infty}(k+1)d_{k+1}t^k,$$

$$x_2''(t) = \frac{-J_0(t)}{t^2} + \frac{2J_0'(t)}{t} + J_0''(t)\log(t) + \sum_{k=1}^{\infty}k(k+1)d_{k+1}t^{k-1},$$

and so

$$Lx_2(t) = tx_2''(t) + x_2'(t) + tx_2(t)$$

$$= 2J_0'(t) + d_1 + \sum_{k=1}^{\infty}[(k+1)^2 d_{k+1} + d_{k-1}]t^k + \log(t)LJ_0(t).$$

Since $LJ_0(t) = 0$ it follows that $Lx_2(t) = 0$ for $t > 0$ if and only if

$$d_1 + \sum_{k=1}^{\infty}[(k+1)^2 d_{k+1} + d_{k-1}]t^k = -2J_0'(t) = -2\sum_{m=1}^{\infty}\frac{(-1)^m 2m}{2^{2m}(m!)^2}t^{2m-1}.$$

Thus

$$d_1 = 0, \quad 2^2 d_2 + d_0 = 1, \quad 3^2 d_3 + d_1 = 0, \ldots,$$

and, since the series for $-2J_0'(t)$ contains only odd powers of t,

$$d_1 = d_3 = d_5 = \cdots = 0.$$

The recursion relation for the other coefficients is

$$(2m)^2 d_{2m} + d_{2m-2} = \frac{(-1)^{m+1}m}{2^{2m-2}(m!)^2}, \quad m = 1, 2, \ldots.$$

The simplest choice for d_0 is $d_0 = 0$, and with this choice

$$d_2 = \frac{1}{2^2}, \quad d_4 = \frac{1}{4^2}\left(-\frac{1}{(2)(2^2)} - \frac{1}{2^2}\right) = -\frac{1}{2^2 4^2}\left(1 + \frac{1}{2}\right),$$

$$d_6 = \frac{1}{6^2}\left[\frac{1}{2^2 4^2}\frac{1}{3} + \frac{1}{2^2 4^2}\left(1 + \frac{1}{2}\right)\right] = \frac{1}{2^2 4^2 6^2}\left(1 + \frac{1}{2} + \frac{1}{3}\right).$$

It can be shown by induction that

$$d_{2m} = \frac{(-1)^{m-1}}{2^{2m}(m!)^2}\left(1 + \frac{1}{2} + \cdots + \frac{1}{m}\right), \quad m = 1, 2, \ldots.$$

The solution thus obtained, usually denoted by K_0, is called a *Bessel function of the second kind of order zero*,

$$K_0(t) = J_0(t)\log(t) + \sum_{m=1}^{\infty}\frac{(-1)^{m-1}}{(m!)^2}\left(1 + \frac{1}{2} + \cdots + \frac{1}{m}\right)\left(\frac{t}{2}\right)^{2m}.$$

The choice $d_0 \neq 0$ in the above calculations would give the solution $x_2 = K_0 + d_0 J_0$.

The pair of functions $\mathbf{X} = (J_0, K_0)$ is a basis for the Bessel equation (6.73) when $\alpha = 0$ for $t > 0$. The linear independence of $\mathbf{X}$ can be easily verified directly. Suppose that, for some constants $c_1, c_2 \in \mathcal{C}$,

$$c_1 J_0(t) + c_2 K_0(t) = c_1 J_0(t) + c_2 J_0(t)\log(t) + c_2 p_2(t) = 0$$

for all $t > 0$. Letting $t \to +0$ we see that $J_0(t) \to J_0(0) = 1$, $p_2(t) \to p_2(0) = 0$, and if $c_2 \neq 0$ then $|c_2 J_0(t)\log(t)| \to +\infty$, which is a contradiction. Hence $c_2 = 0$, and then $c_1 J_0 t(t) = 0$ implies $c_1 J_0(0) = c_1 = 0$.

6.3.5   The Bessel equation, continued

Now consider the Bessel equation of order α, where $\alpha \neq 0$. If in addition $\text{Re}(\alpha) \geq 0$, then $\lambda_1 = \alpha$, $\lambda_2 = -\alpha$ are distinct, with $\text{Re}(\lambda_1) \geq \text{Re}(\lambda_2)$. There is thus a solution x_1 of the form

$$x_1(t) = |t|^\alpha \sum_{k=0}^\infty c_k t^k, \quad c_0 \neq 0,$$

where the series converges for all $t \in \mathcal{R}$. Let us compute this by considering the case $t > 0$. After a short calculation we find that if

$$Lx(t) = t^2 x''(t) + t x'(t) + (t^2 - \alpha^2) x(t),$$

then

$$Lx_1(t) = (0)c_0 t^\alpha + [(\alpha+1)^2 - \alpha^2]c_1 t^{\alpha+1} + t^\alpha \sum_{k=2}^\infty ([(\alpha+k)^2 - \alpha^2]c_k + c_{k-2})t^k = 0.$$

Therefore $c_1 = 0$ and

$$[(\alpha + k)^2 - \alpha^2]c_k + c_{k-2} = 0, \quad k = 2, 3, \ldots,$$

or

$$k(2\alpha + k)c_k + c_{k-2} = 0, \quad k = 2, 3, \ldots.$$

It follows that

$$c_1 = c_3 = c_5 = \cdots = 0,$$

while

$$c_2 = \frac{-c_0}{2(2\alpha + 2)} = \frac{-c_0}{2^2(\alpha + 1)},$$

$$c_4 = \frac{-c_2}{4(2\alpha + 4)} = \frac{c_0}{2^4 2!(\alpha + 1)(\alpha + 2)},$$

$$c_6 = \frac{-c_4}{6(2\alpha + 6)} = \frac{-c_0}{2^6 3!(\alpha + 1)(\alpha + 2)(\alpha + 3)},$$

and in general

$$c_{2m} = \frac{(-1)^m c_0}{2^{2m} m!(\alpha + 1)(\alpha + 2) \cdots (\alpha + m)}, \quad m = 1, 2, 3, \ldots.$$

The solution x_1 is thus given by

$$x_1(t) = c_0 t^\alpha + c_0 t^\alpha \sum_{m=1}^\infty \frac{(-1)^m}{m!(\alpha + 1)(\alpha + 2) \cdots (\alpha + m)} \left(\frac{t}{2}\right)^{2m}. \tag{6.74}$$

For $\alpha = 0$, $c_0 = 1$ this reduces to $J_0(t)$.

One usually chooses

$$c_0 = \frac{1}{2^\alpha \Gamma(\alpha + 1)},$$ (6.75)

where Γ is the *gamma function* defined by

$$\Gamma(z) = \int_0^\infty e^{-t} t^{z-1}\, dt, \quad \text{Re}(z) > 0.$$

This function is an extension of the factorial function to numbers z which are not integers. This can be seen by showing that Γ satisfies the equation

$$\Gamma(z+1) = z\Gamma(z),$$ (6.76)

and observing that

$$\Gamma(1) = \int_0^\infty e^{-t}\, dt = 1.$$

Thus $\Gamma(n+1) = n!$ for all nonnegative integers n.

The proof of (6.76) can be obtained by integrating by parts

$$\Gamma(z+1) = \lim_{T\to+\infty} \int_0^T e^{-t} t^z\, dt$$

$$= \lim_{T\to+\infty}\left[-e^{-t}t^z\big|_0^T + z\int_0^T e^{-t}t^{z-1}\, dt\right] = z\lim_{T\to+\infty}\int_0^T e^{-t}t^{z-1}\, dt = z\Gamma(z).$$

The relation (6.76) can be used to define $\Gamma(z)$ if $\text{Re}(z) < 0$ provided z is not a negative integer. Suppose N is the positive integer such that

$$-N < \text{Re}(z) \le -N + 1.$$

Then $\text{Re}(z+N) > 0$, and we can define $\Gamma(z)$ in terms of $\Gamma(z+N)$ by

$$\Gamma(z) = \frac{\Gamma(z+N)}{z(z+1)\cdots(z+N-1)}, \quad \text{Re}(z) < 0,$$

provided that $z \ne -N + 1$. The gamma function is not defined for $z = 0, -1, -2, \ldots$, but it is natural to define

$$1/\Gamma(k) = 0, \quad k = 0, -1, -2, \ldots,$$

since this is the limiting value of $1/\Gamma(z)$ at these points. The gamma function satisfies (6.76) everywhere it is defined.

Let us return to the solution x_1 in (6.74). Using the c_0 given in (6.75) we obtain a solution which is called the *Bessel function of the first kind of order* α, denoted by J_α. Since

$$\Gamma(m + \alpha + 1) = (m+\alpha)\Gamma(m+\alpha) = (m+\alpha)\cdots(1+\alpha)\Gamma(\alpha+1),$$

it follows that

$$J_\alpha(t) = \left(\frac{t}{2}\right)^\alpha \sum_{m=0}^\infty \frac{(-1)^m}{m!\Gamma(m+\alpha+1)} \left(\frac{t}{2}\right)^{2m}, \quad \mathrm{Re}(\alpha) \geq 0.$$

For $\alpha = 0$ this reduces to the J_0 computed earlier.

The determination of a second independent solution for (6.73) falls into two cases, depending on whether $\lambda_1 - \lambda_2 = 2\alpha$ is a positive integer or not. If 2α is not a positive integer, Theorem 6.9 (i) guarantees a second solution x_2 of the form

$$x_2(t) = t^{-\alpha} \sum_{k=0}^\infty d_k t^k.$$

Observe that our calculations for the root $\lambda_1 = \alpha$ carry over provided α is replaced by $-\alpha$ everywhere, so that

$$J_{-\alpha}(t) = \left(\frac{t}{2}\right)^{-\alpha} \sum_{m=0}^\infty \frac{(-1)^m}{m!\Gamma(m-\alpha+1)} \left(\frac{t}{2}\right)^{2m}$$

gives a second solution in case 2α is not a positive integer.

Since $\Gamma(m - \alpha + 1)$ exists for $m = 0, 1, 2, \ldots$, provided α is not a positive integer, $J_{-\alpha}$ exists in this case, even if $\lambda_1 - \lambda_2 = 2\alpha$ is a positive integer. This corresponds to the rather special situation in Theorem 6.9 where the constant $c = 0$. Therefore, if α is not zero or a positive integer, a basis for the Bessel equation of order α is given by $\mathbf{X} = (J_\alpha, J_{-\alpha})$.

The remaining case is when α is a positive integer $\alpha = n$. Note that in the formula above for $J_{-\alpha}$ with $\alpha = n$ we have

$$\frac{1}{\Gamma(m-n+1)} = 0, \quad m = 0, 1, \ldots, n-1,$$

implying that

$$J_{-n}(t) = (-1)^n J_n(t), \quad n = 1, 2, \ldots.$$

According to Theorem 6.9 a second solution x_2 exists in the form

$$x_2(t) = c J_n(t) \log(t) + t^{-n} \sum_{k=0}^\infty d_k t^k$$

for $t > 0$. By substituting this into the Bessel equation c and the d_k can be computed. The calculation is somewhat tedious and so will be omitted. It turns out that c and d_0 are related by

$$c = \frac{-d_0}{2^{n-1}(n-1)!}.$$

Using $c = 1$ gives a solution which is called a *Bessel function of the second kind of order* n, which is denoted by K_n. It is given by the rather lengthy formula

$$K_n(t) = J_n(t) \log(t) - \frac{1}{2} \left(\frac{t}{2}\right)^{-n} \sum_{k=0}^{n-1} \frac{(n-k-1)!}{k!} \left(\frac{t}{2}\right)^{2k}$$

$$-\frac{1}{2}\left(\frac{t}{2}\right)^n \sum_{m=0}^{\infty} \frac{(-1)^m}{m!(m+n)!}(h_m + h_{m+n})\left(\frac{t}{2}\right)^{2m},$$

where $h_0 = 0$ and

$$h_m = 1 + \frac{1}{2} + \cdots + \frac{1}{m}, \quad m = 1, 2, \ldots.$$

When $n = 0$ this formula reduces to the K_0 we computed earlier, provided the first sum from $k = 0$ to $k = n - 1$ is interpreted to be zero. A basis for the Bessel equation of order n is given by $\mathbf{X} = (J_n, K_n)$.

There are several different Bessel functions of the second kind, each of which is useful for a particular application. All have the form $aJ_n + bK_n$ for some constants a, b. One of the most widely used is the function Y_n, defined by

$$Y_n = \frac{2}{\pi}[(\gamma - \log(2))J_n + K_n],$$

where γ is the *Euler constant* given by

$$\gamma = \lim_{m \to \infty} (h_m - \log(m)) = 6.772\ldots.$$

Therefore

$$Y_n(t) = \frac{2}{\pi}J_n(t)\log\left(\frac{t}{2}\right) - \frac{1}{\pi}\left(\frac{t}{2}\right)^{-n} \sum_{k=0}^{n-1} \frac{(n-k-1)!}{k!}\left(\frac{t}{2}\right)^{2k}$$

$$-\frac{1}{\pi}\left(\frac{t}{2}\right)^n \sum_{m=0}^{\infty} \frac{(-1)^m}{m!(m+n)!}(h_m + h_{m+n} - 2\gamma)\left(\frac{t}{2}\right)^{2m}. \qquad (6.77)$$

With this curious choice of a, b to define Y_n, for large $t > 0$, $J_n(t)$ and $Y_n(t)$ are related to each other in a manner which is similar to the relationship between $\cos(t)$ and $\sin(t)$. In fact, it can be shown that, for large $t > 0$,

$$J_n(t) = \left(\frac{2}{\pi t}\right)^{1/2}\left[\cos\left(t - \frac{\pi}{4} - \frac{n\pi}{2}\right) + j_n(t)\right],$$

$$Y_n(t) = \left(\frac{2}{\pi t}\right)^{1/2}\left[\sin\left(t - \frac{\pi}{4} - \frac{n\pi}{2}\right) + y_n(t)\right],$$

where j_n and y_n are functions which are small for large $t > 0$ in the sense that there is a constant $M > 0$ such that

$$t|j_n(t)| \le M, \quad t|y_n(t)| \le M$$

for all $t \ge T$ for some $T > 0$.

If α is not an integer the function Y_α given by

$$Y_\alpha = \frac{J_\alpha \cos(\alpha\pi) - J_{-\alpha}}{\sin(\alpha\pi)}$$

is a solution of the Bessel equation of order α, and J_α, Y_α is a basis for the solutions. This expression is not defined if $\alpha = n$ is an integer. However we can define Y_n if n is an integer by

$$Y_n(t) = \lim_{\alpha \to n} Y_\alpha(t),$$

and this limit can be evaluated using l'Hôpital's rule. This leads to

$$Y_n(t) = \frac{1}{\pi} \lim_{\alpha \to n} \left[\frac{\partial J_\alpha}{\partial \alpha}(t) - (-1)^n \frac{\partial J_{-\alpha}}{\partial \alpha}(t) \right],$$

which can be shown to give the expression for Y_n in (6.77). In taking these derivatives with respect to α, use is made of the fact that

$$\frac{\Gamma'(m+1)}{\Gamma(m+1)} = h_m - \gamma,$$

which results from the deeper study of the gamma function.

Note that

$$J_{-\alpha} = J_\alpha \cos(\alpha\pi) - Y_\alpha \sin(\alpha\pi)$$

and

$$Y_{-\alpha} = J_\alpha \sin(\alpha\pi) + Y_\alpha \cos(\alpha\pi)$$

for all $\alpha \in \mathcal{C}$, $t > 0$. For $\alpha = n$, an integer, this gives

$$J_{-n} = (-1)^n J_n, \quad Y_{-n} = (-1)^n Y_n.$$

6.4   Infinity as a singular point

It is often necessary to investigate the behavior of solutions to a system

$$\mathbf{A}_1(t)X' + \mathbf{A}_0(t)X = 0, \tag{6.78}$$

or an equation of order n,

$$a_n(t)x^{(n)} + \cdots + a_0(t)x = 0 \tag{6.79}$$

for large values of $|t|$. One way of doing this is to make the substitution $s = 1/t$ and to study the solutions of the resulting equation, called the *equations induced by the substitution* $s = 1/t$, near $s = 0$. We say that ∞ is a regular point for (6.78) or (6.79) or a singular point of a given type if the origin is a regular point or singular point of this type for the corresponding induced equation.

Consider (6.78) and put

$$Y(s) = X\left(\frac{1}{s}\right), \quad \mathbf{B}_1(s) = \mathbf{A}_1\left(\frac{1}{s}\right), \quad \mathbf{B}_0(s) = \mathbf{A}_0\left(\frac{1}{s}\right),$$

where X is a solution of (6.78) for $|t| > r$ for some $r > 0$. We have

$$Y'(s) = -\frac{1}{s^2} X'\left(\frac{1}{s}\right),$$

so that Y satisfies the induced equation

$$\mathbf{B}_1(s)Y' = \frac{\mathbf{B}_0(s)}{s^2} Y \tag{6.80}$$

for $0 < |s| < 1/r$. Now the origin is a regular point for (6.80) if and only if the matrix-valued functions $\mathbf{B}_1(s)$, $\mathbf{B}_0(s)/s^2$, can be defined to be continuous on $|s| < \rho$ for some $\rho > 0$, and $\mathbf{B}_1(0)$ is invertible. Hence ∞ is a regular point for (6.78) if and only if $\mathbf{A}_1(t)$ and $t^2\mathbf{A}_0(t)$ tend to limits,

$$\mathbf{A}_1(t) \to \mathbf{L}_1, \quad t^2\mathbf{A}_0(t) \to \mathbf{L}_0, \quad |t| \to \infty,$$

and $\mathbf{L}_1$ is invertible.

Let us look further at the special case of the equation

$$X' = \mathbf{A}(t)X, \tag{6.81}$$

where $\mathbf{A}$ is continuous for $|t| > r$ for some $r > 0$. Thus ∞ is a regular point for (6.81) if and only if

$$t^2\mathbf{A}(t) \to \mathbf{L} \in M_n(\mathcal{C}), \quad |t| \to \infty.$$

This equation has ∞ as an ordinary point if and only if $\mathbf{A}$ has a power series representation

$$\mathbf{A}(t) = t^{-2} \sum_{k=0}^{\infty} \mathbf{A}_k t^{-k}, \quad \mathbf{A}_k \in M_n(\mathcal{C})$$

which is convergent for $|t| > r$ for some $r > 0$. Infinity is a singular point of the first kind for (6.81) if and only if $\mathbf{A}$ has an expansion

$$\mathbf{A}(t) = t^{-1} \sum_{k=0}^{\infty} \mathbf{A}_k t^{-k}, \quad \mathbf{A}_k \in M_n(\mathcal{C}), \quad \mathbf{A}_0 \neq 0,$$

and it is a singular point of the second kind if and only if $\mathbf{A}$ can be written as

$$\mathbf{A}(t) = t^m \sum_{k=0}^{\infty} \mathbf{A}_k t^{-k}, \quad \mathbf{A}_k \in M_n(\mathcal{C}), \quad \mathbf{A}_0 \neq 0,$$

where m is a nonnegative integer and the series converges for $|t| > r$ for some $r > 0$.

Therefore, if k is an integer, the only possible singular points for the equation

$$t^k X' = \mathbf{A}X, \quad \mathbf{A} \in M_n(\mathcal{C}), \quad \mathbf{A} \neq 0$$

are 0 and ∞. If $k \leq 0$, the origin is an ordinary point and ∞ is a singular point of the second kind. If $k = 1$, both 0 and ∞ are singular points of the first kind. If $k \geq 2$, the origin is a singular point of the second kind and ∞ is an ordinary point.

Now consider a second-order equation of the form

$$x'' + a_1(t)x' + a_0(t)x = 0, \tag{6.82}$$

where a_0, a_1 are continuous complex-valued functions for $|t| > r$ for some $r > 0$. If x is a solution of (6.82) for $|t| > r$, we put

$$y(s) = x(1/s), \quad b_1(s) = a_1(1/s), \quad b_0(s) = a_0(1/s)$$

and

$$x'(1/s) = -s^2 y'(s), \quad x''(1/s) = s^4 y''(s) + 2s^3 y'(s).$$

Thus the induced equation satisfied by y is

$$s^4 y'' + [2s^3 - s^2 b_1(s)]y' + b_0(s)y = 0. \tag{6.83}$$

The origin is a regular point for (6.83) if and only if the functions $2/s - b_1(s)/s^2$, b_0/s^4, can be defined on $|s| < \rho$ for some $\rho > 0$ so as to be continuous there. The origin is an ordinary point for (6.83) if and only if these functions can be defined so as to be analytic at $s = 0$, and this means that b_1, b_0 can be written as

$$b_1(s) = 2s + s^2 \sum_{k=0}^{\infty} c_k s^k, \quad b_0(s) = s^4 \sum_{k=0}^{\infty} d_k s^k, \quad c_k, d_k \in C,$$

where the series converge for $|s| < \rho$ for some $\rho > 0$. Hence ∞ is an ordinary point for (6.82) if and only if a_1, a_0 have expansions

$$a_1(t) = 2t^{-1} + t^{-2} \sum_{k=0}^{\infty} c_k t^{-k}, \quad a_0(t) = t^{-4} \sum_{k=0}^{\infty} d_k t^{-k}, \quad c_k, d_k \in C,$$

where these series converge for $|t| > 1/\rho$.

Infinity is a regular singular point for (6.82) if and only if

$$a_1(t) = t^{-1} \sum_{k=0}^{\infty} c_k t^{-k}, \quad a_0(t) = t^{-2} \sum_{k=0}^{\infty} d_k t^{-k}, \quad c_k, d_k \in C,$$

and these series converge for $|t| > r$ for some $r > 0$, and either $c_0 \neq 2$ or $d_0 \neq 0$.

An example is the Euler equation,

$$t^2 x'' + atx' + bx = 0, \quad a, b \in C.$$

Here the only possible singular points are 0 and ∞. If a and b are not both zero, the origin is a regular singular point, whereas if $a - 2$ and b are not both zero, infinity is a regular singular point. Otherwise 0 and ∞ are ordinary points.

A very important equation having at most three singular points is the *hypergeometric equation*

$$t(1-t)x'' + [\gamma - (\alpha+\beta+1)t]x' - \alpha\beta x = 0, \tag{6.84}$$

where $\alpha, \beta, \gamma \in \mathcal{C}$. The three points $0, 1, \infty$ are either ordinary points or regular singular points, and there are no other possible singular points.

If we let $s = \beta t$, $z(s) = x(s/\beta)$ in (6.84), then z satisfies the equation

$$s(1-s/\beta)z'' + \left[(\gamma - s) - \frac{(\alpha+1)}{\beta}s\right]z' - \alpha z = 0,$$

which has $0, \beta, \infty$ as ordinary points or regular singular points. Letting $\beta \to \infty$ we obtain the equation

$$sz'' + (\gamma - s)z' - \alpha z = 0, \tag{6.85}$$

which has 0 as either an ordinary point or a regular singular point. However ∞ is a singular point for (6.85) which is not a regular singular point. The equation (6.85) is called the *confluent hypergeometric equation*.

6.5   Notes

For the study of solutions to classically important equations with singular points, one would be hard pressed to find a better reference than [27]. Additional material on linear systems with singularities of the second kind can be found in [3]. For a more extensive treatment, one may consult [26]; [8] also has interesting material.

The terminology used for classification of singularities of linear equations is not completely standardized. For instance, [26] uses the terms *regular singular point* and *irregular singular point* for what we have called singularities of the first and second kind. In [3] the terminology regular singular point and irregular singular point is used to describe the solutions of the equations, rather than the coefficients of the equations.

6.6   Exercises

1. Consider the system with a singular point at $\tau = 0$,

$$\mathbf{A}_1 X' + \mathbf{A}_0 X = 0, \quad \mathbf{A}_1 = \begin{pmatrix} 1 & 1 \\ 1 & 1 \end{pmatrix}, \quad \mathbf{A}_0 = \begin{pmatrix} 1 & 0 \\ 0 & 0 \end{pmatrix}. \tag{6.86}$$

(a) Show that if X is a solution of (6.86) which exists at $\tau = 0$, then $X(0) = \xi$ has the form $\xi = \xi_2 E_2$ for some $\xi_2 \in \mathcal{C}$.

(b) If $\xi_2 \in \mathcal{C}$ is arbitrary, show that there is a solution X of (6.86) satisfying $X(0) = \xi_2 E_2$.

2. Consider the system with a singular point at $\tau = 0$,

$$\mathbf{A}_1(t)X' + \mathbf{A}_0(t)X = 0, \quad \mathbf{A}_1(t) = \begin{pmatrix} t & t \\ 0 & 1 \end{pmatrix}, \quad \mathbf{A}_0(t) = \begin{pmatrix} 0 & 0 \\ 1 & 1 \end{pmatrix}. \quad (6.87)$$

(a) Compute a basis $\mathbf{X} = (u, v)$ for (6.87) satisfying $\mathbf{X}(0) = I_2$.

(b) Show that given any $\xi \in \mathcal{C}^2$ there exists a solution X of (6.87) such that $X(0) = \xi$. Even though $\tau = 0$ is a singular point, every initial value problem at 0 is solvable.

3. Consider the system $tX' = \mathbf{A}(t)X$, with a singular point of the first kind at $\tau = 0$. If $\mathbf{X}$ is a solution matrix which exists at $t = 0$, show that $\mathbf{X}(0)$ is not invertible.

4. The system with constant coefficients

$$\frac{dY}{ds} = \mathbf{A}Y, \quad \mathbf{A} \in M_n(\mathcal{C}), \quad s \in \mathcal{R}, \quad (6.88)$$

has a basis $\mathbf{Y}(s) = e^{\mathbf{A}s}$.

(a) Let $X(t) = Y(\log(t))$, $t > 0$. Show that Y satisfies (6.88) if and only if X satisfies

$$t\frac{dX}{dt} = \mathbf{A}X, \quad t > 0. \quad (6.89)$$

Hence a basis for (6.89) is

$$\mathbf{X}(t) = \mathbf{Y}(\log(t)) = e^{\mathbf{A}\log(t)} = t^{\mathbf{A}}.$$

(b) Find a basis for

$$tX' = \mathbf{A}X, \quad \mathbf{A} = \begin{pmatrix} -1 & 1 & 0 \\ 0 & -1 & 0 \\ 0 & 0 & 2 \end{pmatrix}, \quad t > 0.$$

5. (a) If

$$X(t) = Y\left(\frac{t^{r+1}}{r+1}\right), \quad r \in \mathcal{R}, \quad r \neq -1, \quad t > 0,$$

show that Y satisfies

$$\frac{dY}{ds} = \mathbf{A}Y, \quad \mathbf{A} \in M_n(\mathcal{C}), \quad s \in \mathcal{R},$$

if and only if X satisfies

$$\frac{dX}{dt} = t^r \mathbf{A}X, \quad t > 0.$$

(b) Find a basis $\mathbf{X}$ for

$$X' = t^r \mathbf{A}X, \quad \mathbf{A} = \begin{pmatrix} 2 & 0 & 0 \\ 0 & -1 & 1 \\ 0 & 0 & -1 \end{pmatrix}, \quad t > 0, \quad r \neq -1.$$

6. (a) Show that a basis for the system $Y' = \alpha(t)\mathbf{B}Y$, where $\alpha \in C(I, \mathcal{C})$, $\mathbf{B} \in M_n(\mathcal{C})$, is given by $\mathbf{Y}(t) = e^{a(t)\mathbf{B}}$, $t \in I$, where $a(t)$ is any function satisfying $a' = \alpha$, for example,

$$a(t) = \int_\tau^t \alpha(s)\, ds, \quad \tau, t \in I.$$

(b) Consider the system $X' = [\alpha(t)\mathbf{B} + \mathbf{C}(t)]X$, where $\alpha \in C(I, \mathcal{C})$, $\mathbf{B} \in M_n(\mathcal{C})$, $\mathbf{C} \in C(I, M_n(\mathcal{C}))$. If $\mathbf{B}\mathbf{C}(t) = \mathbf{C}(t)\mathbf{B}$ for all $t \in I$, show that a basis is given by

$$\mathbf{X}(t) = e^{a(t)\mathbf{B}}\mathbf{Z}(t), \quad t \in I,$$

where $\mathbf{Z}$ is a basis for the system $Z' = \mathbf{C}(t)Z$. (Hint: Use (a) and Theorem 3.3.)

7. (a) If

$$\mathbf{R} = \begin{pmatrix} \alpha & \beta \\ -\beta & \alpha \end{pmatrix}, \quad \alpha, \beta \in \mathcal{R},$$

show that

$$|t|^{\mathbf{R}} = |t|^\alpha \begin{pmatrix} \cos(\beta \log(|t|)) & \sin(\beta \log(|t|)) \\ -\sin(\beta \log(|t|)) & \cos(\beta \log(|t|)) \end{pmatrix}, \quad t \neq 0.$$

(b) Find a basis for the system

$$tX' = \mathbf{R}X, \quad \mathbf{R} = \begin{pmatrix} -2 & -3 \\ 3 & -2 \end{pmatrix}, \quad t > 0.$$

8. Consider the system

$$X' = \left(\frac{\mathbf{R}}{t} + \mathbf{I}_2\right)X, \quad \mathbf{R} = \begin{pmatrix} 2 & 1 \\ 0 & 1/2 \end{pmatrix}, \quad t > 0. \tag{6.90}$$

(a) For which $\lambda \in \mathcal{C}$ are there solutions X of (6.90) of the form

$$X(t) = t^\lambda P(t), \quad P(t) = \sum_{k=0}^\infty P_k t^k, \quad P_0 \neq 0.$$

(b) For each such λ compute a P_0, and in terms of this compute the successive P_k, $k = 1, 2, \ldots$.

(c) The system (6.90), when written out, is

$$x_1' = \left(\frac{2}{t} + 1\right)x_1 + x_2, \quad x_2' = \left(\frac{1}{2t} + 1\right)x_2,$$

which can be solved directly by first solving the second equation for x_2 and then solving the first equation for x_1. Do this and compare your answer with those of (a), (b).

9. Consider the example which was solved in section 6.2.1,

$$X' = \left(\frac{\mathbf{R}}{t} + 3\mathbf{I}_2\right)x, \quad \mathbf{R} = \frac{1}{6}\begin{pmatrix} 5 & -6 \\ 4 & -6 \end{pmatrix}, \quad t \neq 0. \tag{6.91}$$

(a) Show that by putting $X(t) = e^{3t}Y(t)$, then X satisfies (6.91) if and only if Y satisfies $Y' = (\mathbf{R}/t)Y$. Thus $\mathbf{X}_1(t) = e^{3t}|t|^{\mathbf{R}}$ gives a basis for $t \neq 0$.

(b) Compute a Jordan canonical form $\mathbf{J}$ for $\mathbf{R}$, and find an invertible $\mathbf{Q}$ such that $\mathbf{R} = \mathbf{Q}\mathbf{J}\mathbf{Q}^{-1}$.

(c) Compute $|t|^{\mathbf{R}}$ and thereby $\mathbf{X}_1(t)$.

10. (a) Find a basis $\mathbf{X}$ for the system

$$X' = \left(\frac{\mathbf{R}}{t} + \mathbf{A}t^k\right) X, \quad k \neq -1, \quad t \neq 0,$$

where $\mathbf{R}, \mathbf{A} \in M_n(\mathcal{C})$ commute. (Hint: Use exercise 6.)

(b) Find a basis for the system

$$X' = \left(\frac{\mathbf{R}}{t} + \mathbf{A}t^2\right) X, \quad \mathbf{R} = \begin{pmatrix} 2 & 3 \\ -3 & 2 \end{pmatrix}, \quad \mathbf{A} = \begin{pmatrix} 0 & \pi \\ -\pi & 0 \end{pmatrix}, \quad t \neq 0.$$

(Hint: Use (a).)

11. Compute a basis for the system

$$tX' = \mathbf{B}(t)X, \quad \mathbf{B}(t) = \mathbf{A}(\mathbf{I}_n - \mathbf{A}t)^{-1}, \quad 0 < |t| < 1/|A|,$$

where $\mathbf{A} \in M_n(\mathcal{C})$. (Hint: See exercise 11, Chapter 5.)

12. If $\mathbf{A} = (a_{ij}) \in M_n(\mathcal{C})$, show that $\mathbf{A}$ and its transpose $\mathbf{A}^T = (a_{ji})$ have the same characteristic polynomial and hence have the same eigenvalues.

13. Show that the converse of Theorem 6.4 is valid; that is, if

$$\mathbf{U}\mathbf{A} - \mathbf{B}\mathbf{U} = 0, \quad \mathbf{A}, \mathbf{B} \in M_n(\mathcal{C}),$$

has only the trivial solution $\mathbf{U} = 0$, then $\mathbf{A}$ and $\mathbf{B}$ have no eigenvalue in common. Proceed as follows. Suppose λ is an eigenvalue of $\mathbf{A}$ and $\mathbf{B}$. Then λ is an eigenvalue of $\mathbf{A}^T$ and $\mathbf{B}$ (exercise 12), and there are $\alpha, \beta \in \mathcal{C}^n$, $\alpha \neq 0$, $\beta \neq 0$ such that $\mathbf{A}^T \alpha = \lambda \alpha$, $\mathbf{B}\beta = \lambda\beta$. If $\mathbf{U} = \beta\alpha^T \in M_n(\mathcal{C})$, show that $\mathbf{U} \neq 0$ and $\mathbf{U}\mathbf{A} - \mathbf{B}\mathbf{U} = 0$.

14. Given $\mathbf{A}, \mathbf{B} \in M_n(\mathcal{C})$ the function

$$f : \mathbf{U} \in M_n(\mathcal{C}) \to f(\mathbf{U}) = \mathbf{U}\mathbf{A} - \mathbf{B}\mathbf{U} \in M_n(\mathcal{C})$$

is linear. Show that ν is an eigenvalue of f if and only if $\nu = \lambda - \mu$, where λ is an eigenvalue of $\mathbf{A}$, and μ is an eigenvalue of $\mathbf{B}$. (Hint: Use exercise 13 and Theorem 6.4.)

15. Consider the equation

$$X' = \left[\frac{\mathbf{R}}{t} + \frac{\mathbf{S}}{t-1}\right] X, \quad t \neq 0, 1, \tag{6.92}$$

where

$$\mathbf{R} = \begin{pmatrix} -3 & 1 \\ -1 & -5 \end{pmatrix}, \quad \mathbf{S} = \begin{pmatrix} 1 & 1 \\ -1 & -1 \end{pmatrix}.$$

(a) Show that (6.92) has singular points of the first kind at $\tau_1 = 0$ and at $\tau_2 = 1$.

(b) Compute the eigenvalues of $\mathbf{R}$ and $\mathbf{S}$.

(c) Compute a Jordan canonical form $\mathbf{J}$ for $\mathbf{R}$, and a matrix $\mathbf{Q}$ such that $\mathbf{R} = \mathbf{QJQ}^{-1}$.

(d) Putting $X = \mathbf{Q}Y$ show that X satisfies (6.92) if and only if Y satisfies

$$Y' = \left[\frac{\mathbf{J}}{t} + \frac{\mathbf{K}}{t-1} \right] Y, \quad t \neq 0, 1, \tag{6.93}$$

where $\mathbf{K} = \mathbf{QSQ}^{-1}$. Compute $\mathbf{K}$.

(e) Find a basis Y for (6.93) for $0 < t < 1$, and then compute a basis X for (6.94) for $0 < t < 1$. (Hint: $\mathbf{X} = \mathbf{Q}Y$.)

(f) Show that $\mathbf{X}_1(t) = |t|^{\mathbf{R}} |t-1|^{\mathbf{S}}$ is a basis for (6.92) on any interval not containing 0 or 1. Compute $\mathbf{X}_1$. (Hint: Note that $\mathbf{RS} = \mathbf{SR}$.)

16. The example in section 6.2.4 can be written as

$$x_1' = -\frac{x_1}{t} + \frac{x_2}{t} + 3x_1 - x_2,$$

$$x_2' = -\frac{2x_2}{t} + 3x_2$$

for $t > 0$. Solve this system directly. (Hint: Find x_2 from the second equation, and then solve for x_1.)

17. Let x satisfy the equation

$$t^2 x'' + a(t)tx' + b(t)x = 0, \quad 0 < |t| < \rho, \tag{6.94}$$

where

$$a(t) = \sum_{k=0}^{\infty} a_k t^k, \quad b(t) = \sum_{k=0}^{\infty} b_k t^k, \quad a_k, b_k \in \mathcal{C},$$

and the series converge for $|t| < \rho$, $\rho > 0$. Let

$$\hat{x}(t) = \begin{pmatrix} x(t) \\ tx'(t) \end{pmatrix}, \quad 0 < |t| < \rho.$$

(a) Show that $Z = \hat{x}$ satisfies the equation

$$tZ' = \mathbf{B}(t)Z, \quad \mathbf{B}(t) = \begin{pmatrix} 0 & 1 \\ -b & 1-a \end{pmatrix}, \quad 0 < |t| < \rho, \tag{6.95}$$

and, conversely, if

$$Z = \begin{pmatrix} z_1 \\ z_2 \end{pmatrix}$$

satisfies (6.95) then $Z = \hat{x}$, where x satisfies (6.94).

(b) Show that (6.95) can be written as

$$Z' = \left[\frac{\mathbf{R}}{t} + \mathbf{A}(t) \right] Z, \quad 0 < |t| < \rho, \tag{6.96}$$

where
$$\mathbf{R} = \begin{pmatrix} 0 & 1 \\ -b_0 & 1 - a_0 \end{pmatrix}, \quad \mathbf{A}(t) = \sum_{k=0}^{\infty} \mathbf{A}_k t^k,$$

where the latter series converges for $|t| < \rho$.

18. Find a basis for the solutions of the following equations for $t > 0$:

(a)
$$t^2 x'' + 2t x' - 6x = 0,$$

(b)
$$2t^2 x'' + t x' - x = 0,$$

(c)
$$t^2 x'' + t x' + 4x = 0,$$

(d)
$$t^2 x'' + 5t x' + 4x = 0.$$

(Hint: See exercise 17.)

19. Find the singular points of the following equations, and determine those which are regular singular points:

(a)
$$t^2 x'' + (t + t^2) x' - x = 0,$$

(b)
$$3t^2 x'' + t^2 x' + 2t x = 0,$$

(c)
$$t x'' + 4x = 0,$$

(d)
$$(1 - t^2) x'' - 2t x' + 2x = 0,$$

(e)
$$(t^2 + t - 2)^2 x'' + 3(t + 2) x' + (t - 1)x = 0,$$

(f)
$$t^2 x'' + \sin(t) x' + \cos(t) x = 0.$$

20. Compute the indicial polynomials relative to $\tau = 0$ and their roots for the following equations:

(a)
$$t^2 x'' + (t + t^2) x' - x = 0,$$

(b)
$$t^2 x'' + t x' + (t^2 - 1/4)x = 0,$$

(c)
$$t^2 x'' + (t - 3t^2) x' + e^t x = 0,$$

(d)
$$t^2 x'' + \sin(t) x' + \cos(t) x = 0.$$

21. (a) Show that -1 and 1 are regular singular points for the Legendre equation

$$(1 - t^2)x'' - 2tx' + \alpha(\alpha + 1)x = 0, \quad \alpha \in \mathcal{C}.$$

(b) Find the indicial polynomial and its roots corresponding to the point $\tau = 1$.

22. Find all solutions of the following equations for $t > 0$:

(a)
$$t^2 x'' + 2tx' - 2x = 0,$$

(b)
$$t^3 x''' + 2t^2 x'' - tx + x = 0,$$

(c)
$$t^2 x'' + tx' - 4x = t,$$

(d)
$$t^2 x'' - 5tx' + 9x = t^3.$$

23. Find all solutions of the following equations for $|t| > 0$:

(a)
$$t^2 x'' + tx' + 4t = 1,$$

(b)
$$t^2 x'' - 3tx'' + 5t = 0,$$

(c)
$$t^2 x'' - (2 + i)tx' + 3ix = 0,$$

(d)
$$t^2 x'' + tx' - 4\pi x = t.$$

24. Consider the Euler equation

$$t^2 x'' + atx' + bx = 0, \quad a, b \in \mathcal{C}, \quad t > 0. \tag{6.97}$$

(a) Show that x satisfies (6.97) if and only if $y = x(e^t)$ satisfies

$$y'' + (a - 1)y + by = 0. \tag{6.98}$$

(b) Compute the characteristic polynomial p for (6.98) and the indicial polynomial q for (6.97). How do these compare?

(c) Show that $x(t) = y(\log(t))$.

25. Let L be the Euler differential operator

$$L = t^n D^n + a_{n-1}t^{n-1}D^{n-1} + \cdots + a_1 tD + a_0, \quad D = \frac{d}{dt},$$

where $a_0, \ldots, a_{n-1} \in \mathcal{C}$, and let q be the indicial polynomial of the equation $Lx = 0$.

(a) If $r \in \mathcal{C}$ is such that $q(r) \neq 0$, show that the equation

$$Lx = t^r, \quad t > 0, \tag{6.99}$$

has a solution x of the form $x(t) = ct^r$, and compute $c \in \mathcal{C}$. (Hint: Let $L(ct^r) = cq(r)t^r$.)

(b) Suppose r is a root of q of multiplicity k. Show that (6.99) has a solution x of the form $x(t) = ct^r[\log(t)]^k$, and compute $c \in \mathcal{C}$.

(c) Find all solutions of

$$t^2 x'' - 5tx' + 9t = 2t^3, \quad t > 0.$$

26. Find a basis $\mathbf{X} = (u, v)$ for the following equations for $t > 0$:

(a)
$$2t^2 x'' + (t^2 - t)x' + x = 0,$$

(b)
$$t^2 x'' + (t - t^2)x' + x = 0.$$

27. Find all solutions of the form

$$x(t) = |t|^\lambda \sum_{k=0}^{\infty} p_k t^k, \quad |t| > 0,$$

for the following equations:

(a)
$$3t^2 x'' + 5tx' + 3tx = 0,$$

(b)
$$t^2 x'' + tx' + t^2 x = 0,$$

(c)
$$t^2 x'' + tx' + (t^2 - 1/4)x = 0.$$

28. Consider the following four equations near $\tau = 0$:

(i)
$$2t^2 x'' + (5t + t^2)x' + (t^2 - 2)x = 0,$$

(ii)
$$4t^2 x'' - 4te^t x' + 3\cos(t)x = 0,$$

(iii)
$$(1 - t^2)t^2 x'' + 3(t + t^2)x' + x = 0,$$

(iv)
$$t^2 x'' + 2t^2 x' - 2x = 0.$$

(a) Compute the roots λ_1, λ_2 ($\mathrm{Re}(\lambda_1) \geq \mathrm{Re}(\lambda_2)$)) of the indicial polynomial for each relative to $\tau = 0$.

(b) Describe (do not compute) the nature of a basis $\mathbf{X}$ near $\tau = 0$. In the case of $\lambda_1 = \lambda_2$ determine the first nonzero coefficient in $p_2(t)$ (in the terminology of Theorem 6.9 (ii)), and when $\lambda_1 - \lambda_2$ is a positive integer, determine whether $c = 0$ or not (in the terminology of Theorem 6.10).

29. Consider the Legendre equation

$$(1 - t^2)x'' - 2tx' + \alpha(\alpha + 1)x = 0, \quad \alpha \in \mathcal{C}.$$

(a) For which $\lambda \in \mathcal{C}$ will there exist solutions x of the form

$$x(t) = |t - 1|^\lambda \sum_{k=0}^\infty p_k(t - 1)^k, \quad |t - 1| > 0 \,?$$

(b) Find a solution x of the form given in (a) as follows. Note that $t = (t - 1) + 1$, and express the coefficients in powers of $t - 1$. Show that the p_k satisfy the recursion relation

$$p_{k+1} = \frac{(\alpha - k)(\alpha + k + 1)}{2(k + 1)^2} p_k, \quad k = 0, 1, 2, \ldots.$$

(c) For what values of t does the series found in (b) converge?
(d) Show that there is a polynomial solution if $\alpha = n$, a nonnegative integer.
30. The equation

$$tx'' + (1 - t)x' + \alpha x = 0, \quad \alpha \in \mathcal{C},$$

is called the *Laguerre equation*.
(a) Show that $\tau = 0$ is a regular singular point for this equation.
(b) Compute the indicial polynomial relative to $\tau = 0$ and find its roots.
(c) Find a solution x of the form

$$x(t) = |t|^\lambda \sum_{k=0}^\infty p_k t^k, \quad |t| > 0.$$

(Hint: Show that the p_k satisfy the recursion relation

$$p_{k+1} = \frac{(k - \alpha)}{(k + 1)^2} p_k, \quad k = 0, 1, 2, \ldots.)$$

(d) For what values of t does the series found in (c) converge?
(e) Show the there is a polynomial solution of degree n if $\alpha = n$, a nonnegative integer.
31. Let L_n, for $n = 0, 1, 2, \ldots$, be defined by

$$L_n(t) = e^t D^n (t^n e^{-t}), \quad D = \frac{d}{dt}.$$

(a) Show that L_n is a polynomial of degree n given explicitly by

$$L_n(t) = \sum_{k=0}^\infty (-1)^k \binom{n}{k} \frac{n!}{k!} t^k, \quad \binom{n}{k} = \frac{n!}{k!(n - k)!}.$$

This polynomial is called the *nth Laguerre polynomial.*

(b) Verify that the first five Laguerre polynomials are given by

$$L_0(t) = 1, \quad L_1(t) = 1 - t, \quad L_2(t) = 2 - 4t + t^2,$$

$$L_3(t) = 6 - 18t + 9t^2 - t^3, \quad L_4(t) = 24 - 96t + 72t^2 - 16t^3 + t^4.$$

(c) Show that L_n satisfies the Laguerre equation of exercise 30 if $\alpha = n$.

32. The *Mathieu equation* is given by

$$x'' + (a + b\cos(2t))x = 0, \quad a, b \in \mathcal{R}.$$

(a) Let $s = \cos(2t)$ and $y(\cos(2t)) = x(t)$ or

$$y(s) = x\left(\frac{1}{2}\cos^{-1}(s)\right), \quad 0 \le t \le \pi/2.$$

Show that y satisfies the equation

$$4(1 - s^2)y'' - 4sy' + (a + bs)y = 0. \tag{6.100}$$

(b) Show that 1 and -1 are regular singular points for (6.100), and compute the roots of the indicial polynomials relative to 1 and -1.

33. Using the ratio test, show that the series defining J_0 and K_0 converge for all $t \in \mathcal{R}$.

34. Show that x satisfies the Bessel equation of order 0,

$$t^2 x'' + tx' + t^2 x = 0, \quad t > 0,$$

if and only if

$$y(t) = t^{1/2} x(t)$$

satisfies the equation

$$t^2 y'' + \left(t^2 + \frac{1}{4}\right) y = 0, \quad t > 0.$$

35. (a) If $\lambda > 0$ is such that $J_0(\lambda) = 0$, show that $J_0'(\lambda) \ne 0$. (Hint: Apply the uniqueness result with initial conditions at $\tau = \lambda$.)

(b) Show that on each finite interval $[0, b]$, $b > 0$, there are at most a finite number of zeros of J_0. (Hint: Show that there would be some point $\lambda \in [0, b]$ such that $J_0(\lambda) = 0 = J_0'(\lambda)$ if there were infinitely many zeros on $[0, b]$.)

36. Since $J_0(0) = 1$ and J_0 is continuous, $J_0(t) \ne 0$ in some interval $0 < t < a$, for some $a > 0$. Let $0 < \tau < a$.

(a) Show that there is a second solution y of the Bessel equation of order 0 of the form

$$y(t) = J_0(t) \int_\tau^t \frac{1}{sJ_0^2(s)} \, ds, \quad 0 < t < a.$$

(b) Show that J_0, y are linearly independent on $0 < t < a$.

37. Show that if x is any solution of the Bessel equation of order 0, then $y = x'$ is a solution of the Bessel equation of order 1.

38. (a) If $\lambda > 0$ and $x_\lambda(t) = t^{1/2}J_0(\lambda t)$, show that

$$x_\lambda'' + \frac{1}{4t^2}x_\lambda = -\lambda^2 x_\lambda. \qquad (6.101)$$

(b) If λ, μ are positive constants, show that

$$(\lambda^2 - \mu^2)\int_0^1 x_\lambda(t)x_\mu(t)\ dt = x_\lambda(1)x_\mu'(1) - x_\mu(1)x_\lambda'(1).$$

(Hint: Multiply (6.101) by x_μ, multiply

$$x_\mu'' + \frac{1}{4t^2}x_\mu = -\mu^2 x_\lambda$$

by x_λ, and subtract to obtain

$$(x_\lambda x_\mu' - x_\mu x_\lambda')' = (\lambda^2 - \mu^2)x_\lambda x_\mu.)$$

(c) If $\lambda \neq \mu$ and $J_0(\lambda) = 0$, $J_0(\mu) = 0$, show that

$$\int_0^1 x_\lambda(t)x_\mu(t)\ dt = \int_0^1 tJ_0(\lambda t)J_0(\mu t)\ dt = 0.$$

39. (a) Show that the series defining J_α, $J_{-\alpha}$ converge for all $t \in \mathcal{R}$.

(b) Show that the infinite series involved in the definition of K_n converges for all $t \in \mathcal{R}$.

40. Show that x satisfies the Bessel equation of order α,

$$tx'' + tx' + (t^2 - \alpha^2)x = 0, \quad t > 0,$$

if and only if

$$y = t^{1/2}x(t)$$

satisfies

$$y'' + \left[1 + \frac{1/4 - \alpha^2}{t^2}\right]y = 0, \quad t > 0.$$

41. Show that

$$t^{1/2}J_{1/2}(t) = \frac{\sqrt{2}}{\Gamma(1/2)}\sin(t),$$

$$t^{1/2}J_{-1/2}(t) = \frac{\sqrt{2}}{\Gamma(1/2)}\cos(t).$$

(Hint: See exercise 40. It can be shown that $\Gamma(1/2) = \sqrt{\pi}$.)

42. Show that x satisfies the Bessel equation of order $1/3$,

$$tx'' + tx' + \left(t^2 - \frac{1}{9}\right)x = 0, \quad t > 0,$$

if and only if

$$y = t^{1/2} x(2t^{3/2}/3), \quad t > 0,$$

satisfies

$$y'' + ty = 0, \quad t > 0,$$

which is one version of the Airy equation. Thus a basis for the latter equation is given by $\mathbf{Y}$, where

$$\mathbf{Y}(t) = [t^{1/2} J_{1/3}(2t^{3/2}/3), t^{1/2} J_{-1/3}(2t^{3/2}/3)].$$

43. (a) Show that x satisfies the Bessel equation of order α,

$$tx'' + tx' + (t^2 - \alpha^2)x = 0, \quad t > 0,$$

if and only if

$$y = t^r x(\lambda t^s), \quad t > 0,$$

satisfies the equation

$$t^2 y'' + (1 - 2r)ty' + [\lambda^2 s^2 t^{2s} + (r^2 - \alpha^2 s^2)]y = 0, \quad t > 0. \tag{6.102}$$

Here $r, s, \lambda \in \mathcal{R}$ are such that $s \neq 0$, $\lambda \neq 0$.

(b) Show that in the case of $r = 1/2$, $s = (p+2)/2$, $\alpha = 1/(p+2)$, $\lambda = 2/(p+2)$, the equation (6.102) reduces to

$$y'' + t^p y = 0, \quad t > 0, \tag{6.103}$$

if $p \neq -2$. Thus a basis for this equation consists of Bessel functions.

(c) Solve equation (6.103) when $p = -2$.

44. For α fixed, $\alpha > 0$, and $\lambda > 0$, let $x_\lambda(t) = t^{1/2} J_\alpha(\lambda t)$. Show that x_λ satisfies the equation

$$x_\lambda'' + \frac{1/4 - \alpha^2}{t^2} x_\lambda = -\lambda^2 x_\lambda.$$

45. Use exercise 44 to show that

$$x_\lambda x_\mu'' - x_\mu x_\lambda'' = (x_\lambda x_\mu' - x_\mu x_\lambda')' = (\lambda^2 - \mu^2) x_\lambda x_\mu. \tag{6.104}$$

If λ, μ are positive, show that

$$(\lambda^2 - \mu^2) \int_0^1 x_\lambda(t) x_\mu(t) \, dt = x_\lambda(1) x_\mu'(1) - x_\mu(1) x_\lambda'(1),$$

where x_λ is as in exercise 44.

46. If $\alpha > 0$ and λ, μ are positive zeros of J_α, show that

$$\int_0^1 x_\lambda(t) x_\mu(t) \, dt = \int_0^1 t J_\alpha(\lambda t) J_\alpha(\mu t) \, dt = 0$$

if $\lambda \neq \mu$. (Hint: See exercise 45.)

47. (a) Use the formula for $J_\alpha(t)$ to show that

$$(t^\alpha J_\alpha)'(t) = t^\alpha J_{\alpha-1}(t).$$

(b) Show that

$$(t^{-\alpha} J_\alpha)'(t) = -t^{-\alpha} J_{\alpha+1}(t).$$

48. Show that

$$J_{\alpha-1}(t) - J_{\alpha+1}(t) = 2J_\alpha'(t)$$

and

$$J_{\alpha-1}(t) + J_{\alpha+1}(t) = 2\alpha t^{-1} J_\alpha(t).$$

(Hint: Use exercise 47.)

49. (a) Show that between any two positive zeros of J_α there is a zero of $J_{\alpha+1}$. (Hint: Use exercise 47(b) and Rolle's theorem.)

(b) Show that between any two positive zeros of $J_{\alpha+1}$ there is a zero of J_α. (Hint: Use exercise 47(a) and Rolle's theorem.)

50. (a) Let $\mathbf{R}_1 \neq 0, \mathbf{R}_2 \neq 0 \in M_n(\mathcal{C})$, and suppose τ_1, τ_2 are distinct real numbers. Show that the system

$$x' = \left[\frac{\mathbf{R}_1}{t - \tau_1} + \frac{\mathbf{R}_2}{t - \tau_2} \right] x, \quad t \neq \tau_1, \tau_2,$$

has τ_1, τ_2 as singular points of the first kind and all other $\tau \in \mathcal{R}$ as ordinary points.

(b) Show that ∞ is a singular point of the first kind if $\mathbf{R}_1 + \mathbf{R}_2 \neq 0$ and an ordinary point if $\mathbf{R}_1 + \mathbf{R}_2 = 0$.

(c) Consider the system

$$x' = \left[\sum_{k=1}^{p} (t - \tau_k)^{-1} \mathbf{R}_k \right] x, \quad t \neq \tau_1, \ldots, \tau_p,$$

where the τ_k are distinct real numbers and the $\mathbf{R}_k$ are nonzero matrices, $\mathbf{R}_k \in M_n(\mathcal{C})$. Show that $\tau_1, \ldots, \tau_p$ are singular points of the first kind, and all other $\tau \in \mathcal{R}$ are ordinary points. Show that ∞ is a singular point of the first kind or an ordinary point according to $\mathbf{R}_1 + \cdots + \mathbf{R}_p \neq 0$ or $\mathbf{R}_1 + \cdots + \mathbf{R}_p = 0$.

51. Consider the Legendre equation of order α,

$$(1 - t^2)x'' - 2tx' + \alpha(\alpha + 1)x = 0, \quad \alpha \in \mathcal{C}.$$

(a) Show that ∞ is an ordinary point if $\alpha(\alpha + 1) = 0$.

(b) Show that ∞ is an regular singular point if $\alpha(\alpha + 1) \neq 0$.

(c) Compute the equation induced by the substitution $t = 1/s$, and compute the indicial polynomial, and its roots, of this equation relative to $s = 0$.

52. Find two linearly independent solutions of

$$(1 - t^2)x'' - 2tx' + 2x = 0$$

having the form

$$x(t) = t^{-\lambda} \sum_{k=0}^{\infty} c_k t^{-k}$$

valid for $|t| > 1$. (Hint: See exercise 51 with $\alpha = 1$.)

53. (a) Show that x satisfies the Legendre equation of order p,

$$(1 - t^2)x'' - 2tx' + p(p+1)x = 0,$$

if and only if $y(t) = x(2t - 1)$ satisfies the equation

$$t(1 - t)y'' + (1 - 2t)y' + p(p+1)y = 0. \qquad (6.105)$$

(b) Verify that (6.105) is a hypergeometric equation (6.84) with

$$\alpha = p + 1, \quad \beta = -p, \quad \gamma = 1.$$

54. Classify ∞ as an ordinary point, a regular singular point, or a singular point which is not regular for the following equations:

(a)
$$x'' - tx = 0, \quad \text{(Airy)},$$

(b)
$$x'' - 2tx' + 2\alpha x = 0, \quad \alpha \in C \quad \text{(Hermite)},$$

(c)
$$(1 - t^2)x'' - tx' + \alpha^2 x = 0, \quad \alpha \in C \quad \text{(Chebyshev)},$$

(d)
$$t^2 x'' + tx' + (t^2 - \alpha^2)x = 0, \quad \alpha \in C \quad \text{(Bessel)},$$

(e)
$$tx'' + (1 - t)x' + \alpha x = 0, \quad \alpha \in C \quad \text{(Laguerre)}.$$

55. (a) Compute the indicial polynomial relative to the origin for the hypergeometric equation (6.84).

(b) If γ is not zero or a negative integer, show that x given by $x(t) = F(\alpha, \beta, \gamma; t)$, where F is the *hypergeometric series*

$$F(\alpha, \beta, \gamma; t) = 1 + \sum_{k=1}^{\infty} \frac{(\alpha)_k (\beta)_k}{(\gamma)_k} \frac{t^k}{k!}$$

and

$$(z)_k = z(z + 1)(z + 2) \cdots (z + k - 1), \quad z \in C,$$

satisfies (6.84).

(c) Show that the hypergeometric series is convergent for $|t| < 1$.

(d) Show that if $\gamma \neq 2, 3, \ldots$,

$$y(t) = t^{1-\gamma} F(\alpha - \gamma + 1, \beta - \gamma + 1, 2 - \gamma; t)$$

satisfies the hypergeometric equation for $|t| < 1$.

(e) Show that if γ is not an integer, then $\mathbf{X} = (x, y)$, where x is given in (b), y in (d), is a basis for the hypergeometric equation for $|t| < 1$.

(f) Verify that

$$F(1, 1, 1; t) = (1 - t)^{-1}.$$

(Thus the hypergeometric series reduces to the geometric series in this special case.)

(g) Verify that

$$tF(1, 1, 2; t) = \log(1 + t).$$

Chapter 7

Existence and Uniqueness

7.1   Introduction

The main result of this chapter is a proof of the existence and uniqueness of solutions to initial value problems for linear systems

$$X' = \mathbf{A}(t)X + B(t), \quad X(\tau) = \xi, \quad t \in I. \tag{IVP}$$

Here I is a real interval, $\mathbf{A} \in C(I, M_n(\mathcal{F}))$, and $B \in C(I, \mathcal{F}^n)$, where $\mathcal{F}$ is either $\mathcal{C}$ or $\mathcal{R}$. The point $\tau \in I$ is called an *initial point* and the condition $X(\tau) = \xi$ is called an *initial condition*. A *solution* of (*IVP*) is a differentiable function $X(t)$ on I which satisfies the differential equation and the initial condition; that is,

$$X'(t) = \mathbf{A}(t)X(t) + B(t), \quad t \in I, \quad X(\tau) = \xi. \tag{7.1}$$

If $X(t)$ is a solution of (*IVP*), then integration of (7.1) and the initial condition yield

$$X(t) = \xi + \int_\tau^t [\mathbf{A}(s)X(s) + B(s)] \, ds, \quad t \in I. \tag{7.2}$$

Conversely, if $X \in C(I, \mathcal{F}^n)$ and satisfies (7.2), then the integrand $\mathbf{A}X + B$ is continuous on I, so we can differentiate and obtain (7.1). Putting $t = \tau$ in (7.2) shows that $X(\tau) = \xi$. Thus $X(t)$ is a solution of (*IVP*) if and only if $X \in C(I, \mathcal{F}^n)$ and satisfies the integral equation (7.2).

The right-hand side of the equality in (7.2) can be considered a rule for mapping one continuous function to another. For each $Y \in C(I, \mathcal{F}^n)$ define TY to be that continuous (actually differentiable) function given by

$$TY(t) = \xi + \int_\tau^t [\mathbf{A}(s)Y(s) + B(s)] \, ds, \quad t \in I. \tag{7.3}$$

T is a function which associates another function $TY \in C(I, \mathcal{F}^n)$ with each $Y \in C(I, \mathcal{F}^n)$. To distinguish between the function T and the elements in its domain which are vector-valued functions, T is often called a *transformation* or an *operator* from $C(I, \mathcal{F}^n)$ into $C(I, \mathcal{F}^n)$. In these terms, a solution of (7.2) is an $X \in C(I, \mathcal{F}^n)$ satisfying

$$X(t) = TX(t), \quad t \in I,$$

or simply $X = TX$. Such an X is called a *fixed point* of T in $C(I, \mathcal{F}^n)$.

As a simple example when $n = 1$, $\mathcal{F} = \mathcal{R}$, consider the problem

$$x' = 2tx + 3t, \quad x(1) = -1, \quad t \in \mathcal{R}. \tag{7.4}$$

In this case the integral equation (7.2) becomes

$$x(t) = -1 + \int_1^t [2sx(s) + 3s] \, ds, \quad t \in \mathcal{R},$$

the operator T is given by

$$(Tx)(t) = -1 + \int_1^t [2sx(s) + 3s] \, ds, \quad t \in \mathcal{R},$$

and a solution of (7.4) is an $x \in C(\mathcal{R}, \mathcal{R})$ such that $Tx = x$. The differential equation (7.4) can be solved explicitly to yield

$$x(t) = \frac{-3}{2} + \frac{1}{2e} e^{t^2}, \quad t \in \mathcal{R}.$$

For this example it is easy to verify that

$$(Tx)(t) = -1 + \int_1^t \left[-3s + \frac{s}{e} e^{s^2} + 3s \right] ds = -1 + \frac{1}{e} \int_1^t s e^{s^2} \, ds$$

$$= -1 + \frac{1}{2e} [e^{t^2} - e] = -\frac{3}{2} + \frac{1}{2e} e^{t^2} = x(t).$$

The existence and uniqueness problem for (IVP) has thus been reduced to showing that the integral operator T given by (7.3) has a unique fixed point $X \in C(I, \mathcal{F}^n)$. The question now is how to demonstrate the existence of such a fixed point or, better yet, to give a method for constructing the fixed point. The following example illustrates the general method, which is known as *successive approximations*.

A simple example involving a system of differential equations is

$$X' = \mathbf{A}X, \quad X(0) = \xi, \quad t \in \mathcal{R}, \tag{7.5}$$

where $\mathbf{A} \in M_n(\mathcal{F})$ and $\xi \in \mathcal{F}^n$. The integral equation is now

$$X(t) = \xi + \int_0^t \mathbf{A} \, X(s) \, ds = TX(t), \quad t \in \mathcal{R}. \tag{7.6}$$

We try to construct a sequence of approximations $\{X_m(t)\}$ to the desired function $X(t)$.

For the first approximation, try $X_0(t) = \xi$. Putting this into the right-hand side of (7.6) produces

$$TX_0(t) = \xi + \int_0^t \mathbf{A}\ \xi\ ds = \xi + \mathbf{A}\xi t = (I_n + \mathbf{A}t)\xi.$$

Now take this result and define it to be $X_1(t)$. Thus

$$X_1(t) = TX_0(t) = \xi + \int_0^t \mathbf{A}\ \xi\ ds = (I_n + \mathbf{A}t)\xi.$$

For the next approximation in our sequence, $X_1(t)$ is inserted into the right-hand side of (7.6), giving

$$X_2(t) = \xi + \int_0^t \mathbf{A}\ X_1(s)\ ds = \xi + \int_0^t \mathbf{A}\ [I_n + \mathbf{A}s]\xi\ ds = [I_n + \mathbf{A}t + \mathbf{A}^2 t^2/2]\xi.$$

This procedure is beginning to show some promise. In general, define

$$X_{m+1}(t) = TX_m(t) = \xi + \int_0^t \mathbf{A}\ X_m(s)\ ds, \quad m = 0, 1, 2, \ldots.$$

Then for any $m = 0, 1, 2, \ldots$,

$$X_m(t) = \left[I_n + \mathbf{A}t + \mathbf{A}^2 t^2/2 + \cdots + \frac{(\mathbf{A}t)^m}{m!}\right]\xi.$$

As m increases

$$X_m(t) \to X(t) = \left[\sum_{m=0}^{\infty} \frac{(\mathbf{A}t)^m}{m!}\right]\xi = e^{\mathbf{A}t}\xi, \quad t \in \mathcal{R},$$

which is the solution of (7.5).

7.2   Convergence of successive approximations

The construction of a sequence by using an iteration scheme of the form

$$X_{m+1} = TX_m$$

will often produce a sequence converging to a fixed point of T. Before examining this idea for the general integral equation (7.2), it helps to understand a bit about the convergence of a sequence of continuous functions. We begin by reviewing some of the material developed in section 5.2.

Suppose that I is a compact interval, and that $X(t) \in C(I, \mathcal{F}^n)$. The size of $X(t)$ is measured by

$$\|X\|_\infty = \max_{t \in I} |X(t)|.$$

Since the interval is compact and the function continuous, the maximum will exist. The distance between two functions is given by

$$d(X(t), Y(t)) = \max_{t \in I} |X(t) - Y(t)| = \|X - Y\|_\infty.$$

With this distance function $C(I, \mathcal{F}^n)$ is a metric space, and the usual ideas of convergence can be discussed. The sequence $\{X_m(t)\}$ converges to $X(t)$ if for any $\epsilon > 0$ there is an $N(\epsilon)$ such that $m \geq N(\epsilon)$ implies $d(X_m, X) < \epsilon$, or equivalently that $\|X_m - X\|_\infty < \epsilon$. A sequence $\{X_m\}$ is a *Cauchy sequence* in $C(I, \mathcal{F}^n)$ if for every $\epsilon > 0$ there is an $N(\epsilon)$ such that $k, m \geq N(\epsilon)$ implies $d(X_k, X_m) < \epsilon$, or equivalently $\|X_k(t) - X_m(t)\|_\infty < \epsilon$.

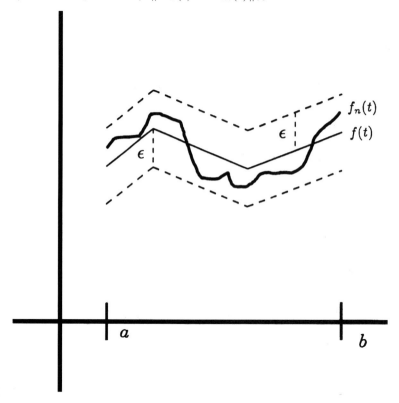

Figure 7.1: Functions satisfying $\|f_n - f\|_\infty < \epsilon$.

Consider for the moment the case in which $\{x_m(t)\}$ is a sequence of continuous scalar-valued functions, $x_m \in C(I, \mathcal{F})$. In this case the convergence just discussed is called uniform convergence. We say the sequence $\{x_m(t)\}$ *converges uniformly* to $x(t) \in C(I, \mathcal{F})$ if for any $\epsilon > 0$ there is an $N(\epsilon)$ such that $m \geq N(\epsilon)$ implies $|x_m(t) - x(t)| < \epsilon$ for all $t \in I$ (see Figure 7.1). A sequence $\{x_m\}$ is a *Cauchy sequence* in $C(I, \mathcal{F})$ if for every $\epsilon > 0$ there is an $N(\epsilon)$ such that $k, m \geq N$ implies $|x_k(t) - x_m(t)| < \epsilon$ for all $t \in I$.

The notion of uniform convergence should be contrasted with the weaker pointwise convergence. The sequence $\{x_m\}$ is said to *converge pointwise* to $x(t)$ (which may or may not be continuous) if for any $t_0 \in I$ and any $\epsilon > 0$ there is an $N(\epsilon, t_0)$ such that $m \geq N(\epsilon, t_0)$ implies $|x_m(t_0) - x(t_0)| < \epsilon$. As exercise 5 demonstrates, every uniformly convergent sequence converges pointwise, but the converse is generally false.

Recall that a *Banach space* is a vector space V with a norm, with the property that every Cauchy sequence in V converges to an element of V. The next result asserts that $C(I, \mathcal{F})$ with the norm $\| \; \|_\infty$ is a Banach space.

Theorem 7.1: *If $I = [a, b] \subset \mathcal{R}$ is compact, then a Cauchy sequence $x_m(t) \in C(I, \mathcal{F})$ converges uniformly to a continuous function $x(t) \in C(I, \mathcal{F})$.*

Proof: Let $x_k \in C(I, \mathcal{F})$, $k = 0, 1, 2, \ldots$, be a Cauchy sequence. Given any $\epsilon > 0$ there is an N such that

$$\|x_k - x_l\|_\infty < \frac{\epsilon}{2}, \quad k, l > N.$$

Hence

$$|x_k(t) - x_l(t)| \leq \|x_k - x_l\|_\infty < \frac{\epsilon}{2}, \quad k, l > N, \tag{7.7}$$

for all $t \in I$. This implies that $\{x_k(t)\}$ is a Cauchy sequence in $\mathcal{F}$ for each $t \in I$. Since $\mathcal{F}$ is complete, for each $t \in I$ there is an $x(t) \in \mathcal{F}$ such that $x_l(t) \to x(t)$ as $l \to \infty$. This defines a function $x(t)$. Letting $l \to \infty$ in (7.7) we see that

$$|x_k(t) - x(t)| \leq \frac{\epsilon}{2}, \quad k > N, \quad \text{all } t \in I,$$

and hence

$$\|x_k - x\|_\infty < \epsilon, \quad k > N,$$

or $\|x_k - x\|_\infty \to 0$ as $k \to \infty$.

It remains to show that x is continuous at each $c \in I$. Given any $\epsilon > 0$ there is an N_1 such that

$$|x_k(t) - x(t)| \leq \|x_k - x\|_\infty < \frac{\epsilon}{3}, \quad k > N_1.$$

Let k be fixed and satisfy $k > N_1$. The function x_k is continuous at c. Thus, for the given $\epsilon > 0$ there is a $\delta > 0$ such that

$$|x_k(t) - x_k(c)| < \frac{\epsilon}{3}, \quad |t - c| < \delta, \quad t \in I.$$

A typical application of the triangle inequality now shows that

$$|x(t) - x(c)| = |x(t) - x_k(t) + x_k(t) - x_k(c) + x_k(c) - x(c)|$$

$$\leq |x(t) - x_k(t)| + |x_k(t) - x_k(c)| + |x_k(c) - x(c)| < \frac{\epsilon}{3} + \frac{\epsilon}{3} + \frac{\epsilon}{3}$$

$$= \epsilon, \quad |t - c| < \delta, \quad t \in I,$$

so that x is continuous at c. $\square$

Corollary 7.2: *If $I = [a, b] \subset \mathcal{R}$ is compact, then $C(I, M_{mn}(\mathcal{F}))$ with the norm $\| \; \|_\infty$ given by*

$$\|\mathbf{A}\|_\infty = \sup\{|\mathbf{A}(t)| \mid t \in I\}, \quad \mathbf{A} \in C(I, M_{mn}(\mathcal{F})),$$

is a Banach space. In particular, $C(I, \mathcal{F}^n)$ with the norm $\| \ \|_\infty$ given by

$$\|X\|_\infty = \sup\{|X(t)| \mid t \in I\}, \quad X \in C(I, \mathcal{F}^n),$$

is a Banach space.

Proof: If $\mathbf{A} = (a_{ij}) \in C(I, M_{mn}(\mathcal{F}))$, then

$$|\mathbf{A}(t)| = \sum_{j=1}^{n} \sum_{i=1}^{m} |a_{ij}(t)|, \quad t \in I,$$

and this implies that

$$\|a_{ij}\|_\infty \le \|\mathbf{A}\|_\infty \le \sum_{j=1}^{n} \sum_{i=1}^{m} \|a_{ij}(t)\|_\infty.$$

If $\mathbf{A}_k = (a_{ij,k})$, $k = 0, 1, 2, \ldots$, is a Cauchy sequence in $C(I, M_{mn}(\mathcal{F}))$, then $\|\mathbf{A}_k - \mathbf{A}_l\|_\infty \to 0$ as $k, l \to \infty$. The inequality

$$\|a_{ij,k} - a_{ij,l}\|_\infty \le \|\mathbf{A}_k - \mathbf{A}_l\|_\infty$$

shows that $\|a_{ij,k} - a_{ij,l}\|_\infty \to 0$ as $k, l \to \infty$ for each i, j, and hence $\{a_{ij,k}\}$ is a Cauchy sequence in $C(I, F)$. Theorem 7.1 implies there is an $a_{ij} \in C(I, F)$ such that for each i, j,

$$\|a_{ij,k} - a_{ij}\|_\infty \to 0, \quad k \to \infty.$$

If $\mathbf{A} = (a_{ij}) \in C(I, M_{mn}(\mathcal{F}))$, then

$$\|\mathbf{A}_k - \mathbf{A}\|_\infty \le \sum_{j=1}^{n} \sum_{i=1}^{m} \|a_{ij,k} - a_{ij}\|_\infty \to 0, \quad k \to \infty. \quad \square$$

Returning to the initial value problem

$$X' = \mathbf{A}(t)X + B(t), \quad X(\tau) = \xi, \quad t \in I, \qquad (IVP)$$

the case where I is compact is considered first.

Theorem 7.3: *If $I = [a, b]$ is a compact interval and $\mathbf{A} \in C(I, M_n(\mathcal{F}))$, $B \in C(I, \mathcal{F}^n)$, then (IVP) has a unique solution X on I for every $\tau \in I$, $\xi \in \mathcal{F}^n$.*

Proof: $X(t)$ is a solution of (IVP) if and only if $X \in C(I, \mathcal{F}^n)$ and satisfies (7.2):

$$X(t) = \xi + \int_\tau^t [\mathbf{A}(s)X(s) + B(s)] \, ds, \quad t \in I.$$

The treatment of the example (7.5) is used as a model. Define

$$X_0(t) = \xi + \int_\tau^t B(s) \, ds, \quad t \in I,$$

and for $m > 0$ define

$$X_{m+1}(t) = \xi + \int_\tau^t [\mathbf{A}(s)X_m(s) + B(s)] \, ds, \quad t \in I.$$

There are four main steps in the proof. First we will establish an estimate for $|X_{m+1}(t) - X_m(t)|$. The second step will be to show that this estimate implies that $\{X_m(t)\}$ is a Cauchy sequence of continuous functions, which has a continuous limit $X(t)$ by Corollary 7.2. The third step will be to check that $X(t)$ satisfies the integral equation (7.2), and the final step is to demonstrate that $X(t)$ is the unique continuous solution of (7.2).

Subtraction of consecutive terms gives

$$X_1(t) - X_0(t) = \int_\tau^t \mathbf{A}(s)X_0(s) \, ds$$

and for any $m > 0$

$$X_{m+1}(t) - X_m(t) = \int_\tau^t \mathbf{A}(s)[X_m(s) - X_{m-1}(s)] \, ds, \quad t \in I.$$

Since the matrix norm $| \ |$ is submultiplicative (inequality (5.7)), we have the inequality

$$|X_1(t) - X_0(t)| \le \left| \int_\tau^t \mathbf{A}(s)X_0(s) \, ds \right| \le \int_\tau^t |\mathbf{A}(s)| \, |X_0(s)| \, ds$$

or

$$|X_1(t) - X_0(t)| \le \|\mathbf{A}\|_\infty \|X_0\|_\infty [t - \tau].$$

An application of the same idea for any $m > 0$ gives us

$$|X_{m+1}(t) - X_m(t)| \le \|\mathbf{A}\|_\infty \int_\tau^t |X_m(s) - X_{m-1}(s)| \, ds, \quad t \in I.$$

An induction argument can be used to show that for all $m \ge 0$ the inequality

$$|X_{m+1}(t) - X_m(t)| \le \|X_0\|_\infty \big[\|\mathbf{A}\|_\infty [t - \tau] \big]^{m+1} / (m+1)! \tag{7.8}$$

is valid. The case $m = 0$ is already established, as is the general estimate

$$|X_{k+1}(t) - X_k(t)| \le \|\mathbf{A}\|_\infty \int_\tau^t |X_k(s) - X_{k-1}(s)| \, ds, \quad t \in I.$$

If the hypothesis (7.8) holds for $m < k$, then the hypothesized estimate may be applied under the integral, yielding

$$|X_{k+1}(t) - X_k(t)| \le \|\mathbf{A}\|_\infty \int_\tau^t \|X_0(s)\|_\infty \big[\|\mathbf{A}\|_\infty [t - \tau] \big]^k / k! \, ds$$

$$\le \|X_0(s)\|_\infty \|\mathbf{A}\|_\infty^{k+1} \int_\tau^t [t - \tau]^k / k! \, ds = \|X_0(s)\|_\infty \|\mathbf{A}\|_\infty^{k+1} [t - \tau]^{k+1} / (k+1)!.$$

This establishes the desired inequality, completing step one.

The next step is to show that the sequence $X_m(t)$ is a Cauchy sequence. Suppose that $k < m$ and let $I = [a, b]$. To establish the desired inequality, note that

$$\|X_m - X_k\|_\infty = \|(X_m - X_{m-1}) + (X_{m-1} - X_{m-2}) + \cdots + (X_{k+1} - X_k)\|_\infty$$

$$\leq \sum_{j=k}^{m-1} \|X_{j+1} - X_j\|_\infty \leq \|X_0\|_\infty \sum_{j=k}^{m-1} (\|\mathbf{A}\|_\infty [b - \tau])^{j+1}/(j+1)!.$$

This last sum is part of the Taylor series for $\exp(\|\mathbf{A}\|_\infty [b - \tau])$. Since this Taylor series converges, we can make $\|X_m - X_k\|_\infty$ less than any $\epsilon > 0$ by taking k sufficiently large. This shows that the sequence $\{X_m(t)\}$ is a Cauchy sequence, which by Corollary 7.2 converges to a continuous function $X(t)$.

The third step is to show that $X(t)$ satisfies the equation

$$X(t) = \xi + \int_\tau^t [\mathbf{A}(s)X(s) + B(s)] \, ds, \quad t \in I.$$

We have

$$X(t) = \lim_{m \to \infty} X_{m+1}(t)$$

$$= \xi + \int_\tau^t B(s) \, ds + \lim_{m \to \infty} \int_\tau^t \mathbf{A}(s)X_m(s) \, ds, \quad t \in I.$$

Now

$$\lim_{m \to \infty} \int_\tau^t \mathbf{A}(s)X_m(s) \, ds$$

$$= \int_\tau^t \mathbf{A}(s)X(s) \, ds + \lim_{m \to \infty} \int_\tau^t \mathbf{A}(s)[X_m(s) - X(s)] \, ds, \quad t \in I.$$

Since $X_m(s) - X(s)$ converges uniformly to 0 and

$$\left\| \int_\tau^t \mathbf{A}(s)[X_m(s) - X(s)] \, ds \right\|_\infty \leq [b - \tau] \|\mathbf{A}\|_\infty \|X_m - X\|_\infty \to 0,$$

it follows that

$$\lim_{m \to \infty} \int_\tau^t \mathbf{A}(s)X_m(s) \, ds = \int_\tau^t \mathbf{A}(s)X(s) \, ds, \quad t \in I,$$

and X satisfies (7.2).

The final point is that this solution is unique. If $X(t)$ and $Y(t)$ both satisfied the integral equation (7.2), then subtraction would yield

$$X(t) - Y(t) = \int_\tau^t \mathbf{A}(s)[X(s) - Y(s)] \, ds.$$

In particular, $X(\tau) - Y(\tau) = 0$. Suppose there is a point $t_1 > \tau$ where $X(t_1) - Y(t_1) \neq 0$. Letting

$$\sigma = \inf\{s \mid X(t) - Y(t) \neq 0, \ s < t < t_1\},$$

we have $X(\sigma) - Y(\sigma) = 0$, but $X(t) - Y(t) \neq 0$ for $\sigma < t < t_1$, Notice too that

$$X(\sigma) - Y(\sigma) = 0 = \int_\tau^\sigma \mathbf{A}(s)[X(s) - Y(s)] \, ds,$$

so that

$$X(t) - Y(t) = \int_\sigma^t \mathbf{A}(s)[X(s) - Y(s)] \, ds, \quad \sigma \le t. \tag{7.9}$$

Let $\| \; \|_{\infty, J}$ denote the sup norm on $J = [\sigma, \sigma_1] \subset I$,

$$\|X\|_{\infty, J} = \sup\{|X(t)| \mid t \in J\}.$$

Taking norms of both sides of (7.9), we get the inequality

$$\|X - Y\|_{\infty, J} \le \|\mathbf{A}\|_\infty \|X - Y\|_{\infty, J}[\sigma_1 - \sigma].$$

If the length of J is positive but sufficiently small, then

$$K = \|\mathbf{A}\|_\infty [\sigma_1 - \sigma] < 1.$$

On such an interval J,

$$\|X - Y\|_{\infty, J} \le K \|X - Y\|_{\infty, J}, \quad K < 1,$$

which implies that

$$\|X - Y\|_{\infty, J} = 0.$$

This contradiction establishes that $X(t_1) - Y(t_1) = 0$ for all $t_1 > \tau \in I$. A similar argument applies for $t_1 < \tau$. $\square$

An alternative proof of the uniqueness of solutions to (7.2) may be based on Gronwall's inequality (see exercise 6(b)).

The existence and uniqueness theorem is next extended to an arbitrary real interval, which may be infinite in length.

Theorem 7.4 (existence and uniqueness theorem for linear systems): *Let I be any real interval and suppose that $\mathbf{A} \in C(I, M_n(\mathcal{F}))$, $B \in C(I, \mathcal{F}^n)$. Given any $\tau \in I$, $\xi \in \mathcal{F}^n$, there exists a unique solution X of (IVP) on I.*

Proof: Given any $t \in I$ let $J = [c, d]$ be any compact subinterval of I such that $\tau, t \in J$. By Theorem 7.3 there is a unique differentiable X_J on J such that

$$X_J'(s) = \mathbf{A}(s)X_J(s) + B(s), \quad X_J(\tau) = \xi, \quad s \in J.$$

Define $X(t) = X_J(t)$. If we choose any other $J_1 = [c_1, c_2] \subset I$ such that $\tau, t \in J_1$, then $J_1 \cap J$ is a compact interval containing τ, t, and the uniqueness result applied to this interval implies that $X_{J_1}(s) = X_J(s)$, $s \in J_1 \cap J$. In particular $X_{J_1}(t) = X_J(t)$, so that our definition of $X(t)$ is independent of the J chosen. Since X is differentiable on I and satisfies

$$X'(t) = \mathbf{A}(t)X(t) + B(t), \quad X(\tau) = \xi, \quad t \in I,$$

it is a solution of (IVP) on I. It is unique, for if Y is also a solution on I, then for any $t \in I$ there is a compact interval J containing τ, t, and the uniqueness result for J implies that $X(t) = Y(t)$. □

Before continuing the theoretical development, we look at another example. Consider the problem where $n = 1$:

$$x' = 3t^2 x, \quad x(0) = 1, \quad t \in \mathcal{R}.$$

The corresponding integral equation is

$$x(t) = 1 + \int_0^t 3s^2 x(s) \, ds = (Tx)(t), \quad t \in \mathcal{R}.$$

If $x_0(t) = 1$, then

$$x_{m+1}(t) = 1 + \int_0^t 3s^2 x_m(s) \, ds, \quad m = 0, 1, 2, \ldots.$$

Thus

$$x_1(t) = 1 + \int_0^t 3s^2 \, ds = 1 + t^3,$$

$$x_2(t) = 1 + \int_0^t 3s^2[1 + s^3] \, ds = 1 + t^3 + t^6/2,$$

$$x_3(t) = 1 + \int_0^t 3s^2[1 + s^3 + s^6/2] \, ds = 1 + t^3 + t^6/2 + t^9/6,$$

and an induction shows that

$$x_m = 1 + t^3 + \frac{(t^3)^2}{2} + \frac{(t^3)^3}{3!} + \cdots + \frac{(t^3)^m}{m!}.$$

We recognize $x_m(t)$ as a partial sum for the series expansion of the function

$$x(t) = e^{t^3}.$$

This series converges to $x(t)$ for all $t \in \mathcal{R}$, and the function $x(t)$ is the solution of the problem.

Taking a second look at the method of proof for Theorem 7.3, it is not difficult to see that any choice of the initial continuous function $X_0(t)$ will result in the same solution $X(t)$. Indeed the same basic inequality would hold

$$|X_{m+1}(t) - X_m(t)| \leq \|\mathbf{A}\|_\infty \int_\tau^t |X_m(s) - X_{m-1}(s)| \, ds, \quad m \geq 1, \quad t \in I.$$

The only distinction arises because of the initial difference $X_1(t) - X_0(t)$. The estimate resulting from the induction argument then becomes

$$|X_{m+1}(t) - X_m(t)| \leq \|X_1 - X_0\|_\infty \left[\|\mathbf{A}\|_\infty [t - \tau]\right]^m /m!. \tag{7.10}$$

The rest of the argument proceeds as before, yielding the unique solution $X(t)$ of (7.2).

If (IVP) is considered on any interval I, it is also possible to estimate the distance between $X_m(t)$ and $X(t)$ on any compact subinterval $J = [a, b] \subset I$ containing τ. For all $k > m$,

$$\|X - X_m\|_{\infty, J} \leq \|X - X_k\|_{\infty, J} + \|X_k - X_m\|_{\infty, J}$$

$$\leq \|X - X_k\|_{\infty, J} + \|(X_k - X_{k-1}) + (X_{k-1} - X_{k-2}) + \cdots + (X_{m+1} - X_m)\|_{\infty, J}$$

and, using the triangle inequality and taking the limit as $k \to \infty$, (7.10) implies that

$$\|X - X_m\|_{\infty, J} \leq \sum_{k=m}^{\infty} \|X_{k+1} - X_k\|_{\infty, J} \tag{7.11}$$

$$\leq \|X_1 - X_0\|_{\infty, J} \sum_{k=m}^{\infty} \left[\|\mathbf{A}\|_{\infty, J} [b - \tau] \right]^m / m!.$$

Of course this last series is again the tail of the series for $\exp(\|\mathbf{A}\|_{\infty, J}[b - \tau])$.

Thus (7.11) implies that $X_m \to X$ in the sup norm on each J. We summarize in the following theorem.

Theorem 7.5: *Define successive approximations by*

$$X_{m+1}(t) = \xi + \int_{\tau}^{t} [\mathbf{A}(s) X_m(s) + B(s)] \, ds, \quad t \in I,$$

where $X_0 \in C(I, \mathcal{F}^n)$ is arbitrary. If $X(t)$ is the solution of (IVP) on I, then $X_m \to X$ uniformly,

$$\|X - X_m\|_{\infty, J} \to 0, \quad k \to \infty,$$

on each compact interval $J \subset I$ containing τ.

7.3 Continuity of solutions

Returning to the situation in Theorem 7.3, where $I = [a, b]$ is a compact interval, the solution $X(t)$ of the initial value problem

$$X' = \mathbf{A}(t) X + B(t), \quad X(\tau) = \xi, \quad t \in I, \tag{IVP}$$

evidently depends on $\tau \in I$, $\xi \in \mathcal{F}^n$, $\mathbf{A} \in C(I, M_n(\mathcal{F}))$, and $B \in C(I, \mathcal{F}^n)$. The main result of this section asserts that for any $t \in I$, the value $X(t)$ is a continuous function of these variables. The analysis of this dependence begins with an estimate for $\|X\|_{\infty}$ which is derived using the method of proof for Theorem 7.3.

It is convenient to begin the successive approximations with

$$X_0(t) = \xi + \int_{\tau}^{t} B(s) \, ds.$$

The solution

$$X(t) = \lim_{k \to \infty} X_k(t)$$

then satisfies the estimate

$$\|X\|_\infty = \|\lim_{k \to \infty} X_k\|_\infty = \left\| \lim_{k \to \infty} \left[X_0(t) + \sum_{m=0}^{k-1} (X_{m+1}(t) - X_m(t)) \right] \right\|_\infty$$

$$\leq \|X_0\|_\infty + \sum_{m=0}^{\infty} \|X_{m+1} - X_m\|_\infty.$$

The inequality (7.8) can now be applied, giving

$$\|X\|_\infty \leq \|X_0\|_\infty + \|X_0\|_\infty \sum_{m=0}^{\infty} \frac{\|\mathbf{A}\|_\infty^{m+1} [b - \tau]^{m+1}}{(m+1)!}$$

$$= \|X_0(t)\|_\infty \exp(\|\mathbf{A}\|_\infty [b - \tau]).$$

Since

$$\|X_0(t)\|_\infty = \left\| \xi + \int_\tau^t B(s)\, ds \right\|_\infty$$

$$\leq |\xi| + |b - a| \|B\|_\infty,$$

the desired estimate for $\|X\|_\infty$ is

$$\|X\|_\infty \leq [|\xi| + |b - a| \|B\|_\infty] \, \exp(\|\mathbf{A}\|_\infty [b - a]). \tag{7.12}$$

The simple estimate (7.12) can be used to show that X is jointly continuous in all these variables. Thus a small change in $t, \mathbf{A}, B, \tau, \xi$ will produce a small change in X. If we denote the solution of (IVP) at t by $X(t, \mathbf{A}, B, \tau, \xi)$, then Theorem 7.6 gives a precise meaning to the statement that

$$X(s, \mathbf{C}, D, \sigma, \eta) \to X(t, \mathbf{A}, B, \tau, \xi),$$

as

$$(s, \mathbf{C}, D, \sigma, \eta) \to (t, \mathbf{A}, B, \tau, \xi);$$

that is, X is continuous at $(t, \mathbf{A}, B, \tau, \xi)$.

Theorem 7.6: *Let I be a compact interval, $\mathbf{A}, \mathbf{C} \in C(I, M_n(\mathcal{F}))$, $B, D \in C(I, \mathcal{F}^n)$, $\tau, \sigma \in I$, $\xi, \eta \in \mathcal{F}^n$. Suppose X is the solution of*

$$X' = \mathbf{A}(t)X + B(t), \quad X(\tau) = \xi, \quad t \in I.$$

Given any $t \in I$ and $\epsilon > 0$, there exists a $\delta > 0$ such that if Y is the solution of

$$Y' = \mathbf{C}(t)Y + D(t), \quad y(\sigma) = \eta, \quad t \in I,$$

and

$$|s - t| < \delta, \quad \|\mathbf{C} - \mathbf{A}\|_\infty < \delta, \quad \|D - B\|_\infty < \delta, \tag{7.13}$$

$$|\sigma - \tau| < \delta, \quad |\eta - \xi| < \delta,$$

then

$$|Y(s) - X(t)| < \epsilon. \tag{7.14}$$

Proof: Subtracting the equations for $X(t)$ and $Y(t)$ gives

$$(Y - X)' = \mathbf{C}(t)(Y - X) + (\mathbf{C}(t) - \mathbf{A}(t))X + D(t) - B(t).$$

Thus if $Z = Y - X$ then Z satisfies the initial value problem

$$Z' = \mathbf{C}(t)Z + E(t), \quad Z(\sigma) = \eta - X(\sigma),$$

where

$$E(t) = (\mathbf{C}(t) - \mathbf{A}(t))X(t) + D(t) - B(t).$$

We may apply the estimate (7.12) to Z and obtain

$$\|Y - X\|_\infty = \|Z\|_\infty \le [|Z(\sigma)| + |b - a| \|E\|_\infty] \ \exp(\|\mathbf{C}\|_\infty [b - a]). \tag{7.15}$$

Let $\epsilon > 0$ be given. First observe that

$$|Y(s) - X(t)| \le |Y(s) - X(s)| + |X(s) - X(t)|$$

$$\le \|Y - X\|_\infty + |X(s) - X(t)|. \tag{7.16}$$

Since X is continuous at t, given any $\epsilon > 0$ there is a $\delta_1 > 0$ such that $|s - t| < \delta_1$ implies

$$|X(s) - X(t)| < \frac{\epsilon}{3}. \tag{7.17}$$

Also

$$|Z(\sigma)| = |\eta - X(\sigma)| \le |\eta - \xi| + |X(\tau) - X(\sigma)|.$$

Since X is continuous at τ, given any $\epsilon > 0$ there is a $\delta_2 > 0$ such that

$$|\eta - \xi| < \delta_2, \quad |\tau - \sigma| < \delta_2$$

implies

$$|Z(\sigma)| \ \exp(\|\mathbf{C}\|_\infty [b - a]) < \frac{\epsilon}{3}. \tag{7.18}$$

Finally, since

$$E(t) = (\mathbf{C}(t) - \mathbf{A}(t))X(t) + D(t) - B(t),$$

there is a δ_3 such that

$$\|\mathbf{C} - \mathbf{A}\|_\infty < \delta_3, \quad \|D - B\|_\infty < \delta_3$$

implies

$$|b - a| \|E\|_\infty \exp(\|\mathbf{C}\|_\infty [b - a]) < \frac{\epsilon}{3}. \tag{7.19}$$

Now choose $\delta > 0$ so that $\delta = \min(\delta_1, \delta_2, \delta_3)$. Then if (7.13) is valid for this δ, (7.14) follows from (7.15)–(7.19). $\square$

7.4 More general linear equations

The initial value problem

$$X' = \mathbf{A}(t)X + B(t), \quad X(\tau) = \xi, \quad t \in I, \qquad (IVP)$$

was solved by using the equivalent integral equation

$$X(t) = \xi + \int_\tau^t [\mathbf{A}(s)X(s) + B(s)]\, ds, \quad t \in I. \qquad (7.20)$$

Now this equation (7.20) makes sense for $\mathbf{A}$ and B which may be discontinuous, and it is often necessary to consider such functions in practice. Then a continuous X satisfying (7.20) would satisfy (IVP) at those t where $\mathbf{A}$, B are continuous.

An important instance of this situation is when $\mathbf{A}$ and B are *piecewise continuous* on a compact interval $I = [a, b]$. This means that $\mathbf{A}$, B are continuous on I except possibly at a finite number of points $t_1, \ldots, t_p$, where they may have simple discontinuities. Thus they have left and right limits at t_j if $a < t_j < b$, right limits at a, and left limits at b. That is, if $a < t_j < b$ there exist $\mathbf{L}_j^+, \mathbf{L}_j^- \in M_n(\mathcal{F})$, and $l_j^+, l_j^- \in \mathcal{F}^n$ such that

$$|\mathbf{A}(t) - \mathbf{L}_j^+| \to 0, \quad |B(t) - l_j^+| \to 0, \quad 0 < t - t_j \to 0,$$

$$|\mathbf{A}(t) - \mathbf{L}_j^-| \to 0, \quad |B(t) - l_j^-| \to 0, \quad 0 < t_j - t \to 0.$$

We also write

$$\mathbf{A}(t) \to \mathbf{A}(t_j\pm) = \mathbf{L}_j^\pm, \quad t \to t_j\pm,$$

$$B(t) \to B(t_j\pm) = l_j^\pm, \quad t \to t_j \pm.$$

Let $PC(I, M_n(\mathcal{F}))$ denote the set of all piecewise continuous $\mathbf{A} : I \to M_n(\mathcal{F})$, and let $PC(I, \mathcal{F}^n)$ be the set of piecewise continuous $B : I \to \mathcal{F}^n$. It is clear that

$$C(I, M_n(\mathcal{F})) \subset PC(I, M_n(\mathcal{F})),$$

$$C(I, \mathcal{F}^n) \subset PC(I, \mathcal{F}^n).$$

All piecewise continuous $\mathbf{A}$, B are bounded on I,

$$\|\mathbf{A}\|_\infty = \sup_{t \in I} |\mathbf{A}(t)| < \infty, \quad \|B\|_\infty = \sup_{t \in I} |B(t)| < \infty.$$

Such $\mathbf{A}$ and B are integrable and the integral equation (7.20) makes sense for any $X \in C(I, \mathcal{F}^n)$. Of course, if $X \in C(I, \mathcal{F}^n)$ satisfies (7.20) in this case, then $X(\tau) = \xi$, and

$$X'(t) = \mathbf{A}(t)X(t) + B(t),$$

for $t \in I \setminus \{t_1, \ldots, t_p\}$. In case $\mathbf{A} \in PC(I, M_n(\mathcal{F}))$, $B \in PC(I, \mathcal{F}^n)$, we are thus motivated to define a *solution* X of (IVP) to be any $X \in C(I, \mathcal{F}^n)$ which satisfies (7.20).

Since for $X \in C(I, \mathcal{F}^n)$

$$(TX)(t) = \xi + \int_{\tau}^{t} [\mathbf{A}(s)X(s) + B(s)] \, ds, \quad t \in I,$$

it is clear that $TX \in C(I, \mathcal{F}^n)$. Then the argument in the proof of Theorem 7.3 can be applied to this new situation, producing a slight generalization.

Theorem 7.7: *If $I = [a, b]$ is a compact interval and $\mathbf{A} \in PC(I, M_n(\mathcal{F}))$, $B \in PC(I, \mathcal{F}^n)$, then the integral equation (7.20) has a unique solution $X \in C(I, \mathcal{F}^n)$, and this satisfies $X(\tau) = \xi$. Moreover*

$$X'(t) = \mathbf{A}(t)X(t) + B(t)$$

at those $t \in I$ where $\mathbf{A}$, B are continuous.

A simple example when $n = 1$ is

$$x' = -3x + b(t), \quad x(0) = 1, \quad t \in [0, 2],$$

where b is a step function with a unit jump at $t = 1$,

$$b(t) = \begin{cases} 0, & 0 \le t < 1, \\ 1, & 1 \le t \le 2. \end{cases}$$

The relevant integral equation is

$$x(t) = 1 + \int_{0}^{t} [-3x(s) + b(s)] \, ds, \quad t \in [0, 2],$$

so that

$$x(t) = \begin{cases} 1 - 3\int_{0}^{t} x(s) \, ds, & 0 \le t < 1, \\ t - 3\int_{0}^{t} x(s) \, ds, & 1 \le t \le 2. \end{cases}$$

The usual variation of parameters formula can be used to solve this problem,

$$x(t) = e^{-3t} + e^{-3t} \int_{0}^{t} e^{3s} b(s) \, ds,$$

yielding

$$x(t) = \begin{cases} e^{-3t}, & 0 \le t < 1, \\ e^{-3t} + \frac{1}{3}[1 - e^{3-3t}], & 1 \le t \le 2. \end{cases}$$

Note that x is indeed continuous at $t = 1$ but that x' is discontinuous there, since

$$x'(1+) = -3e^3 + 1, \quad x'(1-) = -3e^{-3},$$

so that x' has a unit jump at $t = 1$,

$$x'(1+) - x'(1-) = 1.$$

If I is any interval, say that a matrix-valued function $\mathbf{A} : I \to M_{mn}(\mathcal{F})$ is *piecewise continuous* on I if $\mathbf{A} \in PC(J, M_{mn}(\mathcal{F}))$ for every compact subinterval $J = [c, d] \subset I$, and denote the set of all such $\mathbf{A}$ as $PC(I, M_{mn}(\mathcal{F}))$. The proof of Theorem 7.4 carries over to give the following result.

Theorem 7.8: *Let I be any real interval and suppose $\mathbf{A} \in PC(I, M_n(\mathcal{F}))$, $B \in PC(I, \mathcal{F}^n)$. Given any $\tau \in I$, $\xi \in \mathcal{F}^n$, there exists a unique solution $X \in C(I, \mathcal{F}^n)$ of the integral equation (7.20) on I and this satisfies $X(\tau) = \xi$ and*

$$X'(t) = \mathbf{A}(t)X(t) + B(t)$$

at those $t \in I$ where $\mathbf{A}$, B are continuous.

Since the solution of the integral equation (7.20) is obtained as the limit of successive approximations, the estimate given in Theorem 7.3 is valid for the more general case considered in Theorem 7.7. Moreover, the result of Theorem 7.6 also carries over to this case, so that the solution X is a continuous function of t, $\mathbf{A}$, B, τ, ξ.

7.5   Estimates for second-order equations

The idea of recasting a differential equation as an integral equation can often be used to extract information about the solutions of the differential equation. In order to make explicit use of special features of a class of differential equations, the form of the integral equation may be somewhat different from that obtained by integration of first-order linear systems. This idea may be applied to the second-order scalar equations with parameter λ,

$$x'' + \lambda x = p(t)x, \quad x(0) = a, \quad x'(0) = b, \quad \lambda > 0, \quad t \geq 0. \tag{7.21}$$

This equation arises in a variety of problems in physics.

If the function $p(t)$ vanishes for $t > T$, then for large t the solution of the initial value problem will be a function of the form

$$x(t) = A \cos(\sqrt{\lambda}t) + B \sin(\sqrt{\lambda}t), \quad t > T. \tag{7.22}$$

It is reasonable to expect that if $p(t)$ exhibits sufficiently rapid decay as $t \to \infty$, then the solution of the initial value problem will "look like" one of the solutions (7.22). The problem is to make this idea precise.

For the moment we focus on the nonhomogeneous equation

$$x'' + \lambda x = f(t), \quad x(0) = a, \quad x'(0) = b, \quad t \in \mathcal{R}.$$

Since the homogeneous equation

$$x'' + \lambda x = 0$$

has a basis of solutions

$$x_1(t) = \cos(\sqrt{\lambda}t), \quad x_2(t) = \sin(\sqrt{\lambda}t)/\sqrt{\lambda}$$

satisfying

$$x_1(0) = 1, \quad x_1'(0) = 0,$$
$$x_2(0) = 0, \quad x_2'(0) = 1,$$

the variation of parameters method, specialized to second-order scalar equations as in Chapter 2, exercise 25, gives the following formula for $x(t)$:

$$x(t) = a \ \cos(\sqrt{\lambda}t) + b \ \frac{\sin(\sqrt{\lambda}t)}{\sqrt{\lambda}}$$

$$+ \int_0^t \left[\cos(\sqrt{\lambda}s) \frac{\sin(\sqrt{\lambda}t)}{\sqrt{\lambda}} - \cos(\sqrt{\lambda}t) \frac{\sin(\sqrt{\lambda}s)}{\sqrt{\lambda}} \right] f(s) \ ds.$$

Applying the trigonometric identity

$$\sin(s)\cos(t) - \cos(s)\sin(t) = \sin(s - t), \tag{7.23}$$

the formula for $x(t)$ assumes the simpler appearance

$$x(t) = a \ \cos(\sqrt{\lambda}t) + b \ \frac{\sin(\sqrt{\lambda}t)}{\sqrt{\lambda}} + \int_0^t \frac{\sin(\sqrt{\lambda}[t - s])}{\sqrt{\lambda}} f(s) \ ds. \tag{7.24}$$

Two ideas now come into play. The first idea is to use (7.24) to obtain an integral equation for the solution of (7.21). Formally, the solution $x(t)$ of (7.21) should satisfy

$$x(t) = a \ \cos(\sqrt{\lambda}t) + b \ \frac{\sin(\sqrt{\lambda}t)}{\sqrt{\lambda}} + \int_0^t \frac{\sin(\sqrt{\lambda}[t - s])}{\sqrt{\lambda}} p(s)x(s) \ ds. \tag{7.25}$$

One can check by differentiation that if $x(t)$ satisfies (7.25), then $x(t)$ satisfies the initial value problem (7.21). What needs to be shown is that there is a solution to this integral equation.

Theorem 7.9: *If $p(t)$ is continuous, then the integral equation (7.25) has a unique continuous solution satisfying (7.21). For fixed a, b, the solution $x(t, \lambda)$ is bounded in any set of the form*

$$(t, \lambda) \in [0, T] \times [\lambda_0, \infty), \quad T > 0, \quad \lambda_0 > 0.$$

Proof: We sketch the ideas of the proof, which are much the same as the arguments for Theorem 7.3. The remaining details are left as an exercise.

First, set up the iteration scheme

$$x_0(t) = a \ \cos(\sqrt{\lambda}t) + b \ \frac{\sin(\sqrt{\lambda}t)}{\sqrt{\lambda}}$$

and, for $m \geq 0$,

$$x_{m+1}(t) = x_0(t) + \int_0^t \frac{\sin(\sqrt{\lambda}[t - s])}{\sqrt{\lambda}} p(s)x_m(s) \ ds.$$

Observe that for $m \geq 1$

$$|x_{m+1}(t) - x_m(t)| \leq \frac{1}{\sqrt{\lambda}} \int_0^t |p(s)| \, |x_m(s) - x_{m-1}(s)| \, ds.$$

Let $P(t) = \max_{s \leq t} |p(s)|$. By induction we establish the inequalities

$$|x_m(t) - x_{m-1}(t)| \leq \|x_0\|_\infty \Big(\frac{P(t)}{\sqrt{\lambda}}\Big)^m \frac{t^m}{m!}, \quad m \geq 1.$$

Thus for $k < m$,

$$|x_m(t) - x_k(t)| = |[x_m(t) - x_{m-1}(t)] + [x_{m-1}(t) - x_{m-2}(t)] + \cdots + [x_{k+1}(t) - x_k(t)]|$$

$$\leq |x_m(t) - x_{m-1}(t)| + |x_{m-1}(t) - x_{m-2}(t)| + \cdots + |x_{k+1}(t) - x_k(t)|$$

$$\leq \|x_0\|_\infty \sum_{j=k+1}^m \Big(\frac{P(t)}{\sqrt{\lambda}}\Big)^j \frac{t^j}{j!}.$$

This is again a portion of the convergent series for $\exp(P(t)t/\sqrt{\lambda})$. It follows that $x_m(t)$ is a Cauchy sequence and hence converges uniformly on any compact interval to a continuous function $x(t)$, which is a solution of the integral equation (7.23).

Suppose that $T > 0$ and $\lambda_0 > 0$. Since $x(t, \lambda) = \lim x_m(t)$, the estimates above yield

$$|x(t, \lambda)| = \lim_{m \to \infty} |x_m(t)| \leq |x_0| + \sum_{m=0}^\infty |x_{m+1}(t) - x_m(t)|$$

$$\leq \|x_0\|_\infty \exp(TP(T)/\sqrt{\lambda_0}),$$

which establishes the bound on $|x(t, \lambda)|$. $\square$

So far no serious constraints have been imposed on $p(t)$, but it should be "small" for large t if $x(t)$ is to look like a trigonometric function. The smallness is part of the second idea. The representation (7.24) is awkward if we want to know about the behavior of the solution $x(t)$ for large values of t. At least formally,

$$\int_0^t \frac{\sin(\sqrt{\lambda}[t - s])}{\sqrt{\lambda}} p(s) x(s) \, ds$$

$$= \int_0^\infty \frac{\sin(\sqrt{\lambda}[t - s])}{\sqrt{\lambda}} p(s) x(s) \, ds - \int_t^\infty \frac{\sin(\sqrt{\lambda}[t - s])}{\sqrt{\lambda}} p(s) x(s) \, ds.$$

Moreover, after unwinding $\sin(\sqrt{\lambda}[t-s])$ with the trigonometric identity (7.23), we see that formally

$$\int_0^\infty \frac{\sin(\sqrt{\lambda}[t - s])}{\sqrt{\lambda}} p(s) x(s) \, ds = A_1 \cos(\sqrt{\lambda}t) + B_1 \frac{\sin(\sqrt{\lambda}t)}{\sqrt{\lambda}}.$$

This suggests rewriting (7.24) as

$$x(t) = A \, \cos(\sqrt{\lambda}t) + B\frac{\sin(\sqrt{\lambda}t)}{\sqrt{\lambda}} - \int_t^\infty \frac{\sin(\sqrt{\lambda}[t-s])}{\sqrt{\lambda}}p(s)x(s) \, ds. \quad (7.26)$$

There are two problems with this idea so far, both connected to the existence of the integral

$$\int_0^\infty \frac{\sin(\sqrt{\lambda}[t-s])}{\sqrt{\lambda}}p(s)x(s) \, ds.$$

The first problem is that nothing has been said about the growth of $p(t)$, and the second is that the growth of $x(t)$ is not well understood. Let's try to take care of both at once.

Theorem 7.10: *Suppose that $p(t)$ is continuous and*

$$\int_0^\infty |p(t)| \, dt < \infty.$$

Then the solution $x(t)$ of (7.21) can be written in the form

$$x(t) = A \, \cos(\sqrt{\lambda}t) + B\frac{\sin(\sqrt{\lambda}t)}{\sqrt{\lambda}} + \Xi(t),$$

where $\Xi(t) \to 0$ as $t \to \infty$. In particular, every solution $x(t)$ is bounded for all $t \geq 0$.

Proof: Following the previous pattern, an iteration scheme is set up, this time with

$$x_0(t) = A \, \cos(\sqrt{\lambda}t) + B\frac{\sin(\sqrt{\lambda}t)}{\sqrt{\lambda}}$$

and

$$x_{m+1}(t) = x_0(t) - \int_t^\infty \frac{\sin(\sqrt{\lambda}[t-s])}{\sqrt{\lambda}}p(s)x_m(s) \, ds.$$

The critical estimate again comes from

$$|x_{m+1}(t) - x_m(t)| \leq \int_t^\infty \left| \frac{\sin(\sqrt{\lambda}[t-s])}{\sqrt{\lambda}}p(s)[x_m(s) - x_{m-1}(s)] \right| \, ds$$

$$\leq \frac{1}{\sqrt{\lambda}} \int_t^\infty |p(s)| \, |x_m(s) - x_{m-1}(s)| \, ds.$$

Looking for an induction hypothesis we evaluate

$$|x_1(t) - x_0(t)| \leq \frac{1}{\sqrt{\lambda}} \int_t^\infty |p(s)| \, |x_0(s)| \, ds.$$

Now for each fixed $\lambda > 0$, the function $x_0(s)$ is bounded, satisfying

$$|x_0(s)| \leq |A| + \frac{|B|}{\sqrt{\lambda}}.$$

Thus

$$|x_1(t) - x_0(t)| \leq \frac{\|x_0\|_\infty}{\sqrt{\lambda}} \int_t^\infty |p(s)| \, ds.$$

Let

$$K(t) = \frac{1}{\sqrt{\lambda}} \int_t^\infty |p(s)| \, ds.$$

By assumption $K(t) \to 0$ as $t \to \infty$. In particular for T large enough

$$K(t) = \frac{1}{\sqrt{\lambda}} \int_t^\infty |p(s)| \, ds \leq 1/2, \quad t \geq T.$$

The induction hypothesis

$$|x_m(t) - x_{m-1}(t)| \leq \|x_0\| 2^{-m}, \quad t > T,$$

is now easily verified. The estimate

$$\sum_{m=1}^\infty |x_m(t) - x_{m-1}(t)| \leq \|x_0\| \sum_{m=1}^\infty 2^{-m} = \|x_0\|$$

implies that the sequence $x_m(t)$ converges uniformly to a continuous function $x(t)$ as long as $t \geq T$.

The calculation needed to show that $x(t)$ satisfies the equation (7.21) for $t \geq T$ is left as an exercise. But what happens for $t < T$? The answer is that nothing unusual happens. Suppose we select two linearly independent solutions $x(t), y(t)$ of (7.24) by taking $A = 1, B = 0$ for $x(t)$ and $A = 0, B = 1$ for $y(t)$. These solutions are still linearly independent on the interval $[T, T+1]$. But there are two solutions satisfying some initial conditions at $t = 0$ which agree with these solutions on $[T, T + 1]$. It follows that any solution of the initial value problem (7.21) can be represented as a solution of (7.26). □

7.6   Notes

A brief discussion of the history of existence theorems for ordinary differential equations can be found in [12, pp. 717–721].

The ideas introduced in section 7.5 appear in a variety of contexts. A number of applications may be found in [3, e.g., p. 92 (Theorem 8.1)] and various exercises at the end of Chapter 3, as well as p. 255 (exercise 4) [3]. Similar methods for equations with analytic coefficients are used in [8, section 5.6].

7.7   Exercises

1. Treat the initial value problem

$$x' = x - \frac{t^2}{2}, \quad x(0) = 1,$$

with the successive approximation scheme, computing the first four iterates explicitly. Try the distinct initial functions: (i) $x_0(t) = 1$, (ii) $x_0(t) = t$.

2. Consider the equation

$$X' = \mathbf{A}X, \quad X(0) = \xi, \quad \mathbf{A} = \begin{pmatrix} 0 & 1 \\ -1 & 0 \end{pmatrix}.$$

Following the treatment of equation (7.5), compute the first few successive approximations explicitly.

3. Build a sequence of continuous functions $\{f_n(t)\}$ which converge pointwise on $I = [-1, 1]$ to the discontinuous function

$$f(t) = 1/t \quad \text{if} \quad 0 < t \le 1 \quad \text{and} \quad f(t) = 0 \quad \text{if} \quad -1 \le t \le 0.$$

4. Prove that the sequence of functions $f_n(x) = x + x^2 \sin(n^2 x)/n$ converges uniformly to $f(x) = x$ on the interval $I = [0, 1]$.

5. Prove that if the sequence of functions $f_n(t)$ converges uniformly to $f(t)$, then $f_n(t)$ converges pointwise to $f(t)$. Give an example that shows that the converse is generally false.

6. The following inequalities are variations on a theme. They can be used to give bounds on the growth of solutions to differential equations which have been recast as integral equations. Parts (b) and (c) are usually referred to as *Gronwall's inequality*. Throughout the exercise $v(t), u(t)$ are continuous real-valued functions.

(a) Suppose that $v(t) > 0$ is continuously differentiable. Show that if

$$v'(t) \le v(t)u(t), \quad t \ge a,$$

then

$$v(t) \le v(a) \exp\left(\int_a^t u(s) \, ds \right).$$

(Hint: What is $[\log(v(t))]'$?)

(b) Suppose that $v(t)$ is continuous, that $C \in \mathcal{R}$, and that $C, u(t), v(t) \ge 0$. Show that

$$v(t) \le C + \int_a^t u(s)v(s) \, ds \quad \text{implies} \quad v(t) \le C \exp\left(\int_a^t u(s) \, ds \right).$$

(Hint: Define

$$r(t) = C + \int_a^t u(s)v(s) \, ds$$

and show that $r'(t) \le u(t)r(t)$.)

(c) Suppose that $v(t), w(t)$ are continuous and that $u(t) \ge 0$. Show that

$$v(t) \le w(t) + \int_a^t u(s)v(s) \, ds$$

implies

$$v(t) \le w(t) + \int_a^t u(s)w(s) \exp\left(\int_s^t u(z)\,dz\right)\,ds.$$

(Hint: Define

$$r(t) = \int_a^t u(s)v(s)\,ds;$$

what can be said about $r' - u(t)r$?)

7. Use Gronwall's inequality to give an alternate proof of uniqueness in Theorem 7.3.

8. Suppose that we have an initial value problem of the form

$$X' = -X + \mathbf{A}(t)X, \quad X(0) = \xi, \quad t \in \mathcal{R}.$$

(a) Show that the equation can be rewritten as

$$[e^t X(t)]' = e^t \mathbf{A}(t)X(t).$$

(b) Show that the equivalent integral equation is

$$e^t X(t) = \xi + \int_0^t \mathbf{A}(s)e^s X(s)\,ds.$$

(c) This integral equation implies that the solution $X(t)$ satisfies the inequality

$$|e^t X(t)| \le |\xi| + \int_0^t |\mathbf{A}(s)||e^s X(s)|\,ds.$$

Use Gronwall's inequality to show that if

$$\int_0^t |\mathbf{A}(s)|\,ds \le K + (1 - \epsilon)t, \quad K, \epsilon > 0,$$

then

$$\lim_{t \to \infty} |X(t)| = 0.$$

(d) Try to generalize this result.

9. Let V be a metric space with the metric d. Recall that V is said to be complete if every Cauchy sequence $\{x_k\}$, $x_k \in V$, $k \in Z^+$ has a limit $x \in V$. A function $f : V \to V$ is said to be a *contraction* if there exists a constant K, $0 \le K < 1$, such that

$$d(f(x), f(y)) \le Kd(x, y) \quad \text{for all } x, y \in V.$$

A point $x \in V$ is said to be a *fixed point* of f if $f(x) = x$.

(a) With the hypotheses of Theorem 7.3, show that if the length of I is sufficiently small, then the transformation

$$TX(t) = \xi + \int_\tau^t \mathbf{A}(s)X(s)\,ds + \int_\tau^t B(s)\,ds$$

is a contraction on $C(I, \mathcal{F}^n)$ with the metric $d(X(t), Y(t)) = \|X - Y\|_\infty$.

(b) Prove that

if V is a nonempty complete metric space and $f : V \to V$ is a contraction, then f has a unique fixed point $x = f(x)$.

(Hint: Starting with any point $x_0 \in V$, define a sequence of iterates. Show that the sequence converges by estimating $\sum_m d(x_m, x_{m+1})$.)

10. Suppose that

$$a(t) = \begin{array}{ll} 0, & 0 \le t < 1, \\ 1, & 1 \le t < 2, \\ 0, & 2 \le t < \infty. \end{array}$$

(a) Find the solution of the initial value problem

$$x' = a(t)x, \quad x(0) = 1, \quad t \in \mathcal{R}.$$

(b) Find the solution of the initial value problem

$$x'' = a(t)x, \quad x(0) = c_1, \quad x'(0) = c_2, \quad t \in \mathcal{R}.$$

(c) Consider the initial value problem

$$x'' = b(t)x, \quad x(0) = c_1, \quad x'(0) = c_2, \quad t \in I,$$

where $b(t)$ is piecewise continuous on I. Show that the solution $x(t)$ has a continuous derivative on I and a second derivative satisfying the equation wherever $b(t)$ is continuous.

11. Verify that the equation

$$x'' + \lambda x = \frac{2x}{(t+1)^2}, \quad x(0) = a, \quad x'(0) = b, \quad \lambda > 0, \quad t \ge 0, \qquad (7.27)$$

satisfies the hypotheses of Theorem 7.10. Verify that the functions

$$x_1(t) = \frac{\sin(\sqrt{\lambda}[t+1])}{\sqrt{\lambda}[t+1]} - \cos(\sqrt{\lambda}[t+1]), \quad x_2(t) = \sin(\sqrt{\lambda}[t+1]) + \frac{\cos(\sqrt{\lambda}[t+1])}{\sqrt{\lambda}[t+1]}$$

give a basis for (7.27). Express the values of the coefficients $A(\lambda)$, $B(\lambda)$ in the statement of Theorem 7.10 in terms of the initial data $x(0), x'(0)$ of a solution to (7.27).

12. Fill in the missing details in the proof of Theorem 7.9, including the reason why the solution is unique.

13. At the end of the proof of Theorem 7.10 the solution $x(t)$ of (7.26), which existed for $t > T$, was extended to all $t \ge 0$ using a solution of the equation (7.21) on the interval $[0, T+1]$. Show that this extended function satisfies (7.26) on $[0, \infty)$ and that it satisfies the equation (7.21) for $t > 0$.

Chapter 8

Eigenvalue Problems

8.1 Introduction

The problems studied so far have concerned the existence, uniqueness, and properties of solutions to differential equations satisfying a condition at one point, the initial point. This chapter introduces *boundary value problems*, in which solutions of differential equations are subject to restrictions at both endpoints of an interval. These problems are introduced with a brief look at the classical physical problem of heat conduction.

Consider a uniform rod of length l, and let $u(x,t)$ denote the temperature of a cross section of the rod at position x, $0 \le x \le l$, and time $t \ge 0$. Suppose that initially the temperature at position x has the value $f(x)$. The ends of the rod are held at temperature 0 for all $t \ge 0$. The evolution of the temperature function is typically modeled with a partial differential equation

$$\frac{\partial u}{\partial t} = k \frac{\partial^2 u}{\partial x^2}, \quad 0 \le x \le l, \quad t \ge 0,$$

called, naturally, the *heat equation*. Here k is a positive constant called the diffusivity, which depends on the material in the rod. The problem is to determine the temperature $u(x,t)$ for all $0 < x < l$ and $t > 0$. The desired solution u satisfies the following *initial-boundary value problem*:

$$\frac{\partial u}{\partial t} = k \frac{\partial^2 u}{\partial x^2}, \quad 0 \le x \le l, \quad t \ge 0, \qquad (I - BVP)$$

$$u(0,t) = u(l,t) = 0, \quad t \ge 0,$$

$$u(x,0) = f(x), \quad 0 \le x \le l.$$

Solutions for this partial differential equation can be found with the method of separation of variables, which involves searching for solutions of the form

$$u(x,t) = v(x)w(t),$$

where v is a function of x alone and w is a function of t alone. Focusing on the boundary conditions and ignoring for the moment the temperature at time $t = 0$, insert $u(x, t) = v(x)w(t)$ into the differential equation

$$\frac{\partial u}{\partial t} = k\frac{\partial^2 u}{\partial x^2}, \quad u(0, t) = u(l, t) = 0, \quad 0 \le x \le l, \quad t \ge 0. \tag{8.1}$$

This process yields

$$v(x)\frac{\partial w(t)}{\partial t} = k\frac{\partial^2 v(x)}{\partial x^2}w(t).$$

Wherever u is not 0 we can divide by $u = vw$ to obtain

$$\frac{1}{k}\frac{1}{w(t)}\frac{\partial w(t)}{\partial t} = \frac{1}{v(x)}\frac{\partial^2 v(x)}{\partial x^2}.$$

The left side of the equality is a constant function of x, whereas the right side is a constant function of t, so that

$$\frac{1}{k}\frac{1}{w(t)}\frac{\partial w(t)}{\partial t} = \frac{1}{v(x)}\frac{\partial^2 v(x)}{\partial x^2} = -\lambda,$$

where λ is a constant. Thus v and w satisfy ordinary differential equations

$$w' + \lambda k w = 0, \quad v'' + \lambda v = 0. \tag{8.2}$$

The boundary conditions for u result in

$$u(0, t) = v(0)w(t) = 0, \quad u(l, t) = v(l)w(t) = 0, \quad t \ge 0. \tag{8.3}$$

Since the trivial solution $u(x, t) = 0$, $0 \le x \le l$, $t \ge 0$, of (8.1) is not very interesting, assume that $w(t) \ne 0$ for some $t \ge 0$. Then (8.3) implies that $v(0) = v(l) = 0$, and v satisfies the following *boundary value problem*:

$$v'' + \lambda v = 0, \quad v(0) = v(l) = 0. \tag{BVP}$$

For most choices of the constant λ, (BVP) has only the trivial solution. For example, if $\lambda = 0$ all solutions of $v'' = 0$ are given by

$$v(x) = c_1 + c_2 x,$$

where c_1, c_2 are constants; the boundary conditions imply that

$$v(0) = c_1 = 0, \quad v(l)/l = c_2 = 0.$$

A value of λ for which (BVP) has a nontrivial solution is called an *eigenvalue* for (BVP). A nontrivial solution v of (BVP) for an eigenvalue λ is called an *eigenfunction* of (BVP) for λ. Thus (BVP) can be viewed as the problem of computing all the admissible values of λ, the eigenvalues, and their corresponding eigenfunctions. For this reason the boundary value problem (BVP)

containing the parameter λ is often called an *eigenvalue problem (EVP)*. In this particular case, (BVP) is so simple that we can determine all its eigenvalues and eigenfunctions explicitly.

First note that every eigenvalue λ must be a positive real number. If v is an eigenfunction for λ then $-v'' = \lambda v$. Multiplying both sides of this equality by $\overline{v}$, integrating from 0 to l, and using integration by parts, we obtain

$$\lambda \int_0^l |v(x)|^2 \, dx = -\int_0^l v''(x)\overline{v}(x) \, dx$$

$$= v'(0)\overline{v}(0) - v'(l)\overline{v}(l) + \int_0^l |v'(x)|^2 \, dx$$

$$= \int_0^l |v'(x)|^2 \, dx \geq 0.$$

Since v is nontrivial,

$$\int_0^l |v(x)|^2 \, dx > 0$$

and hence $\lambda \geq 0$. The case $\lambda = 0$ has already been excluded, so $\lambda > 0$.

Writing $\lambda = \mu^2$, with $\mu > 0$, our differential equation becomes

$$v'' + \mu^2 v = 0, \tag{8.4}$$

and a basis for the solutions is given by $\cos(\mu x), \sin(\mu x)$. The most general solution v of (8.4) is

$$v(x) = c_1 \cos(\mu x) + c_2 \sin(\mu x),$$

where c_1, c_2 are constants, and this satisfies the boundary conditions $v(0) = v(l) = 0$ if and only if

$$v(0) = c_1 = 0, \quad v(l) = c_1 \cos(\mu l) + c_2 \sin(\mu l) = c_2 \sin(\mu l) = 0.$$

Since $v(x) = c_2 \sin(\mu x)$ is nontrivial, $c_2 \neq 0$, and therefore

$$\sin(\mu l) = 0 \tag{8.5}$$

is the condition which determines the possible values of μ. Now (8.5) is true for $\mu > 0$ if and only if

$$\mu l = n\pi, \quad n = 1, 2, 3, \ldots,$$

or

$$\mu = \frac{n\pi}{l}, \quad n = 1, 2, 3, \ldots.$$

Thus the eigenvalues of (BVP) are

$$\lambda_n = \frac{n^2 \pi^2}{l^2}, \quad n = 1, 2, 3, \ldots,$$

and an eigenfunction for λ_n is

$$v_n(x) = \sin\left(\frac{n\pi}{l}x\right).$$

For $\lambda = \lambda_n$ the equation for w in (8.2) has for a basis the solution

$$w_n(t) = e^{-\lambda_n kt} = \exp\left(-\frac{n^2\pi^2 k}{l^2}t\right).$$

Thus for each $n = 1, 2, 3, \ldots$, the equation (8.1) has a solution

$$u_n(x, t) = v_n(x)w_n(t) = \exp\left(-\frac{n^2\pi^2 k}{l^2}t\right)\sin\left(\frac{n\pi}{l}x\right).$$

It is easy to check that if U_1, U_2 are solutions of (8.1), then any linear combination $cU_1 + dU_2$ is again a solution; the solutions of (8.1) form a vector space. Thus any finite linear combination of the u_n will again satisfy (8.1).

In order to satisfy the initial condition $u(x, 0) = f(x)$, consider the formal infinite sum

$$u(x, t) = \sum_{n=1}^{\infty} c_n u_n(x, t) = \sum_{n=1}^{\infty} c_n \exp\left(-\frac{n^2\pi^2 k}{l^2}t\right)\sin\left(\frac{n\pi}{l}x\right). \tag{8.6}$$

The initial condition will be met if

$$u(x, 0) = f(x) = \sum_{n=1}^{\infty} c_n \sin\left(\frac{n\pi}{l}x\right). \tag{8.7}$$

The problem of expanding a "general" function f in a series of eigenfunctions of (BVP), that is, computing the c_n in terms of f, thus arises in a natural way. If we can do that, then u given by (8.6) will be a solution of $(I - BVP)$ provided the termwise differentiation of the series in (8.6) can be justified.

If $f(x)$ is a finite linear combination of the $\sin(n\pi x/l)$, then the solution is easy. For example, if

$$f(x) = 3\sin\left(\frac{\pi x}{l}\right) - 43\sin\left(\frac{5\pi x}{l}\right),$$

then we can take $c_1 = 3$, $c_5 = -43$, and all other $c_n = 0$. Since the sums are finite, no convergence questions arise. A solution of $(I - BVP)$ will then be

$$u(x, t) = 3\exp\left(-\frac{\pi^2 k}{l^2}t\right)\sin\left(\frac{\pi x}{l}\right) - 43\exp\left(-\frac{25\pi^2 k}{l^2}t\right)\sin\left(\frac{5\pi x}{l}\right).$$

These formulas are greatly simplified if $l = \pi$ and $k = 1$ when $(I - BVP)$ has a formal solution

$$u(x, t) = \sum_{n=1}^{\infty} c_n e^{-n^2 t}\sin(nx),$$

where

$$f(x) = \sum_{n=1}^{\infty} c_n \sin(nx).$$

In the case of

$$f(x) = 3 \sin\left(\frac{\pi x}{l}\right) - 43 \sin\left(\frac{\pi x}{l}\right),$$

for example,

$$u(x,t) = 3 \exp(-t) \sin(x) - 43 \exp(-25t) \sin(5x)$$

gives a solution. Exercise 1 indicates how the use of suitable units of position and time transforms $(I - BVP)$ into the special one with $l = \pi$, $k = 1$.

The methods sketched above work for a very general class of second-order linear equations with linear boundary conditions. The purpose of this chapter is to investigate these boundary value problems in detail.

8.2   Inner products

A discussion of eigenvalue problems is greatly simplified by working in the context of inner product spaces. Inner products are generalizations of the familiar dot product in $\mathcal{R}^n$ or $\mathcal{C}^n$.

Recall that if $\mathcal{F}$ is either $\mathcal{C}$ or $\mathcal{R}$, the usual dot product or *inner product* on $\mathcal{F}^n$ is defined by

$$\langle X, Y \rangle = x_1 \overline{y_1} + \cdots + x_n \overline{y_n}$$

for each $X = (x_1, \ldots, x_n)$, $Y = (y_1, \ldots, y_n) \in \mathcal{F}^n$. For all $X, Y \in \mathcal{F}^n$, and $\alpha, \beta \in \mathcal{F}$, the inner product has the properties

(a) $\langle X, X \rangle \geq 0,$

(b) $\langle X, X \rangle = 0$ if and only if $X = 0,$

(c) $\langle \alpha X + \beta Y, Z \rangle = \alpha \langle X, Z \rangle + \beta \langle Y, Z \rangle,$

(d) $\langle X, Y \rangle = \overline{\langle Y, X \rangle}.$

This inner product has an associated norm, $\| \; \|_2$ defined by $\|X\|_2^2 = \langle X, X \rangle$. The *Schwarz inequality* is $|\langle X, Y \rangle| \leq \|X\|_2 \|Y\|_2$ for all $X, Y \in \mathcal{F}^n$. Clearly, since $\|X\|_2 \leq \|X\|_1 = |X|$, this implies that $|\langle X, Y \rangle| \leq |X||Y|$ for all $X, Y \in \mathcal{F}^n$.

There is a certain interplay between matrices and the usual inner product on $\mathcal{F}^n$. If $\mathbf{A} = (a_{ij}) \in M_{mn}(\mathcal{F})$, its *adjoint*, or *conjugate transpose*, is the matrix $\mathbf{A}^* = (\overline{a}_{ji}) \in M_{nm}(\mathcal{F})$. The map $\mathbf{A} \rightarrow \mathbf{A}^*$ has the properties

(i) $\mathbf{A}^{**} = \mathbf{A},$

(ii) $$(\mathbf{A} + \mathbf{B})^* = \mathbf{A}^* + \mathbf{B}^*,$$

(iii) $$(\alpha\mathbf{A})^* = \overline{\alpha}\mathbf{A}^*,$$

(iv) $$(\mathbf{AB})^* = \mathbf{B}^*\mathbf{A}^*.$$

Notice that if $X \in \mathcal{F}^n$ and $Y \in \mathcal{F}^m$, then

$$\langle \mathbf{A}X, Y \rangle = \sum_{i=1}^{m}\left(\sum_{j=1}^{n} a_{ij}x_j\right)\overline{y_i} = \sum_{i=1}^{m}\sum_{j=1}^{n} a_{ij}x_j\overline{y_i} \tag{8.8}$$

$$= \sum_{j=1}^{n}\left(\sum_{i=1}^{m}\overline{\overline{a_{ij}}y_i}\right)x_j = \langle X, \mathbf{A}^*Y \rangle.$$

A matrix $\mathbf{A} \in M_n(\mathcal{F})$ is *selfadjoint* if $\mathbf{A}^* = \mathbf{A}$. When $\mathcal{F}^n = \mathcal{R}^n$ the adjoint of a matrix is simply the transpose, and the selfadjoint matrices are real symmetric matrices. Eigenvalues of selfadjoint matrices are real, and eigenvectors X, Y, corresponding to distinct eigenvalues $\lambda_1 \neq \lambda_2$ are *orthogonal*, that is, $\langle X, Y \rangle = 0$. These facts follow from simple algebraic manipulations. If

$$\mathbf{A}X = \lambda_1 X, \quad X \neq 0, \quad \mathbf{A}^* = \mathbf{A},$$

then, using 8.8 and (d),

$$\lambda_1\langle X, X \rangle = \langle \mathbf{A}X, X \rangle = \langle X, \mathbf{A}^*X \rangle = \langle X, \mathbf{A}X \rangle = \overline{\langle \mathbf{A}X, X \rangle}.$$

Since $\langle X, X \rangle > 0$, it follows that $\lambda_1 = \overline{\lambda_1}$, or $\lambda_1 \in \mathcal{R}$. On the other hand, if $\mathbf{A}Y = \lambda_2 Y$,

$$\lambda_1\langle X, Y \rangle = \langle \mathbf{A}X, Y \rangle = \langle X, \mathbf{A}Y \rangle = \lambda_2\langle X, Y \rangle.$$

Since $\lambda_1 - \lambda_2 \neq 0$, we must have $\langle X, Y \rangle = 0$. In fact, if $\mathbf{A}$ is a selfadjoint matrix, then $\mathcal{F}^n$ has a basis of n orthogonal eigenvectors of $\mathbf{A}$. (See exercise 14.)

Functions analogous to the dot product arise in more general settings. For any vector space $\mathcal{V}$ over $\mathcal{F}$ a function

$$\langle \, , \, \rangle : \mathcal{V} \times \mathcal{V} \to \mathcal{F}$$

satisfying (a)–(d) above is called an *inner product* on $\mathcal{V}$. The norm associated to an inner product is $\|X\|^2 = \langle X, X \rangle$, and the *Schwarz inequality*

$$|\langle X, Y \rangle| \leq \|X\|\|Y\|$$

is valid in any inner product space (see exercise 7).

Two elements X, Y of an inner product space are said to be *orthogonal* if $\langle X, Y \rangle = 0$. A finite or infinite sequence $\{X_k\}$ is orthogonal if $\langle X_j, X_k \rangle = 0$ for $j \neq k$ and is *orthonormal* if, in addition, $\langle X_k, X_k \rangle = 1$ for all k.

An important case for us is $\mathcal{V} = C(I, \mathcal{C})$, the continuous complex-valued functions on I, with the inner product given by

$$\langle x, y \rangle = \int_a^b x(t) \overline{y}(t) \, dt, \quad x, y \in C(I, \mathcal{C}),$$

and the norm

$$\|x\|_2 = \langle x, x \rangle^{1/2} = \left(\int_a^b |x(t)|^2 \, dt \right)^{1/2}.$$

As we will see in this and the next chapter, there is a strong analogy between the analysis of $(I\text{-}BVP)$ and that of the linear system

$$X' = \mathbf{A}X, \quad X \in \mathcal{F}^n, \quad \mathbf{A} \in M_n(\mathcal{F}), \quad \mathbf{A}^* = \mathbf{A}.$$

In developing this analogy, it will be helpful to think of the mapping $v(x) \to kv''(x)$ as the analogue of a selfadjoint matrix on the vector space $C(I, \mathcal{C})$. In developing this analogy, the boundary conditions $v(0) = 0 = v(l)$ will play a subtle, but crucial, role.

8.3   Boundary conditions and operators

A large part of the theory of eigenvalue problems for differential equations parallels the eigenvalue theory for matrices. The role of the matrix is played by a linear differential operator L. Our focus is on the class of second-order operators on $I = [a, b]$ given by

$$Lx = -(px')' + qx, \quad ' = \frac{d}{dt},$$

where $p \in C^1(I, \mathcal{R})$ and $q \in C(I, \mathcal{R})$. The leading coefficient $p(t)$ is assumed to be nonvanishing on I; for convenience we will assume that $p(t) > 0$. By using the product rule, the action of L on a function $x(t)$ can also be expressed as

$$Lx = -px'' - p'x' + qx.$$

In marked contrast to the theory of matrices, questions about the domain of definition of the operator L can involve surprising subtleties, with apparently minor changes in the domain leading to dramatic differences in the behavior of the operator. To keep the discussion of these issues as straightforward as possible, the domain of L is defined to be the set

$$D_1 = C^2(I, \mathcal{C}).$$

Boundary value problems like the one encountered in the heat conduction problem of section 8.1 involve additional constraints on the domain of the operator. In that case L was given by $Lx = -x''$, where $p(t) = 1$, $q(t) = 0$

on I. The boundary value problem (BVP), in terms of this special L, can be written as

$$Lx = \lambda x, \quad x(0) = x(l) = 0,$$

on $I = [0, l]$. Thus functions $x(t)$, already in D_1, are required to satisfy the additional conditions $x(0) = x(l) = 0$.

With the operators L as described, both the domain and the range of the operator are subspaces of the continuous complex-valued functions on I. (Real-valued functions are then a special case.) $C(I, \mathcal{C})$ will be considered as a vector space over $\mathcal{C}$ with the inner product given by

$$\langle x, y \rangle = \int_a^b x(t)\overline{y}(t)\, dt, \quad x, y \in C(I, \mathcal{C}),$$

and the norm

$$\|x\|_2 = \langle x, x \rangle^{1/2} = \left(\int_a^b |x(t)|^2\, dt \right)^{1/2}.$$

Fundamental to the study of boundary value problems associated with L are the *Lagrange identity* and *Green's formula*.

Theorem 8.1: *If* $\{x, y\}(t)$ *is defined by*

$$\{x, y\}(t) = p(t)[\overline{y}'(t)x(t) - \overline{y}(t)x'(t)], \quad x, y \in D_1,$$

then

$$\overline{y}(Lx) - (\overline{Ly})x = \{x, y\}', \quad x, y \in D_1. \qquad \text{(Lagrange identity)}$$

Proof: Using the fact that p and q are real valued, $p = \overline{p}$, $q = \overline{q}$, a direct computation shows that

$$\overline{y}(Lx) - (\overline{Ly}x) = \overline{y}[-(px')' + qx] - [-(p\overline{y}')' + q\overline{y}]x$$

$$= (p\overline{y}')'x - \overline{y}(px')' = [(p\overline{y}')x - \overline{y}(px')]'$$

$$= [p(\overline{y}'x - \overline{y}x')]' = \{x, y\}'. \quad \square$$

Integrating the Lagrange identity from a to b yields the following result.

Theorem 8.2: *For* $x, y \in D_1$,

$$\langle Lx, y \rangle - \langle x, Ly \rangle = \{x, y\}(b) - \{x, y\}(a). \qquad \text{(Green's formula)}$$

Although the left side of Green's formula involves integration of functions on I, the right-hand side only involves data from the boundary of the interval I. This data can be considered in $\mathcal{C}^4$ if

$$\hat{x} = \begin{pmatrix} \tilde{x}(a) \\ \tilde{x}(b) \end{pmatrix} = \begin{pmatrix} x(a) \\ x'(a) \\ x(b) \\ x'(b) \end{pmatrix}, \quad \hat{y} = \begin{pmatrix} \tilde{y}(a) \\ \tilde{y}(b) \end{pmatrix} = \begin{pmatrix} y(a) \\ y'(a) \\ y(b) \\ y'(b) \end{pmatrix}.$$

The eigenvalue problem above involved boundary conditions $x(0) = 0 = x(l)$. More generally, L will be restricted by two real linear boundary conditions at the endpoints a and b of I,

$$a_{11}x(a) + a_{12}x'(a) + b_{11}x(b) + b_{12}x'(b) = 0, \qquad (BC)$$

$$a_{21}x(a) + a_{22}x'(a) + b_{21}x(b) + b_{22}x'(b) = 0,$$

where the a_{ij} and b_{ij} are real constants. If

$$\mathbf{B}^a = \begin{pmatrix} a_{11} & a_{12} \\ a_{21} & a_{22} \end{pmatrix}, \quad \mathbf{B}^b = \begin{pmatrix} b_{11} & b_{12} \\ b_{21} & b_{22} \end{pmatrix},$$

then these boundary conditions may be written as

$$\mathbf{B}^a \tilde{x}(a) + \mathbf{B}^b \tilde{x}(b) = 0.$$

In terms of the *boundary matrix*

$$\mathbf{B} = [\mathbf{B}^a, \mathbf{B}^b] = \begin{pmatrix} a_{11} & a_{12} & b_{11} & b_{12} \\ a_{21} & a_{22} & b_{21} & b_{22} \end{pmatrix},$$

the boundary conditions (BC) may be written simply as

$$\mathbf{B}\hat{x} = 0.$$

To avoid trivial or equivalent boundary conditions, assume that neither boundary condition is a constant multiple of the other. This is equivalent to assuming that the two rows of $\mathbf{B}$ are linearly independent, or rank$(\mathbf{B}) = 2$.

For a given $\mathbf{B} \in M_{24}(\mathcal{R})$ define a *boundary operator* β by

$$\beta(x) = \mathbf{B}\hat{x} = \mathbf{B}^a \tilde{x}(a) + \mathbf{B}^b \tilde{x}(b), \quad x \in D_1.$$

This β is a linear function from $D_1 = C^2(I, \mathcal{C})$ into $\mathcal{C}^2$. Denote by L_β the operator L restricted to the set D_β of $x \in D_1$ satisfying the boundary conditions,

$$D_\beta = \{x \in D_1 \mid \beta(x) = 0\}.$$

Thus

$$L_\beta x = Lx, \quad x \in D_\beta.$$

The domain D_β is a vector space over $\mathcal{C}$, and L_β is a linear operator from D_β into $C(I, \mathcal{C})$.

For any β, those $x \in D_1$ satisfying $\hat{x} = 0$ clearly satisfy $\beta(x) = 0$. These x are denoted by

$$D_0 = \{x \in D_1 \mid \hat{x} = 0\};$$

in all cases $D_0 \subset D_\beta \subset D_1$.

As an example, the boundary conditions $x(a) = 0 = x(b)$ give rise to

$$\mathbf{B}^a = \begin{pmatrix} 1 & 0 \\ 0 & 0 \end{pmatrix}, \quad \mathbf{B}^b = \begin{pmatrix} 0 & 0 \\ 1 & 0 \end{pmatrix},$$

$$\mathbf{B} = \begin{pmatrix} 1 & 0 & 0 & 0 \\ 0 & 0 & 1 & 0 \end{pmatrix},$$

$$D_\beta = \{x \in D_1 | x(a) = 0 = x(b)\}.$$

An operator L_β such that

$$\langle Lx, y \rangle = \langle x, Ly \rangle, \quad x, y \in D_\beta, \tag{8.9}$$

or

$$\langle L_\beta x, y \rangle = \langle x, L_\beta y \rangle,$$

is said to be a *selfadjoint differential operator*. This particularly important class of operators will occupy much of our attention. Naturally, the boundary conditions defining a selfadjoint L_β are called *selfadjoint boundary conditions*, and the corresponding operator β is called a *selfadjoint boundary operator*.

Notice the similarity between this terminology and the terminology used for $n \times n$ matrices. In that setting, a selfadjoint matrix

$$\mathbf{A} = (a_{ij}) = (\bar{a}_{ji}) = \mathbf{A}^*$$

exhibits the characteristic property

$$\langle \mathbf{A}x, y \rangle = \langle x, \mathbf{A}^* y \rangle = \langle x, \mathbf{A}y \rangle, \quad x, y \in \mathcal{C}^n.$$

This equation is analogous to (8.9).

The operator L_β on $I = [0, 1]$ given by

$$Lx = -x'', \quad x(0) = 0, \quad x(1) = 0,$$

is selfadjoint, as Green's formula shows. Another example occurs with the same L but with *periodic boundary conditions*

$$Lx = -x'', \quad x(0) - x(1) = 0, \quad x'(0) - x'(1) = 0.$$

We emphasize that L_β being selfadjoint depends on both L and β. Boundary conditions may be selfadjoint for one L but not for another L. For example, if $Lx = -x''$ on $I = [0, 1]$, the conditions $\tilde{x}(0) = \tilde{x}(1)$ give a selfadjoint problem, while if $Lx = -(px')'$, where $p(t) = t + 1$ on $I = [0, 1]$, then the same conditions $\tilde{x}(0) = \tilde{x}(1)$ do not give a selfadjoint L_β. For this example, integration by parts twice gives

$$\langle L_\beta x, y \rangle = -\int_0^1 ([t+1]x')' \bar{y}$$

$$= -[t+1]x'\overline{y}\Big|_0^1 + x[t+1]\overline{y'}\Big|_0^1 - \int_0^1 x\overline{([t+1]y')'}.$$

The periodic boundary conditions lead to the reduction

$$-[t+1]x'\overline{y}\Big|_0^1 + x[t+1]\overline{y'}\Big|_0^1 = -x'(1)\overline{y(1)} + x(1)\overline{y'(1)}.$$

The particular choices $x(t) = \cos(2\pi t)$ and $y(t) = \sin(2\pi t)$, which are periodic with period 1, give

$$\langle L_\beta \cos(2\pi t), \sin(2\pi t)\rangle = 2\pi + \langle \cos(2\pi t), L_\beta \sin(2\pi t)\rangle.$$

Thus the desired identity $\langle L_\beta x, y\rangle = \langle x, L_\beta y\rangle$ does not hold for all $x, y \in D_\beta$.

An important class of selfadjoint boundary conditions is the *separated conditions*, where one condition only involves the endpoint a, and the other only b. These have the form

$$a_{11}x(a) + a_{12}x'(a) = 0,$$

$$b_{21}x(b) + b_{22}x'(b) = 0,$$

where not both a_{11}, a_{12} are 0 and not both b_{21}, b_{22} are 0. When the boundary conditions are separated, L_β is called a *Sturm–Liouville differential operator*. Sturm and Liouville were two investigators who made fundamental contributions (in the 1830s) to our understanding of such operators. In these cases the selfadjoint condition (8.9) can be verified directly for every L of the form $Lx = -(px')' + qx$.

Observe that the conditions $x(a) = x(b) = 0$, which we met in (BVP) of section 8.1, are separated ones, but the periodic boundary conditions $\tilde{x}(a) = \tilde{x}(b)$ are not separated.

8.4 Eigenvalues

8.4.1 Selfadjoint eigenvalue problems

Let L_β be a fixed selfadjoint differential operator,

$$L_\beta x = Lx, \quad x \in D_\beta = \{x \in D_1 \mid \beta(x) = \mathbf{B}\hat{x} = 0\}.$$

A number $\lambda \in \mathcal{C}$ is an *eigenvalue* of L_β if there is an $x \in D_\beta$ with $\|x\|_2 > 0$ such that $Lx = \lambda x$. Such a nontrivial x is an *eigenfunction* of L_β for the eigenvalue λ. The problem of finding the eigenvalues and eigenfunctions of L_β is called a *selfadjoint eigenvalue problem* for L_β, abbreviated as (EVP),

$$Lx = \lambda x, \quad \beta(x) = \mathbf{B}\hat{x} = 0. \qquad (EVP)$$

The set $\sigma(L_\beta)$ of all eigenvalues of L_β is called the *spectrum* of L_β, and if $\lambda \in \sigma(L_\beta)$ the *eigenspace* of L_β for λ is the set of all solutions x of (EVP), including the zero function. That is,

$$\mathcal{E}(L_\beta, \lambda) = N(L_\beta - \lambda I),$$

where I is the identity operator on $C(I, \mathcal{C})$. The eigenspace $\mathcal{E}(L_\beta, \lambda)$ is a vector space over $\mathcal{C}$. The *(geometric) multiplicity* of $\lambda \in \sigma(L_\beta)$ is just $\dim(\mathcal{E}(L_\beta, \lambda))$. Since L has order 2, the multiplicity of an eigenvalue λ is either one or two. We say λ is a *simple eigenvalue* in case the multiplicity is one.

As an example, reconsider (BVP) from section 8.1:

$$Lx = -x'' = \lambda x, \quad x(0) = 0 = x(\pi),$$

which has eigenvalues

$$\lambda_n = n^2, \quad n = 1, 2, \ldots.$$

In this case each λ_n is simple, with $\mathcal{E}(L_\beta, \lambda_n)$ having a basis $x_n(t) = \sin(nt)$.

Another example is provided by the same L, but with periodic boundary conditions on $I = [-\pi, \pi]$, namely $\tilde{x}(-\pi) = \tilde{x}(\pi)$. Here the problem is

$$Lx = -x'' = \lambda x, \quad x(-\pi) = x(\pi), \quad x'(-\pi) = x'(\pi). \tag{8.10}$$

Green's formula directly implies that this problem is selfadjoint. If $\lambda \in \mathcal{R}$ is an eigenvalue for (8.10) then $\lambda \geq 0$. This follows from an integration by parts

$$\lambda \langle x, x \rangle = \langle Lx, x \rangle = -\int_{-\pi}^{\pi} \overline{x}(t) x''(t) \, dt$$

$$= \overline{x}(-\pi) x'(-\pi) - \overline{x}(\pi) x'(\pi) + \int_{-\pi}^{\pi} |x'(t)|^2 \, dt$$

$$= \int_{-\pi}^{\pi} |x'(t)|^2 \, dt \geq 0,$$

since $\overline{x}(-\pi) x'(-\pi) = \overline{x}(\pi) x'(\pi)$ for those x satisfying the boundary conditions. For this problem $\lambda = 0$ is a simple eigenvalue with eigenfunction $u_0(t) = 1$ for $t \in I = [-\pi, \pi]$. For $\lambda > 0$ we can write $\lambda = \mu^2$, $\mu > 0$, and all solutions x of $-x'' = \mu^2 x$ have the form

$$x(t) = c_1 \cos(\mu t) + c_2 \sin(\mu t), \quad c_1, c_2 \in \mathcal{C}.$$

A nontrivial such x (not both c_1, c_2 are 0) satisfies the boundary conditions if and only if $\sin(\mu \pi) = 0$, or $\mu = n$, a positive integer. Thus the set of eigenvalues for L_β defined by (8.10) is

$$\sigma(L_\beta) = \{\lambda_n = n^2 \mid n = 0, 1, 2, \ldots\}.$$

For $n \neq 0$ there are two linearly independent eigenfunctions u_n, v_n, where

$$u_n(t) = \cos(nt), \quad v_n(t) = \sin(nt),$$

so that each $\lambda_n = n^2 > 0$ has multiplicity two.

Some general results about selfadjoint L_β are contained in the following theorem.

Theorem 8.3: *If L_β is selfadjoint, then*

(i) *each eigenvalue λ of L_β is real,*

(ii) *eigenfunctions corresponding to distinct eigenvalues are orthogonal,*

(iii) *for $\lambda \in \sigma(L_\beta)$, a basis for $\mathcal{E}(L_\beta, \lambda)$ can be chosen to be real valued.*

Proof: The statement (i) follows directly from the selfadjointness condition (8.9). If $Lx = \lambda x$, $x \in D_\beta$, $\|x\|_2 > 0$, then

$$0 = \langle Lx, x \rangle - \langle x, Lx \rangle = \langle \lambda x, x \rangle - \langle x, \lambda x \rangle$$

$$= \lambda \langle x, x \rangle - \overline{\lambda} \langle x, x \rangle = (\lambda - \overline{\lambda}) \|x\|_2^2,$$

so that $\lambda = \overline{\lambda} \in \mathcal{R}$.

If $y \in D_\beta$, $\|y\|_2 > 0$ is a second eigenfunction with eigenvalue $\nu \in R$, $\nu \neq \lambda$, then (8.9) gives

$$0 = \langle Lx, y \rangle - \langle x, Ly \rangle = \langle \lambda x, y \rangle - \langle x, \nu y \rangle = (\lambda - \nu) \langle x, y \rangle,$$

and hence $\langle x, y \rangle = 0$, which is (ii).

For (iii), write $x = u + iv$, where u, v are real valued. Since the coefficients p, q in L are real valued, Lu, Lv are real valued, and

$$Lx = Lu + iLv = \lambda(u + iv)$$

implies that $Lu = \lambda u$, $Lv = \lambda v$. Also, since $\mathbf{B}$ has real elements, $\beta(u), \beta(v) \in \mathcal{R}^2$, and from

$$\beta(x) = \beta(u) + i\beta(v) = 0$$

we obtain $\beta(u) = \beta(v) = 0$. Thus if $x \in \mathcal{E}(L_\beta, \lambda)$ then $u, v \in \mathcal{E}(L_\beta, \lambda)$. Since $\|x\|_2^2 = \|u\|_2^2 + \|v\|_2^2$, and $\|x\|_2 > 0$, either $\|u\|_2 > 0$ or $\|v\|_2 > 0$, or both. If $\|u\|_2 > 0$ (respectively, $\|v\|_2 > 0$), then u (respectively v) is a real-valued eigenfunction. Thus if λ is simple we have a real-valued basis for $\mathcal{E}(L_\beta, \lambda)$.

If $\dim(\mathcal{E}(L_\beta, \lambda)) = 2$, let x, y be a basis for $\mathcal{E}(L_\beta, \lambda)$, with $x = u + iv$, $y = w + iz$, where u, v, w, z are real valued. As before, $u, v, w, z \in \mathcal{E}(L_\beta, \lambda)$. If the real vector space spanned by u, v, w, z has dimension less than two, one easily sees $\dim(\mathcal{E}(L_\beta, \lambda)) < 2$ as a complex vector space, a contradiction. Hence $\mathcal{E}(L_\beta, \lambda)$ has a real-valued basis. $\square$

For Sturm–Liouville operators L_β we can say more.

Theorem 8.4: *Suppose L_β is selfadjoint with separated boundary conditions. Then each eigenvalue λ of L_β is simple.*

Proof: Let the boundary conditions defining L_β be given by

$$a_{11} x(a) + a_{12} x'(a) = 0, \quad b_{21} x(b) + b_{22} x'(b) = 0.$$

Suppose that u, v are eigenfunctions for λ. The boundary condition at a says that the vectors $(u(a), u'(a)), (v(a), v'(a)) \in \mathcal{R}^2$ are both orthogonal to the nonzero vector (a_{11}, a_{12}), hence they lie in a one-dimensional subspace of $\mathcal{R}^2$.

If $\mathbf{X} = (u, v)$, then the Wronskian $W_{\mathbf{X}}$ at a is

$$W_{\mathbf{X}}(a) = \det \begin{pmatrix} u(a) & v(a) \\ u'(a) & v'(a) \end{pmatrix}.$$

Since the columns are linearly dependent, $W_{\mathbf{X}}(a) = 0$. By Theorem 2.10, $W_{\mathbf{X}}(t) = 0$ for all $t \in I$, so u, v are linearly dependent. This shows that $\dim(\mathcal{E}(L_\beta, \lambda)) = 1$. $\square$

We have considered in detail two examples of selfadjoint problems:

$$Lx = -x'' = \lambda x, \quad x(0) = 0 = x(\pi), \quad I = [0, \pi]; \tag{I}$$

$$Lx = -x'' = \lambda x, \quad \tilde{x}(-\pi) = \tilde{x}(\pi), \quad I = [-\pi, \pi]. \tag{II}$$

If L_β is the selfadjoint operator defined by (I) or (II), then the boundary operators β, where $\beta(x) = \mathbf{B}\hat{x}$, are determined by

$$(I) \qquad\qquad\qquad \mathbf{B} = \begin{pmatrix} 1 & 0 & 0 & 0 \\ 0 & 0 & 1 & 0 \end{pmatrix},$$

$$(II) \qquad\qquad\qquad \mathbf{B} = \begin{pmatrix} 1 & 0 & -1 & 0 \\ 0 & 1 & 0 & -1 \end{pmatrix},$$

respectively. The spectra $\sigma(L_\beta)$ for these examples are

$$(I) \qquad\qquad\qquad \sigma(L_\beta) = \{n^2 \mid n = 1, 2, 3, \ldots\},$$

$$(II) \qquad\qquad\qquad \sigma(L_\beta) = \{n^2 \mid n = 0, 1, 2, \ldots\}.$$

In case (I) each eigenvalue $\lambda_n = n^2$ is simple, and a corresponding eigenfunction v_n is given by

$$(I) \qquad\qquad\qquad v_n(t) = \sin(nt).$$

Now

$$\|v_n\|_2^2 = \langle v_n, v_n \rangle = \int_0^\pi \sin^2(nt)\, dt = \pi/2, \quad n = 1, 2, 3, \ldots.$$

If $x_n = v_n / \|v_n\|_2$, then x_n is an eigenfunction for λ_n with unit norm $\|x_n\|_2 = 1$, and

$$x_n(t) = \sqrt{\frac{2}{\pi}} \sin(nt), \quad n = 1, 2, 3, \ldots.$$

The x_n form an *orthonormal set*

$$\langle x_m, x_n \rangle = \delta_{mn},$$

where $\delta_{nn} = 1$ and $\delta_{mn} = 0$ if $m \neq n$. This is due to the fact that the x_n are eigenfunctions corresponding to distinct eigenvalues; see Theorem 8.3(ii).

In case (II) only λ_0 is simple; the other $\lambda_n = n^2$, $n = 1, 2, 3, \ldots$, have multiplicity two. Corresponding eigenfunctions are

(II) $u_0(t) = 1$, $u_n(t) = \cos(nt)$, $v_n(t) = \sin(nt)$, $n = 1, 2, 3, \ldots$.

Because of Theorem 8.3(ii) we have

$$\langle u_0, v_n \rangle = 0, \quad n \neq 0, \quad \langle u_m, u_n \rangle = \langle v_m, v_n \rangle = 0, \quad m \neq n.$$

Moreover,

$$\langle u_n, v_n \rangle = \int_{-\pi}^{\pi} \sin(nt) \cos(nt) \, dt = \frac{1}{2n} \sin^2(nt)|_{-\pi}^{\pi} = 0.$$

Thus the eigenfunctions in case (II) form a mutually orthogonal set. They can be normalized by dividing by their norms. Since

$$\|u_0\|_2^2 = \int_{-\pi}^{\pi} 1 \, dt = 2\pi,$$

$$\|u_n\|_2^2 = \int_{-\pi}^{\pi} \cos^2(nt) \, dt = \pi, \quad \|v_n\|_2^2 = \int_{-\pi}^{\pi} \sin^2(nt) \, dt = \pi,$$

these normalized eigenfunctions can be written as

$$x_1 = \frac{1}{\sqrt{2\pi}},$$

$$x_{2k} = \frac{1}{\sqrt{\pi}} \cos(nt), \quad x_{2k+1} = \frac{1}{\sqrt{\pi}} \sin(nt), \quad k = 1, 2, 3, \ldots .$$

The corresponding eigenvalues are then

$$\lambda_1 = 0, \quad \lambda_{2k} = \lambda_{2k+1} = k^2, \quad k = 1, 2, 3, \ldots .$$

In both of these selfadjoint cases there are an infinite number of eigenvalues λ_n which are bounded below and tend to $+\infty$ as $n \to \infty$. As we will see a bit later, the properties described above for these examples are typical of the more general selfadjoint problem

$$Lx = -(px')' + qx = \lambda x, \quad \beta(x) = \mathbf{B}\hat{x} = 0, \quad I = [a, b], \quad p(t) > 0, \quad (EVP)$$

and the corresponding selfadjoint operator L_β.

If L_β is not selfadjoint, the behavior can be quite different. For example, if

$$Lx = -x'', \quad \mathbf{B} = \begin{pmatrix} 2 & 0 & -1 & 0 \\ 0 & 2 & 0 & 1 \end{pmatrix}, \quad I = [0, \pi],$$

the corresponding nonselfadjoint L_β has no eigenvalues. For the same L and I, the boundary matrix

$$\mathbf{B} = \begin{pmatrix} 1 & 0 & -1 & 0 \\ 0 & 1 & 0 & 1 \end{pmatrix},$$

determines a nonselfadjoint L_β such that every $\lambda \in C$ is an eigenvalue.

In our examples (I) and (II) we showed directly that $\langle Lx, x \rangle \geq 0$. If the inequality

$$\langle Lx, x \rangle \geq m\langle x, x \rangle, \quad x \in D_\beta, \tag{8.11}$$

is valid for some $m > -\infty$, we say L_β is *bounded below* by m. In this case, the spectrum $\sigma(L_\beta)$ is bounded below by m, for if x is an eigenfunction of L_β with eigenvalue λ, we have

$$\langle Lx, x \rangle = \lambda \langle x, x \rangle \geq m \langle x, x \rangle,$$

or $\lambda \geq m$.

For some selfadjoint L_β it may not be apparent that (8.11) is valid for some $m > -\infty$. For example, let $Lx = -x''$ on $I = [0, 1]$, with

$$\mathbf{B} = \begin{pmatrix} 1 & 1 & 0 & 0 \\ 0 & 0 & 0 & 1 \end{pmatrix}.$$

This leads to the selfadjoint eigenvalue problem

$$Lx = -x'' = \lambda x, \quad x(0) + x'(0) = 0, \quad x'(1) = 0. \tag{III}$$

An integration by parts shows that

$$\langle Lx, x \rangle = -|x(0)|^2 + \int_0^1 |x'(t)|^2 \, dt, \quad x \in D_\beta, \tag{8.12}$$

and an inequality of the form (8.11) is not so clear.

Define the operator L_0 to agree with L on the smaller domain D_0,

$$L_0 x = Lx, \quad x \in D_0.$$

Say that L_0 is bounded below by m if

$$\langle Lx, x \rangle \geq m \langle x, x \rangle, \quad x \in D_0 = \{x \in D_1 \mid \hat{x} = 0\}.$$

Since $D_0 \subset D_\beta$, it is certainly true that if L_β is bounded below by m, then L_0 is bounded below by m. In the opposite direction the following result is true.

Theorem 8.5: *If L_0 is bounded below by $m > -\infty$,*

$$\langle Lx, x \rangle \geq m \langle x, x \rangle, \quad x \in D_0,$$

then for any selfadjoint L_β there are at most two linearly independent eigenfunctions of L_β with eigenvalues λ satisfying $\lambda < m$.

Proof: Suppose that x_1, x_2, x_3 are three linearly independent eigenfunctions, which by Theorem 8.3 may be assumed real, with eigenvalues $\lambda_1, \lambda_2, \lambda_3$ all less than m. Changing these vectors if necessary with the Gram–Schmidt process, we can assume that the vectors x_j are orthonormal. We consider the boundary values $\hat{x}_1, \hat{x}_2, \hat{x}_3 \in \mathcal{R}^4$. Since $x_j \in D_\beta$, that is, the boundary values are orthogonal in $\mathcal{R}^4$ to two independent vectors, the boundary values are already constrained to lie in a two-dimensional subspace of $\mathcal{R}^4$. Consequently, a nontrivial linear combination $w = c_1 x_1 + c_2 x_2 + c_3 x_3$ of these vectors satisfies $\hat{w} = 0$, or $w \in D_0$. Now we consider

$$\langle Lw, w \rangle = (c_1 \lambda_1 x_1 + c_2 \lambda_2 x_2 + c_3 \lambda_3 x_3, c_1 x_1 + c_2 x_2 + c_3 x_3)$$

$$= \lambda_1 |c_1|^2 + \lambda_2 |c_2|^2 + \lambda_3 |c_3|^2$$

$$< m[|c_1|^2 + |c_2|^2 + |c_3|^2] = m(w, w).$$

This inequality contradicts the hypothesis. $\square$

It is simple to use integration by parts to obtain information about (Lx, x) for $x \in D_0$. When $x \in D_0$, the boundary terms vanish, yielding

$$\langle Lx, x \rangle = \int_a^b p(t)|x'(t)|^2 \, dt + \int_a^b q(t)|x(t)|^2 \, dt, \quad x \in D_0.$$

Letting

$$q_0 = \min\{q(t) \mid t \in I\} > -\infty, \quad p_0 = \min\{p(t) \mid t \in I\} > 0,$$

it follows that

$$\langle Lx, x \rangle \geq p_0 \langle x', x' \rangle + q_0 \langle x, x \rangle \geq q_0 \langle x, x \rangle, \quad x \in D_0.$$

An application of Theorem 8.5 proves

Corollary 8.6: *Every selfadjoint L_β has at most two linearly independent eigenfunctions with eigenvalues λ satisfying $\lambda < q_0$.*

As an example consider the selfadjoint problem (III), with corresponding operator L_β. Since (8.12) implies that

$$\langle Lx, x \rangle = \int_0^1 |x'(t)|^2 \, dt \geq 0, \quad x \in D_0,$$

the operator L_0 is bounded below by 0. It is easy to check that $\lambda = 0$ is not an eigenvalue of L_β. Can there exist negative eigenvalues? To answer this, let $\lambda = -\mu^2$, $\mu > 0$. The general solution of $-x'' = -\mu^2 x$ is given by

$$x(t) = c_1 \cosh(\mu t) + c_2 \sinh(\mu t), \quad c_1, c_2 \in \mathcal{C},$$

and

$$x'(t) = \mu[c_1 \sinh(\mu t) + c_2 \cosh(\mu t)].$$

Satisfying the boundary conditions means that

$$x(0) + x'(0) = c_1 + c_2 \mu = 0,$$

$$x'(1) = \mu[c_1 \sinh(\mu) + c_2 \cosh(\mu)] = 0.$$

These conditions are satisfied for c_1, c_2 not both 0 if and only if

$$0 = \det \begin{pmatrix} 1 & \mu \\ \mu \, \sinh(\mu) & \mu \, \cosh(\mu) \end{pmatrix} = \mu[\cosh(\mu) - \mu \sinh(\mu)]$$

or $\coth(\mu) = \mu$.

A sketch of the graphs of $y = \coth(\mu)$ and $y = \mu$ (see Figure 8.1) shows that there is just one $\mu_1 > 0$, where these graphs cross. This is approximately given by $\mu_1 \simeq 1.20$, and hence $\lambda_1 = \mu_1^2 \simeq -1.44$ is the only negative eigenvalue. Since the boundary conditions are separated, all eigenvalues of L_β are simple. An eigenfunction x_1 for λ_1 is given by

$$x_1(t) = \sinh(\mu_1 t) - \mu_1 \cosh(\mu_1 t).$$

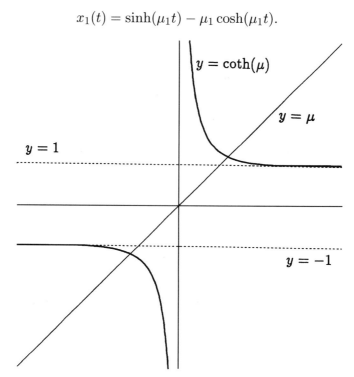

Figure 8.1: Locating a negative eigenvalue.

8.4.2   Eigenvalue asymptotics

It is natural to ask if the eigenvalues for selfadjoint problems are in some sense similar to eigenvalue sequences of examples such as (I), where explicit computations can be made. This question is most easily addressed for Sturm–Liouville problems. We will focus our attention primarily on problems having the form

$$-x'' + q(t)x = \lambda x, \quad x(0) = 0 = x(\pi). \tag{8.13}$$

A set S of real numbers is said to be *discrete* if for every $s \in S$ there is an open interval U such that $U \cap S = s$. Our first result is elementary and applies immediately to all Sturm–Liouville problems.

Lemma 8.7: *The eigenvalues of a Sturm–Liouville problem form a discrete set.*

Proof: The Sturm–Liouville problem has the form

$$Lx = -(px')' + qx = \lambda x, \quad a_{11}x(a) + a_{12}x'(a) = 0, \quad b_{21}x(b) + b_{22}x'(b) = 0.$$

In the two-dimensional vector space of solutions to the equation there is only a one-dimensional space satisfying the boundary condition at a, and this space has a basis given by those solutions $x_1(t, \lambda)$ satisfying the initial condition

$$x_1(a, \lambda) = -a_{12}, \quad x_1'(a, \lambda) = a_{11}.$$

The function $x_1(t, \lambda)$ is real valued for λ real. If λ_j is an eigenvalue for this Sturm–Liouville problem, then the function

$$r(\lambda) = \int_0^\pi x_1(t, \lambda_j)\overline{x_1(t, \lambda)}\ dt = \int_0^\pi x_1(t, \lambda_j)x_1(t, \lambda)\ dt$$

is real valued if λ is real. It follows from Theorem 7.6 (see exercise 7) that for any $\epsilon > 0$ there is a $\delta > 0$ such that $|\lambda - \lambda_j| < \delta$ implies that $|x_1(t, \lambda) - x_1(t, \lambda_j)| < \epsilon$ for all $t \in [a, b]$. Thus $r(\lambda)$ is continuous, and since $r(\lambda_j) > 0$, it must remain positive for λ in some open neighborhood U of λ_j.

If λ_i is another eigenvalue, distinct from λ_j, then Theorem 8.3 says that the eigenfunctions are orthogonal, or $r(\lambda_i) = 0$. Thus no $\lambda \in U$ except for λ_j can be an eigenvalue. $\square$

Our next goal is to describe the locations of the eigenvalues for problem (8.13). The integral equation (7.25) is an effective tool for addressing this problem.

Theorem 7.9 established that a solution of the related initial value problem

$$-x'' + q(t)x = \lambda x, \quad x(0) = a, \quad x'(0) = b$$

can be written as a solution of the integral equation

$$x(t) = a\ \cos(\sqrt{\lambda}t) + b\ \frac{\sin(\sqrt{\lambda}t)}{\sqrt{\lambda}} + \int_0^t \frac{\sin(\sqrt{\lambda}[t-s])}{\sqrt{\lambda}}q(s)x(s)\ ds.$$

Notice that the boundary condition $x(0) = 0$ implies that $a = 0$. The basic existence and uniqueness result for second-order equations implies that any nontrivial eigenfunction for (8.13) must satisfy $x'(0) \neq 0$. In searching for an eigenfunction there is no loss of generality in assuming that $x'(0) = 1$. With this choice the integral equation for $x(t)$ reduces to the form

$$x(t, \lambda) = \frac{\sin(\sqrt{\lambda}t)}{\sqrt{\lambda}} + \int_0^t \frac{\sin(\sqrt{\lambda}[t-s])}{\sqrt{\lambda}}q(s)x(s, \lambda)\ ds. \qquad (8.14)$$

Here the dependence of x on λ is displayed, since this dependence will now play an important role.

As a consequence of Theorem 7.9, $x(t, \lambda)$ is bounded by some constant C for all (t, λ) satisfying $0 \leq t \leq \pi$ and $\lambda > \lambda_0$ for any $\lambda_0 > 0$. When such an estimate is inserted into (8.14), the result is that there is a constant C such that

$$|x(t, \lambda)| \leq C/\sqrt{\lambda}, \quad \lambda > \lambda_0 > 0.$$

Putting this estimate for $|x(t, \lambda)|$ into (8.14) yields

$$x(t, \lambda) = \frac{\sin(\sqrt{\lambda} t)}{\sqrt{\lambda}} + O\left(\frac{1}{\lambda}\right).$$

A solution $x(t, \lambda)$ of the differential equation will also satisfy the boundary condition at $t = \pi$ if and only if

$$x(\pi, \lambda) = 0 = \frac{\sin(\sqrt{\lambda}\pi)}{\sqrt{\lambda}} + \int_0^\pi \frac{\sin(\sqrt{\lambda}[\pi - s])}{\sqrt{\lambda}} q(s) x(s, \lambda) \, ds$$

$$= \frac{\sin(\sqrt{\lambda}\pi)}{\sqrt{\lambda}} + O\left(\frac{1}{\lambda}\right). \tag{8.15}$$

It is possible to use (8.15) to describe the location of eigenvalues of (8.14) when λ is large.

Theorem 8.8: *If $q(t) \in C([0, \pi], \mathcal{R})$ then there is a constant K_1 such that any eigenvalue λ of (8.13) satisfies*

$$|\lambda - n^2| \leq K_1$$

for some positive integer n. Conversely, there is a constant K_2 such that for every positive integer n there is an eigenvalue λ of (8.13) satisfying

$$|\lambda - n^2| \leq K_2.$$

Proof: The estimate for $x(t, \lambda)$ means that there is a constant K such that

$$|\sqrt{\lambda}\, x(\pi, \lambda) - \sin(\sqrt{\lambda}\pi)| \leq K|\lambda|^{-1/2}, \quad \lambda > \lambda_0 > 0.$$

At any eigenvalue $\lambda = \lambda_j$ we have $x(\pi, \lambda_j) = 0$, implying

$$|\sin(\sqrt{\lambda}\pi)| \leq K|\lambda|^{-1/2}.$$

But $|\sin(\sqrt{\lambda}\pi)|$ is only small when $\sqrt{\lambda}$ is near the integers. In fact (see Figure 8.2 and exercise 18), the inequality

$$2\, dist(\sqrt{\lambda}, \mathcal{Z}) \leq |\sin(\sqrt{\lambda}\pi)|, \quad \mathcal{Z} = \{0, \pm 1, \pm 2, \ldots\}$$

is satisfied. If λ_j is an eigenvalue, there is a positive integer n and a constant C such that

$$|\sqrt{\lambda_j} - n| \leq C|\lambda_j|^{-1/2}.$$

Rephrasing,

$$\sqrt{\lambda_j} - C|\lambda_j|^{-1/2} \le n \le \sqrt{\lambda_j} + C|\lambda_j|^{-1/2}$$

or

$$\sqrt{\lambda_j}[1 - C/|\lambda_j|] \le n \le \sqrt{\lambda_j}[1 + C/|\lambda_j|].$$

Squaring gives

$$\lambda_j[1 - 2C/|\lambda_j| + C^2/|\lambda_j|^2] \le n^2 \le \sqrt{\lambda_j}[1 + 2C/|\lambda_j| + C^2/|\lambda_j|^2],$$

which means that

$$|\lambda_j - n^2| \le 2C + C^2/|\lambda_j|,$$

and for $|\lambda_j|$ sufficiently large

$$|\lambda_j - n^2| \le 3C.$$

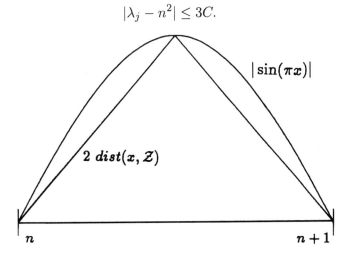

Figure 8.2: Estimating $|\sin(\pi x)|$ from below.

For the converse, consider what happens to $x(\pi, \lambda)$ when $\sqrt{\lambda} = n \pm C/n$, $C > 0$. Then

$$x(\pi, \lambda) = \frac{1}{\sqrt{\lambda}}[\sin(\sqrt{\lambda}\pi) + O(|\lambda^{-1/2}|)]$$

$$= \frac{1}{\sqrt{\lambda}}[\cos(n\pi)\sin(\pm C\pi/n) + O(|\lambda^{-1/2}|)]$$

$$= \frac{1}{\sqrt{\lambda}}[(-1)^n \sin(\pm C\pi/n) + O(|\lambda^{-1/2}|)].$$

For n sufficiently large

$$\sin(C\pi/n) \ge C/n, \quad \sin(-C\pi/n) \le -C/n.$$

There is another constant $C_1 > 0$ such that the $O(|\lambda|^{-1/2})$ terms have magnitude smaller than C_1/n. If we pick $C > 2C_1$ then $x(\pi, \lambda)$ changes sign between $n \pm C/n$. Since $x(\pi, \lambda)$ is continuous, it must vanish when

$$n[1 - C/n^2] < \sqrt{\lambda_j} < n[1 + C/n^2],$$

or there is a $C_2 > 0$ such that for n sufficiently large,

$$n^2 - C_2 < \lambda_j < n^2 + C_2. \quad \square$$

8.5   Nonhomogeneous boundary value problems

8.5.1   Nonhomogeneous problems

When $\mathbf{A}$ is an $n \times n$ matrix, the equation $(\mathbf{A} - \lambda I)X = Y$ has a unique solution X for every $Y \in \mathcal{F}^n$ if and only if λ is not an eigenvalue of $\mathbf{A}$. To find a corresponding result for a selfadjoint operator L_β on an interval $I = [a, b]$, consider the associated *nonhomogeneous problem*

$$Lx = -(px')' + qx = \lambda x + f, \quad \mathbf{B}\hat{x} = 0, \tag{NH}$$

where $\lambda \in \mathcal{C}$ and $f \in C(I, \mathcal{C})$. Since λ may be complex and f complex-valued, solutions of (NH) will in general be complex valued. It turns out that (NH) has a solution for every $f \in C(I, \mathcal{C})$ if and only if λ is not an eigenvalue of L_β. Before demonstrating this, we take a short detour to introduce some notation.

Let $\mathbf{X}(t, \lambda) = (x_1(t, \lambda), x_2(t, \lambda))$ be a basis for the solutions of the homogeneous equation $Lx = \lambda x$ on I. Then

$$\hat{\mathbf{X}}(\lambda) = (\hat{x}_1(\lambda), \hat{x}_2(\lambda))$$

is a 4×2 matrix

$$\hat{\mathbf{X}}(\lambda) = \begin{pmatrix} x_1(a, \lambda) & x_2(a, \lambda) \\ x_1'(a, \lambda) & x_2'(a, \lambda) \\ x_1(b, \lambda) & x_2(b, \lambda) \\ x_1'(b, \lambda) & x_2'(b, \lambda) \end{pmatrix},$$

and $\mathbf{B}\hat{\mathbf{X}}(\lambda)$ is the 2×2 matrix

$$\mathbf{B}\hat{\mathbf{X}}(\lambda) = (\mathbf{B}\hat{x}_1(\lambda), \mathbf{B}\hat{x}_2(\lambda)).$$

For a fixed $\lambda \in \mathcal{C}$, every solution x of $Lx = \lambda x$ has the form $x = \mathbf{X}(\lambda)C$, where $C \in \mathcal{C}^2$. This solution satisfies the boundary condition $\mathbf{B}\hat{x} = 0$ if and only if

$$\mathbf{B}\hat{x} = \mathbf{B}\hat{\mathbf{X}}(\lambda)C = 0, \tag{8.16}$$

and x is nontrivial (an eigenfunction of L_β for λ) if and only if (8.16) has a nontrivial solution $C \in \mathcal{C}^2$. The homogeneous system (8.16) of two linear equations for the components c_1, c_2 of C has a nontrivial solution $c \in \mathcal{C}^2$ if and only if

$$\det(\mathbf{B}\hat{\mathbf{X}}(\lambda)) = 0. \tag{8.17}$$

Hence λ is an eigenvalue of L_β if and only if (8.17) is satisfied.

To illustrate these ideas, consider the earlier example (III) on $I = [0, 1]$,

$$Lx = -x'' = \lambda x, \quad x(0) + x'(0) = 0, \quad x'(1) = 0. \tag{III}$$

The boundary matrix is

$$\mathbf{B} = \begin{pmatrix} 1 & 1 & 0 & 0 \\ 0 & 0 & 0 & 1 \end{pmatrix}.$$

Suppose $\lambda \neq 0$ and put $\lambda = \nu^2$, where $\nu \in \mathcal{C}$. A basis for the solutions of $-x'' = \nu^2 x$ is given by

$$\mathbf{X}(t, \nu^2) = (\cos(\nu t), \sin(\nu t)),$$

where

$$\cos(\nu t) = \frac{e^{i\nu t} + e^{-i\nu t}}{2}, \quad \sin(\nu t) = \frac{e^{i\nu t} - e^{-i\nu t}}{2i}.$$

Then

$$\mathbf{X}'(t, \nu^2) = (-\nu \sin(\nu t), \nu \cos(\nu t)),$$

$$\tilde{\mathbf{X}}(t, \nu^2) = \begin{pmatrix} \mathbf{X}(t, \nu^2) \\ \mathbf{X}'(t, \nu^2) \end{pmatrix} = \begin{pmatrix} \cos(\nu t) & \sin(\nu t) \\ -\nu \sin(\nu t) & \nu \cos(\nu t) \end{pmatrix},$$

and

$$\hat{\mathbf{X}}(\nu^2) = \begin{pmatrix} \tilde{\mathbf{X}}(0, \nu^2) \\ \tilde{\mathbf{X}}(1, \nu^2) \end{pmatrix} = \begin{pmatrix} 1 & 0 \\ 0 & \nu \\ \cos(\nu) & \sin(\nu) \\ -\nu \sin(\nu) & \nu \cos(\nu) \end{pmatrix}.$$

Therefore,

$$\mathbf{B}\hat{\mathbf{X}}(\nu^2) = \begin{pmatrix} 1 & \nu \\ -\nu \sin(\nu) & \nu \cos(\nu) \end{pmatrix}$$

and

$$\det(\mathbf{B}\hat{\mathbf{X}}(\nu^2)) = \nu \cos(\nu) + \nu^2 \sin(\nu).$$

It follows that $\lambda = \nu^2 \neq 0$ is an eigenvalue of L_β if and only if

$$\cot(\nu) + \nu = 0. \tag{8.18}$$

For $\lambda = 0$, a short calculation using the basis given by $\mathbf{X}(t, 0) = (1, t)$ shows that

$$\mathbf{B}\hat{\mathbf{X}}(0) = \begin{pmatrix} 1 & 1 \\ 0 & 1 \end{pmatrix},$$

and, since $\det(\mathbf{B}\hat{\mathbf{X}}(0)) = 1 \neq 0$, we see that $0 \notin \sigma(L_\beta)$.

Since any eigenvalue λ of L_β is real, either $\lambda > 0$, $\lambda = \nu^2$ for some $\nu > 0$, which satisfies (8.18), or $\lambda = -\mu^2$ ($\nu = i\mu$) for some $\mu > 0$. From the relations

$$\cos(i\mu) = \cosh(\mu), \quad \sin(i\mu) = i \sinh(\mu),$$

and (8.18), the negative eigenvalues $\lambda = -\mu^2$ of L_β satisfy $\coth(\mu) - \mu = 0$.

The first step toward understanding (NH) is the following necessary condition for a solution.

Theorem 8.9: *If (NH) has a solution* x, *then* $\langle f, y \rangle = 0$ *for all* $y \in N(L_\beta - \lambda I)$.

Proof: Let x satisfy

$$Lx - \lambda x = f, \quad x \in D_\beta,$$

and let

$$Ly - \lambda y = 0, \quad y \in D_\beta.$$

Either y is the zero function, in which case $\langle f, y \rangle = 0$ is clearly satisfied, or y is an eigenfunction of L_β for λ, and $\lambda \in \mathcal{R}$. In the latter case

$$\langle f, y \rangle = \langle Lx - \lambda x, y \rangle = \langle x, Ly - \lambda y \rangle = 0. \quad \square$$

We are now in a position to demonstrate the result mentioned at the beginning of this section.

Theorem 8.10: *The nonhomogeneous problem* (NH) *has a solution for every* $f \in C(I, \mathcal{C})$ *if and only if* $\lambda \notin \sigma(L_\beta)$, *in which case the solution is unique.*

Proof: Theorem 8.9 says that if (NH) has a solution, then $\langle f, y \rangle = 0$ for all $y \in N(L_\beta - \lambda I)$. If $y \neq 0$ then $\langle y, y \rangle > 0$, so (NH) cannot have a solution in case $f = y$. Thus $\lambda \in \sigma(L_\beta)$ implies there are functions $f \in C(I, \mathcal{C})$ such that (NH) is not solvable.

Conversely, assume $\lambda \notin \sigma(L_\beta)$. Let u be a particular solution of $Lx = \lambda x + f$, which can be constructed using the variation of parameters method (see section 2.6). The general solution x of $Lx = \lambda x + f$ has the form

$$x = u + \mathbf{X}(\lambda)C, \tag{8.19}$$

where $\mathbf{X}(\lambda) = (x_1(\lambda), x_2(\lambda))$ is a basis for the solutions of $Lx = \lambda x$ and $C \in \mathcal{C}^2$. Such an x will satisfy the boundary conditions if and only if

$$\mathbf{B}\hat{x} = \mathbf{B}\hat{u} + \mathbf{B}\widehat{\mathbf{X}(\lambda)}C = 0$$

or

$$\mathbf{B}\widehat{\mathbf{X}(\lambda)}C = -\mathbf{B}\hat{u}. \tag{8.20}$$

Now $\lambda \notin \sigma(L_\beta)$ implies that $\det(\mathbf{B}\widehat{\mathbf{X}(\lambda)}) \neq 0$, and thus the 2×2 matrix $\mathbf{B}\widehat{\mathbf{X}(\lambda)}$ is invertible. Hence (8.20) has a unique solution $C \in \mathcal{C}^2$ given by

$$C = -(\mathbf{B}\widehat{\mathbf{X}(\lambda)})^{-1}\mathbf{B}\hat{u},$$

and with this C the x given by (8.19) is a solution of (NH). The uniqueness of C implies the uniqueness of the solution x. $\square$

If λ is an eigenvalue of L_β, those solutions of (NH) which do exist are not unique. Indeed, if v is a solution and $y \in \mathcal{E}(L_\beta, \lambda)$, then $x = v + y$ is also a solution, for

$$Lx - \lambda x = Lv - \lambda v + Ly - \lambda y = f + 0 = f,$$

$$\mathbf{B}\hat{x} = \mathbf{B}\hat{v} + \mathbf{B}\hat{y} = 0 + 0 = 0.$$

In this case the solutions of (NH) form an affine space

$$v + \mathcal{E}(L_\beta, \lambda) = \{x = v + y | y \in \mathcal{E}(L_\beta, \lambda)\},$$

where v is a particular solution of (NH).

The next result gives necessary and sufficient conditions for the nonhomogeneous problem (NH) to have a solution when λ is an eigenvalue of L_β. In this case uniqueness is assured if the solution is further restricted. The proof is omitted.

Theorem 8.11: *If $\lambda \in \sigma(L_\beta)$, the nonhomogeneous problem (NH) has a solution if and only if $f \in \mathcal{E}(L_\beta, \lambda)^\perp$. If $\lambda \in \sigma(L_\beta)$ and $f \in \mathcal{E}(L_\beta, \lambda)^\perp$, then (NH) has a unique solution x satisfying $x \in \mathcal{E}(L_\beta, \lambda)^\perp$.*

8.5.2   Green's functions

According to Theorem 8.10, if $\lambda \notin \sigma(L_\beta)$ there exists a unique solution x of the nonhomogeneous problem

$$Lx = -(px')' + qx = \lambda x + f, \quad \mathbf{B}\hat{x} = 0, \qquad (NH)$$

for every $f \in C(I, \mathcal{C})$. This solution may be written as

$$x(t, \lambda) = \int_a^b g(t, s, \lambda) f(s) \, ds,$$

where the *kernel* g, defined on the square

$$I \times I = \{(t, s) \in \mathcal{R}^2 | t, s \in I\},$$

depends only on the selfadjoint homogeneous problem

$$Lx = \lambda x, \quad \mathbf{B}\hat{x} = 0, \qquad (H)$$

that is, on $L_\beta - \lambda I$. This g is independent of f and is called a *Green's function* for $L_\beta - \lambda I$. To know g is to know the solution of (NH) for an arbitrary $f \in C(I, \mathcal{C})$. For simplicity, the case $\lambda = 0 \notin \sigma(L_\beta)$ is considered. Then (NH) is

$$Lx = -(px')' + qx = f, \quad \mathbf{B}\hat{x} = 0, \quad f \in C(I, \mathcal{C}). \qquad (8.21)$$

Let $\mathbf{X} = (x_1, x_2)$ be a basis for the solutions of $Lx = 0$ on I. The variation of parameters method (Theorem 2.12) yields a particular solution u of $Lx = f$; from Chapter 2, exercise 19, that solution satisfying $u(a) = u'(a) = 0$ is

$$u(t) = \int_a^t \frac{x_1(t)x_2(s) - x_1(s)x_2(t)}{p(s)W_{\mathbf{X}}(s)} f(s) \, ds.$$

Here $W_{\mathbf{X}} = \det(\tilde{\mathbf{X}})$ is the Wronskian of the basis $\mathbf{X}$. We remark that the formula for $W_{\mathbf{X}}$ given in Theorem 2.10 shows that $p(s)W_{\mathbf{X}}(s)$ is a constant, independent of $s \in I$.

Letting

$$k(s) = \frac{1}{p(s)W_{\mathbf{X}}(s)} \begin{pmatrix} x_2(s) \\ -x_1(s) \end{pmatrix},$$

u may be written as

$$u(t) = \mathbf{X}(t) \int_a^t k(s)f(s)\,ds.$$

Note that

$$\mathbf{X}(t)k(t) = 0, \quad \mathbf{X}'(t)k(t) = \frac{-1}{p(t)}.$$

We have

$$u'(t) = \mathbf{X}'(t) \int_a^t k(s)f(s)\,ds,$$

so that

$$\tilde{u}(t) = \begin{pmatrix} u(t) \\ u'(t) \end{pmatrix} = \begin{pmatrix} \mathbf{X}(t) \\ \mathbf{X}'(t) \end{pmatrix} \int_a^t k(s)f(s)\,ds = \tilde{\mathbf{X}}(t) \int_a^t k(s)f(s)\,ds$$

and

$$\tilde{u}(a) = 0, \quad \tilde{u}(b) = \tilde{\mathbf{X}}(b) \int_a^b k(s)f(s)\,ds.$$

Every solution x of $Lx = f$ has the form $x = u + \mathbf{X}C$, where $C \in \mathcal{C}^2$, and such a solution satisfies the boundary conditions $\mathbf{B}\hat{x} = 0$ if and only if

$$C = -(\mathbf{B}\hat{\mathbf{X}})^{-1}\mathbf{B}\hat{u},$$

so that

$$x = u - \mathbf{X}(\mathbf{B}\hat{\mathbf{X}})^{-1}\mathbf{B}\hat{u}.$$

Now recalling that $\mathbf{B} = (\mathbf{B}^a, \mathbf{B}^b)$, where $\mathbf{B}^a, \mathbf{B}^b$ are 2×2 matrices, we have

$$\mathbf{B}\hat{u} = \mathbf{B}^a\tilde{u}(a) + \mathbf{B}^b\tilde{u}(b) = \mathbf{B}^b\tilde{\mathbf{X}}(b) \int_a^b k(s)f(s)\,ds.$$

Thus

$$x(t) = \mathbf{X}(t) \int_a^t k(s)f(s)\,ds - \mathbf{X}(t)(\mathbf{B}\hat{\mathbf{X}})^{-1}\mathbf{B}^b\tilde{\mathbf{X}}(b) \int_a^b k(s)f(s)\,ds$$

or

$$x(t) = \int_a^b g(t,s)f(s)\,ds, \tag{8.22}$$

where

$$g(t,s) = \begin{cases} \mathbf{X}(t)k(s) - \mathbf{X}(t)(\mathbf{B}\hat{\mathbf{X}})^{-1}\mathbf{B}^b\tilde{\mathbf{X}}(b)k(s), & s \le t, \\ \\ -\mathbf{X}(t)(\mathbf{B}\hat{\mathbf{X}})^{-1}\mathbf{B}^b\tilde{\mathbf{X}}(b)k(s), & s > t. \end{cases}$$

Making use of the identity

$$\mathbf{X}(t)k(s) = \mathbf{X}(t)(\mathbf{B}\hat{\mathbf{X}})^{-1}[\mathbf{B}^a\tilde{\mathbf{X}}(a) + \mathbf{B}^b\tilde{\mathbf{X}}(b)]k(s),$$

$g(t,s)$ can be written in the more symmetric form

$$g(t,s) = \left\{ \begin{array}{ll} \mathbf{X}(t)(\mathbf{B}\hat{\mathbf{X}})^{-1}\mathbf{B}^a\tilde{\mathbf{X}}(a)k(s), & s \leq t, \\ \\ -\mathbf{X}(t)(\mathbf{B}\hat{\mathbf{X}})^{-1}\mathbf{B}^b\tilde{\mathbf{X}}(b)k(s), & s > t. \end{array} \right\} \tag{8.23}$$

With $0 \notin \sigma(L_\beta)$ and this g, the Green function for L_β, the solution x of (8.21), is given by (8.22).

To obtain the Green function for $L_\beta - \lambda I$ in the general case $\lambda \notin \sigma(L_\beta)$ the above argument is altered by replacing $\mathbf{X}$ by a basis $\mathbf{X}(\lambda) = (x_1(\lambda), x_2(\lambda))$ for the solutions of $Lx = \lambda x$ on I and $k(s)$ by

$$k(s,\lambda) = \frac{1}{p(s)W_{\mathbf{X}(\lambda)}(s)} \left(\begin{array}{c} x_2(s,\lambda) \\ -x_1(s,\lambda) \end{array} \right).$$

Then we have the following result.

Theorem 8.12: *If $\lambda \notin \sigma(L_\beta)$, the nonhomogeneous problem*

$$Lx = \lambda x + f, \quad \mathbf{B}\hat{x} = 0, \quad f \in C(I,\mathcal{C}), \tag{NH}$$

has a unique solution x given by

$$x(t,\lambda) = \int_a^b g(t,s,\lambda)f(s)\ ds,$$

where

$$g(t,s,\lambda) = \left\{ \begin{array}{ll} \mathbf{X}(t,\lambda)(\mathbf{B}\widehat{\mathbf{X}(\lambda)})^{-1}\mathbf{B}^a\tilde{\mathbf{X}}(a,\lambda)k(s,\lambda), & s \leq t, \\ \\ -\mathbf{X}(t,\lambda)(\mathbf{B}\widehat{\mathbf{X}(\lambda)})^{-1}\mathbf{B}^b\tilde{\mathbf{X}}(b,\lambda)k(s,\lambda), & s > t. \end{array} \right\} \tag{8.24}$$

As a particular case consider

$$Lx = -x'' = \lambda x + f, \quad x(0) = x(\pi) = 0, \quad I = [0,\pi], \tag{8.25}$$

the nonhomogeneous problem corresponding to our example (I), where $\lambda = 0$ is not an eigenvalue. It is a nice exercise to check that the Green function for L_β is given by

$$g(t,s) = \left\{ \begin{array}{ll} s(\pi - t)/\pi, & s \leq t, \\ \\ t(\pi - s)/\pi, & s > t. \end{array} \right\} \tag{8.26}$$

This can be done by choosing $\mathbf{X}(t) = (1,t)$ and substituting in the formula (8.23) for g with $\lambda = 0$, or g can be computed by calculating the solution of (8.25) directly.

The next result characterizes the Green function g, which is defined on the square $I \times I$. Let T_1, T_2 be the triangles

$$T_1 = \{(t,s) \in I \times I | s < t\},$$

$$T_2 = \{(t,s) \in I \times I | s > t\}.$$

Theorem 8.13: *Let g be the Green function for $L_\beta - \lambda I$ as given by (8.24). Then, for fixed $\lambda \notin \sigma(L_\beta)$,*
    (i) *g is continuous on the square $I \times I$;*
    (ii) *$\partial g/\partial t$ is continuous on each of the triangles T_1, T_2;*
    (iii) *$\partial g/\partial t(s+0, s, \lambda) - \partial g/\partial t(s-0, s, \lambda) = -1/p(s)$;*
    (iv) *if $v_s(t) = g(t, s, \lambda)$, then $Lv_s = 0$, for $t \neq s$;*
    (v) *$\mathbf{B}\hat{v}_s = 0$.*
There is only one function g on $I \times I$ satisfying (i) – (v).
Proof: Write

$$g(t, s, \lambda) = \left\{ \begin{array}{ll} \mathbf{X}(t, \lambda)g_a(s, \lambda), & s \leq t, \\[2mm] -\mathbf{X}(t, \lambda)g_b(s, \lambda), & s > t, \end{array} \right\}$$

where

$$g_a(s, \lambda) = (\mathbf{B}\widehat{\mathbf{X}(\lambda)})^{-1}\mathbf{B}^a\tilde{\mathbf{X}}(a, \lambda)k(s, \lambda),$$

$$g_b(s, \lambda) = (\mathbf{B}\widehat{\mathbf{X}(\lambda)})^{-1}\mathbf{B}^b\tilde{\mathbf{X}}(b, \lambda)k(s, \lambda).$$

Observe that

$$g_a(s, \lambda) + g_b(s, \lambda) = k(s, \lambda).$$

Clearly g and $\partial g/\partial t$ are continuous on T_1 and T_2, for $\mathbf{X}(\lambda), \mathbf{X}'(\lambda), k(\lambda)$ are. The only possible discontinuity of g can occur along the line $s = t$. Using the notation

$$g(s \pm 0, s, \lambda) = \lim_{t \to s\pm} g(t, s, \lambda),$$

we have

$$g(s+0, s, \lambda) = \lim_{t \to s+0} g(t, s, \lambda) = \mathbf{X}(s, \lambda)g_a(s, \lambda),$$

$$g(s-0, s, \lambda) = \lim_{t \to s-0} g(t, s, \lambda) = -\mathbf{X}(s, \lambda)g_b(s, \lambda).$$

This means

$$g(s+0, s, \lambda) - g(s-0, s, \lambda) = \mathbf{X}(s, \lambda)[g_a(s, \lambda) + g_b(s, \lambda)] = \mathbf{X}(s, \lambda)k(s, \lambda) = 0,$$

so that g is continuous on $I \times I$. As to $\partial g/\partial t$,

$$\frac{\partial g}{\partial t}(s+0, s, \lambda) - \frac{\partial g}{\partial t}(s-0, s, \lambda)$$

$$= -\mathbf{X}'(s, \lambda)[g_a(s, \lambda) + g_b(s, \lambda)] = \mathbf{X}'(s, \lambda)k(s, \lambda) = \frac{-1}{p(s)},$$

which proves (iii). Since $L\mathbf{X}(\lambda) = 0$ it follows that $Lv_s = 0$ for $t \neq s$. Finally, v_s satisfies the boundary conditions (v) for

$$\mathbf{B}\hat{v}_s = \mathbf{B}^a\tilde{v}_s(a) + \mathbf{B}^b\tilde{v}_s(b) = \mathbf{B}^a\tilde{g}(a, s, \lambda) + \mathbf{B}^b\tilde{g}(b, s, \lambda)$$

$$= -\mathbf{B}^a\tilde{\mathbf{X}}(a, \lambda)g_b(s, \lambda) + \mathbf{B}^b\tilde{\mathbf{X}}(b, \lambda)g_a(s, \lambda)$$
$$= -[\mathbf{B}^a\tilde{\mathbf{X}}(a, \lambda) + \mathbf{B}^b\tilde{\mathbf{X}}(b, \lambda)]g_b(s, \lambda) + \mathbf{B}^b\tilde{\mathbf{X}}(b, \lambda)[g_a(s, \lambda) + g_b(s, \lambda)]$$
$$= -(\widehat{\mathbf{B}\mathbf{X}(\lambda)})g_b(s, \lambda) + \mathbf{B}^b\tilde{\mathbf{X}}(b, \lambda)k(s, \lambda) = 0,$$

using the definition of $g_b(s, \lambda)$.

Now to show that (i)–(v) characterize g uniquely, suppose h is another function on $I \times I$ satisfying (i)–(v). If

$$r(t, s, \lambda) = h(t, s, \lambda) - g(t, s, \lambda),$$

then from (i)–(iii) we see that $r, \partial r/\partial t$ are continuous on $I \times I$, and (iv) then shows that $r_s(t) = r(t, s, \lambda)$ satisfies

$$Lr_s = -pr_s'' - p'r_s' + qr_s, \quad t \neq s.$$

This equation can be used to define r_s'' to be continuous for $t = s$ via

$$r_s''(s) = \frac{-1}{p(s)}[p'(s)r_s'(s) - q(s)r_s(s)].$$

Then $Lr_s = 0$ for all $t \in I$. The condition (v) implies that $\mathbf{B}\hat{r}_s = 0$. Thus r_s satisfies

$$Lr_s = \lambda r_s, \quad \mathbf{B}\hat{r}_s = 0,$$

or $r_s \in N(L_\beta - \lambda I) = \{0\}$, and hence $h(t, s, \lambda) = g(t, s, \lambda)$ on $I \times I$. $\square$

As a simple application of Theorem 8.13, consider the nonhomogeneous problem

$$Lx = -x'' = \lambda x + f, \quad x(0) + x'(0) = 0, \quad x'(1) = 0, \quad I = [0, 1],$$

which corresponds to our example (III). We know that $\lambda = 0$ is not an eigenvalue for the homogeneous problem. Let $\mathbf{X}$ be the basis for the solutions of $-x'' = 0$ given by $\mathbf{X}(t) = [1, t]$. The condition (iv) in Theorem 8.13 implies that for $\lambda = 0$ the Green function for L_β has the form

$$g(t, s) = \left\{ \begin{array}{ll} c_1 + c_2 t, & s \leq t, \\ d_1 + d_2 t, & s > t, \end{array} \right\}$$

where c_1, c_2, d_1, d_2 may depend on s. The conditions (i), (iii), (v) imply that

$$c_1 + c_2 s - d_1 - d_2 s = 0, \quad c_2 - d_2 = -1, \quad d_1 + d_2 = 0, \quad c_2 = 0,$$

and these equations have the solution

$$c_1 = s - 1, \quad c_2 = 0, \quad d_1 = -1, \quad d_2 = 1.$$

Thus

$$g(t, s) = \left\{ \begin{array}{ll} s - 1, & s \leq t, \\ t - 1, & s > t. \end{array} \right\}$$

For a slightly more complicated illustration, consider the problem

$$Lx = -x'' = \lambda x + f, \quad x(0) = x(\pi) = 0,$$

and suppose $\lambda \notin \sigma(L_\beta)$ and $\lambda \neq 0$. Then the Green function g for $L_\beta - \lambda I$ exists. Applying Theorem 8.13, with the basis $\mathbf{X}(\lambda)$ for the solutions of $-x'' = \lambda x$ given by

$$\mathbf{X}(t, \nu^2) = (\cos(\nu t), \sin(\nu t)), \quad \lambda = \nu^2, \quad \nu \in \mathcal{C},$$

it can be shown that

$$g(t, s, \nu^2) = \left\{ \begin{array}{ll} \sin(\nu s)\sin(\nu[\pi - t])/[\nu \, \sin(\nu\pi)], & s \leq t, \\ \sin(\nu t)\sin(\nu[\pi - s])/[\nu \, \sin(\nu\pi)], & s > t. \end{array} \right\} \tag{8.27}$$

Observe that as $\nu \to 0$,

$$g(t, s, \nu^2) \to \left\{ \begin{array}{ll} s(\pi - t)/\pi, & s \leq t, \\ t(\pi - s)/\pi, & s > t, \end{array} \right\}$$

which is $g(t, s, 0)$, as pointed out previously.

Let us return to the general nonhomogeneous problem

$$Lx = \lambda x + f, \quad \mathbf{B}\hat{x} = 0, \quad f \in C(I, \mathcal{C}). \tag{NH}$$

From Theorem 8.12, if $\lambda \notin \sigma(L_\beta)$, (NH) has a unique solution x given by

$$x(t, \lambda) = \int_a^b g(t, s, \lambda)f(s) \, ds,$$

where g is the Green function for $L_\beta - \lambda I$, which is characterized in Theorem 8.13. Define $G(\lambda) : C(I, \mathcal{C}) \to D_\beta$, the *Green operator* for λ, by

$$G(\lambda)f(t) = \int_a^b g(t, s, \lambda)f(s) \, ds, \quad f \in C(I, \mathcal{C}).$$

The equation

$$(L_\beta - \lambda I)G(\lambda)f = f, \quad f \in C(I, \mathcal{C}),$$

just expressed the fact that $x = G(\lambda)f$ solves (NH). The uniqueness of this solution shows that $x = G(\lambda)f$, or

$$G(\lambda)(L_\beta - \lambda I)x = x, \quad x \in D_\beta. \tag{8.28}$$

Suppose that $f = (L_\beta - \lambda I)x$. Then $f \in C(I, \mathcal{C})$, $G(\lambda)f \in D_\beta$, and

$$(L_\beta - \lambda I)G(\lambda)f = f = (L_\beta - \lambda I)x, \tag{8.29}$$

so that $x = G(\lambda)f$, which is (8.28). The relations (8.28) and (8.29) show that the Green operator

$$G(\lambda) : f \in C(I, \mathcal{C}) \rightarrow G(\lambda)f \in D_\beta$$

is the *inverse* of the differential operator $L_\beta - \lambda I$; that is, $G(\lambda) = (L_\beta - \lambda I)^{-1}$.

8.6   Notes

The study of selfadjoint operators on a Hilbert space is a major topic in mathematics, particularly because of the applications to quantum mechanics. In a sense, the selfadjoint problems studied in this and the following chapter are among the most elementary. In many of the physically motivated examples, the interval I is not compact. This has a substantial impact on the description of the spectrum of the operator. The book [21], particularly Chapter 8, serves as a modern introduction to more general problems. The reference [3] also has an extensive treatment of boundary value problems.

8.7   Exercises

1.   By making use of the chain rule, express equation (8.1) in the new variables

$$\tau = at + b, \quad \xi = cx + d,$$

where a, b, c, d are real constants. What choices of a, b, c, d will convert (8.1) to the special case $k = 1$, $l = \pi$?

2. Many of the differential equations

$$-(p(t)x'(t))' + q(t)x(t) = \lambda x(t) \tag{8.30}$$

are related by elementary transformations.

(a) Using the chain rule, rewrite the equation (8.30) using the variable

$$s = \int_0^t \frac{1}{\sqrt{p(u)}}\, du$$

to get a new equation of the form

$$-x''(s) + a(t(s))x'(s) + b(t(s))x(s) = \lambda x(s). \tag{8.31}$$

What are the new coefficients $a(t), b(t)$? (The function $t(s)$ is implicitly defined.)

(b) Make the substitution

$$x(s) = y(s)w(s), \quad w(s) = \exp\left(\frac{1}{2}\int_0^s a(u)\ du\right)$$

to reduce the equation (8.31) to the *Liouville normal form*

$$-y'' + c(s)y = \lambda y.$$

(c) Express c in terms of the original coefficients p, q.

3. Using the development in section 8.1 as a model, carry through the computation of eigenvalues and eigenfunctions arising in the heat equation with insulated boundary conditions,

$$\frac{\partial u}{\partial t} = k\frac{\partial^2 u}{\partial x^2}, \quad \frac{\partial u}{\partial x}(0, t) = 0 = \frac{\partial u}{\partial x}(l, t), \quad u(x, 0) = f(x),$$

$$0 \le x \le l, \quad t \ge 0.$$

4. Using the development in section 8.1, and particularly (8.6), (8.7), as a model, find the formal solution $u(x, t)$ of the time-dependent *Schrödinger equation*

$$i\frac{\partial u}{\partial t} = k\frac{\partial^2 u}{\partial x^2}, \quad u(0, t) = 0 = u(\pi, t), \quad u(x, 0) = f(x),$$

$$0 \le x \le \pi, \quad t \in \mathcal{R}.$$

5. Let $X, Y \in \mathcal{C}^3$ be given by $X = (i, -i, 1)$, $Y = (1 + i, -2, i)$. Compute

$$\langle X, Y \rangle.$$

Verify that

$$|\langle X, Y \rangle| \le \|X\|_2 \|Y\|_2.$$

6. Verify that the usual dot product on C^n has properties (a)–(d) of section 8.2.

7. Prove the Schwarz inequality for any inner product space $\mathcal{V}$ using the following outline. For $\alpha \in \mathcal{C}$ and $X, Y \in \mathcal{V}$,

$$\langle X - \alpha Y, X - \alpha Y \rangle \ge 0.$$

Show this implies

$$\langle X, X \rangle + \langle Y, Y \rangle \ge 2\mathrm{Re}(\alpha \langle Y, X \rangle).$$

Now take $\langle X, X \rangle = 1 = \langle Y, Y \rangle$ and write $\langle Y, X \rangle = |\langle Y, X \rangle| e^{i\theta}$ for $\theta \in \mathcal{R}$. Use $\alpha = e^{-i\theta}$ to establish the Schwarz inequality in this special case. Finally, apply the special case to $W = X/\|X\|$ and $Z = Y/\|Y\|$.

8. Suppose that
$$f(t) \bullet g(t) = \langle f(t), g(t) \rangle_{\mathcal{R}^n}$$

and that
$$f : (a, b) \to \mathcal{R}^n$$

is differentiable. Show that
$$[f(t) \bullet f(t)]' = 2[f'(t) \bullet f(t)].$$

What happens if $\mathcal{R}^n$ is replaced by $\mathcal{C}^n$?

9. (a) If
$$\|\mathbf{A}\|_2 = \left[\sum_{j=1}^n \sum_{i=1}^m |a_{ij}|^2 \right]^{1/2}, \quad \mathbf{A} \in M_{mn}(\mathcal{F}),$$

show that $\| \ \|$ is a norm on $M_{mn}(\mathcal{F})$.

(b) Show that $\|\mathbf{A}\|_2^2 = \mathrm{tr}(\mathbf{A}^*\mathbf{A})$, where $\mathrm{tr}(\mathbf{C})$ denotes the *trace* of $\mathbf{C} = (c_{ij}) \in M_n(\mathcal{F})$,
$$\mathrm{tr}(\mathbf{C}) = \sum_{i=1}^n c_{ii}.$$

(c) Show that
$$\|\mathbf{A}\|_2 \leq |\mathbf{A}| \leq \sqrt{mn}\|\mathbf{A}\|_2$$

for all $\mathbf{A} \in M_{mn}(\mathcal{F})$.

10. For $f, g \in C(I, \mathcal{C})$, where $I = [0, 1]$, let
$$\langle f, g \rangle = \int_0^1 f(t)\overline{g}(t) \ dt.$$

Show that $\langle \ , \ \rangle$ is an inner product on $C(I, \mathcal{C})$.

11. For $f, g \in C(I, \mathcal{C}^n)$, where $I = [0, 1]$, let
$$\langle f, g \rangle = \int_0^1 g^*(t)f(t) \ dt.$$

(a) Show that $\langle \ , \ \rangle$ is an inner product on $C(I, \mathcal{C}^n)$.

(b) If $\|f\|_2 = \langle f, f \rangle^{1/2}$, $f \in C(I, \mathcal{C}^n)$, show that $\| \ \|_2$ is a norm on $C(I, \mathcal{C}^n)$.

(c) Show that $\|f\|_2 \leq \|f\|_\infty$, $f \in C(I, \mathcal{C}^n)$.

(d) Are $\| \ \|_\infty$ and $\| \ \|_2$ equivalent norms on $C(I, \mathcal{C}^n)$?

12. Let $\mathcal{V}$ be a vector space with inner product $(\ , \ )$. For a set $S \subset \mathcal{V}$, define
$$S^\perp = \{X \in \mathcal{V} \mid \langle X, Y \rangle = 0 \quad \text{for all} \quad Y \in S\}.$$

(a) Show that $S^\perp$ is a subspace of $\mathcal{V}$.

(b) If $V = \mathcal{F}^n$ and S is a subspace of dimension k, what is the dimension of $S^\perp$?

13. Suppose that $\mathbf{A}$ is an $n \times n$ matrix, and $\langle X, Y \rangle$ denotes the usual inner product on $\mathcal{F}^n$. Using the equation

$$\langle \mathbf{A}X, Y \rangle = \langle X, \mathbf{A}^*Y \rangle, \quad X, Y \in \mathcal{F}^n,$$

show that

$$N(\mathbf{A}) = Ran(\mathbf{A}^*)^\perp$$

and consequently that

$$N(\mathbf{A})^\perp = Ran(\mathbf{A}^*).$$

(See exercise 12.)

14. Suppose that V is a finite-dimensional vector space over $\mathcal{F}$ with inner product $\langle\,,\,\rangle$. Say that a linear transformation $\mathbf{A} : V \to V$ is selfadjoint if

$$\langle \mathbf{A}X, Y \rangle = \langle X, \mathbf{A}Y \rangle \quad \text{for all} \quad X, Y \in V.$$

(a) If Z is an eigenvector for $\mathbf{A}$, show that $\mathbf{A}$ is selfadjoint on $Z^\perp$ with the inner product $\langle\,,\,\rangle$ inherited from V. (See exercise 12.)

(b) Extend part (a) to show that V has a basis of orthogonal eigenvectors of $\mathbf{A}$.

15. Show directly from Green's formula that the real separated boundary conditions are always selfadjoint.

16. Let $Lx = (px')' + q(x)$ on $I = [0, 1]$. If $p(1) = p(0)$, show that the boundary conditions

$$x(1) = e^{i\theta} x(0), \quad x'(1) = e^{i\theta} x'(0), \quad \theta \in \mathcal{R},$$

are selfadjoint.

17. (a) In the proof of Lemma 8.7 we asserted that Theorem 7.6 implied that for any $\epsilon > 0$ there is a $\delta > 0$ such that $|\lambda - \lambda_j| < \delta$ implies that $|x_1(t, \lambda) - x_1(t, \lambda_j)| < \epsilon$ for all $t \in [a, b]$. Provide the details for this argument.

(b) Why does this imply the positivity of the integral

$$\int_0^\pi x_1(t, \lambda_j) \overline{x_1(t, \lambda)} \, dt$$

for λ in some real open neighborhood of λ_j?

18. Show analytically that

$$\sin(\pi t) \geq 2t, \quad 0 \leq t \leq 1/2.$$

Use this result, together with the symmetry and periodicity of the sine function, to establish the claim

$$|\sin(\sqrt{\lambda}\pi)| \geq 2 dist(\sqrt{\lambda}, \mathcal{Z})$$

in the proof of Theorem 8.8.

19. Consider a pair of equations

$$x'' + q_1(t)x = 0, \quad y'' + q_2(t)y = 0, \quad q_1(t) > q_2(t).$$

Suppose that t_1 and t_2 are consecutive zeros of y; that is, $y(t_1) = 0 = y(t_2)$ and $y(t) \neq 0$ for $t_1 < t < t_2$.

(a) Show that the derivatives of y at t_1 and t_2 have opposite sign; more precisely $y'(t_1)y'(t_2) < 0$.

(b) If $W(t) = [x(t)y'(t) - x'(t)y(t)]$, observe that

$$W(t)' = [x(t)y'(t) - x'(t)y(t)]' = x(t)y''(t) - x''(t)y(t) = [q_1(t) - q_2(t)]x(t)y(t).$$

Integrating from t_1 to t_2 we get

$$W(t_2) - W(t_1) = \int_{t_1}^{t_2} [q_1(t) - q_2(t)]x(t)y(t) \, dt.$$

On the other hand,

$$W(t_2) = x(t_2)y'(t_2), \quad W(t_1) = x(t_1)y'(t_1).$$

Show that x cannot be positive over the entire interval $[t_1, t_2]$, so that x vanishes at least once between successive zeros of y, a result known as the *Sturm comparison theorem*.

20. Use exercise 19 to show that if $\alpha > 0$ the Bessel function J_α has infinitely many positive zeros as follows. By Chapter 6 exercise 40, the function $y(t) = t^{1/2} J_\alpha(t)$ satisfies

$$y'' + b(t)y = 0, \quad b(t) = 1 + \frac{1/4 - \alpha^2}{t^2}, \quad t > 0.$$

For all large enough t we have $b(t) > 1/4$. Compare this equation with

$$y'' + \frac{1}{4}y = 0,$$

which is satisfied by $\sin(t/2)$.

21. Modify the argument used for Theorem 8.8 to estimate the locations of the eigenvalues for the problem

$$-x'' + q(t)x = \lambda x, \quad x'(0) = 0 = x'(\pi).$$

22. Show that the Green function for the problem (8.25) is given by (8.26).

23. (a) Using the particular Green function given in (8.27), compute $(L_\beta - \lambda)^{-1} \sin(\omega s)$. What happens as $\lambda = \nu^2$ approaches an eigenvalue of L_β?

(b) Suppose that $f \in C(I, \mathcal{C})$, with norm $\|f\|_\infty = \max_{t \in I} |f(t)| = 1$. Estimate $\|(L_\beta - \lambda)^{-1}f\|_\infty$ as a function of λ.

Chapter 9

Eigenfunction Expansions

9.1 Introduction

In the course of trying to solve the heat equation in section 8.1, we were faced with the problem of expanding a function f in a series of eigenfunctions of a particular selfadjoint eigenvalue problem. This chapter is devoted to eigenfunction expansion results which can be proved for the general selfadjoint problem

$$Lx = -(px')' + qx = \lambda x, \quad \beta(x) = \mathbf{B}\hat{x} = 0, \quad I = [a, b],$$

and the corresponding operator L_β.

Since the vector space $C(I, \mathcal{C})$ is not finite dimensional, the ability to expand a general function f in a series of eigenfunctions will require an infinite sequence of linearly independent eigenfunctions. In addition to the question of the existence of a sequence of independent eigenfunctions, several other questions arise immediately:

    (i) which functions have an eigenfunction expansion?

    (ii) does the series converge?

    (iii) if the series converges, is its sum equal to f?

To begin, suppose that $\{x_n\}$ is an orthonormal sequence $\langle x_m, x_n \rangle = \delta_{mn}$ in an inner product space $\mathcal{V}$, which might be $\mathcal{F}^n$ with its usual inner product, or $C(I, \mathcal{C})$ with the inner product

$$\langle x, y \rangle = \int_a^b x(t)\overline{y(t)} \; dt.$$

If $f \in \mathcal{V}$ is a linear combination of the x_n,

$$f = \sum_{n=1}^{N} a_n x_n, \quad a_n \in \mathcal{C},$$

then the coefficients a_n may be identified by taking the inner product of both sides with x_m. Since the sequence $\{x_n\}$ is orthonormal,

$$\langle f, x_m \rangle = \sum_{n=1}^{N} a_n \langle x_n, x_m \rangle = a_m.$$

Turning to the case where $V = C(I, \mathcal{C})$, the *Fourier series* of $f \in C(I, \mathcal{C})$ with respect to $\{x_n\}$ is the series

$$\sum_{n=1}^{\infty} \langle f, x_n \rangle x_n,$$

and $\langle f, x_n \rangle$ is called the *nth Fourier coefficient* of f with respect to $\{x_n\}$.

As an example, consider the orthonormal eigenfunctions x_n for the selfadjoint problem

$$-x'' = \lambda x, \quad x(0) = x(\pi) = 0, \quad I = [0, \pi], \tag{9.1}$$

namely,

$$x_n(t) = \sqrt{2/\pi} \sin(nt), \quad n = 1, 2, 3, \ldots.$$

If $f(t) = t$, $t \in I$, then an integration shows that

$$\langle f, x_n \rangle = \sqrt{2/\pi} \int_0^\pi t \, \sin(nt) \, dt = -\frac{\sqrt{2\pi}}{n} \cos(n\pi) = \sqrt{2\pi} \frac{(-1)^{n+1}}{n},$$

and hence the Fourier series for this f is

$$\sum_{n=1}^{\infty} \sqrt{2\pi} \frac{(-1)^{n+1}}{n} \sqrt{2/\pi} \sin(nt) = 2 \sum_{n=1}^{\infty} \frac{(-1)^{n+1}}{n} \sin(nt).$$

Note that the sum of this series for $t = \pi$ is 0, which is not $f(\pi) = \pi$, so that question (iii) may have a negative answer. It can be shown that for this f the series converges to $f(t)$ for $0 \le t < \pi$, although we will not stop to show this.

The first main goal for this chapter is to establish the following result on the existence of eigenvalues for a selfadjoint differential operator L_β.

Theorem 9.1: *A selfadjoint L_β has an infinite sequence of eigenvalues λ_n, $n = 1, 2, 3, \ldots$,*

$$-\infty < \lambda_1 \le \lambda_2 \le \cdots \le \lambda_n \le \cdots, \quad \lambda_n \to \infty, \quad n \to \infty,$$

with corresponding eigenfunctions x_n which form an orthonormal set,

$$\langle x_m, x_n \rangle = \delta_{mn}.$$

To simplify the exposition, consider the case when $\lambda = 0$ is not an eigenvalue for L_β. It will be easy to handle the general situation once this case is understood (exercise 6). Let $G = G(0)$ be the Green operator

$$Gf(t) = \int_a^b g(t, s) f(s) \, ds, \quad f \in C(I, \mathcal{C}).$$

From the discussion at the end of Chapter 8 we know that G is the inverse of L_β in the sense that

$$L_\beta G f = f, \quad f \in C(I, \mathcal{C}),$$

$$GL_\beta x = x, \quad x \in D_\beta.$$

Suppose that $L_\beta x = f$ and $L_\beta y = h$, where f, h may be any functions in $C(I, \mathcal{C})$. Since L_β is selfadjoint,

$$\langle Gf, h \rangle = \langle x, L_\beta y \rangle = \langle L_\beta x, y \rangle = \langle f, y \rangle = \langle f, Gh \rangle, \tag{9.2}$$

and G is also called *selfadjoint*.

The basic results concerning eigenvalues and eigenfunctions for L_β are more readily established by considering the Green operator G. A number $\mu \in \mathcal{C}$ is an *eigenvalue* of G if there is a nontrivial $f \in C(I, \mathcal{C})$ such that

$$Gf = \mu f,$$

and such an f is called an *eigenfunction* of G for μ. Since G is the inverse of L_β, it is not surprising that the eigenvalues of G should be the inverses of those for L_β.

Theorem 9.2: *If* $\dim(N(L_\beta)) = 0$, λ *is an eigenvalue of* L_β *with eigenfunction* x *if and only if* $\mu = 1/\lambda$ *is an eigenvalue of* G *with eigenfunction* x.

Proof: Suppose $L_\beta x = \lambda x$, $x \in D_\beta$, $x \neq 0$. Then $GL_\beta x = \lambda Gx = x$, or $Gx = \mu x$, $\mu = 1/\lambda$. Conversely, let $Gx = \mu x$ for some $x \in C(I, \mathcal{C})$, $x \neq 0$. Then $\mu \neq 0$, for if $\mu = 0$ then $Gx = 0$ and $0 = L_\beta Gx = x$, a contradiction. $Gx \in D_\beta$ implies that $x = \lambda Gx \in D_\beta$, $\lambda = 1/\mu$, and $L_\beta x = \lambda L_\beta Gx = \lambda x$. $\square$

A helpful roadmap for this chapter can be drawn by considering the Green operator for example (9.1) as a generalization of a selfadjoint matrix. A selfadjoint matrix $\mathbf{A} \in M_n(\mathcal{F})$ has a finite set of real eigenvalues, and there is an orthonormal basis $\{X_n\}$ of eigenfunctions. The Green operator for (9.1) has a sequence of eigenvalues $\mu_n = 1/n^2$, $\mu_n \to 0$, and there is a corresponding orthonormal sequence $\{x_n(t)\}$ of eigenfunctions.

In the matrix case eigenvalues can be found by looking at the roots of the characteristic polynomial. Since this tool is not available for the Green operator, some other means of finding eigenvalues is needed. By using the orthonormal eigenfunctions we can actually identify two expressions which reveal the value of the eigenvalue of largest magnitude. The following computations are meant as a formal guide, so for now questions about convergence of series will be ignored. First, if $f(t) = \sum a_n x_n(t)$, then

$$\langle f, f \rangle = \left\langle \sum_m a_m x_m(t), \sum_n a_n x_n(t) \right\rangle = \sum_{m,n} a_m \overline{a_n} \langle x_m(t), x_n(t) \rangle.$$

Since the sequence $\{x_n\}$ is orthonormal, i.e., $\langle x_m, x_n \rangle = \delta_{m,n}$, we have

$$\langle f, f \rangle = \sum_n |a_n|^2.$$

Taking $\|f\|^2 = \langle f, f \rangle = 1$, it follows that

$$\langle Gf, Gf \rangle = \left\langle G \sum a_m x_m(t), G \sum a_n x_n(t) \right\rangle = \left\langle \sum_m \mu_m a_m x_m(t), \sum_n \mu_n a_n x_n(t) \right\rangle$$

$$= \sum_{m,n} \mu_m a_m \overline{\mu_n a_n} \langle x_m(t), x_n(t) \rangle = \sum_n |\mu_n a_n|^2.$$

Since $\sum_n |a_n|^2 = 1$,

$$\|Gf\| \leq \sup_n |\mu_n|.$$

If it happens that $|\mu_1| = \sup_n |\mu_n|$, as in the case (9.1), then the choice $f(t) = x_1(t)$ gives

$$\sup_f \|Gf\| = |\mu_1|, \quad \|f\| = 1.$$

A similar calculation reveals that

$$|\langle Gf, f \rangle| = \left| \sum_n \mu_n |a_n|^2 \right| \leq \sup_n |\mu_n|$$

and

$$\sup_f |(Gf, f)| = |\mu_1|, \quad \|f\| = 1.$$

These inequalities motivate the following plan. We will try to find a single eigenfunction for G by considering a sequence f_n such that $\|f_n\| = 1$ and

$$\lim_n \langle Gf_n, f_n \rangle = \sup \langle Gf, f \rangle, \quad \|f\| = 1.$$

A compactness argument will show that some subsequence of the sequence $\{Gf_n\}$ converges to an eigenfunction of G. The magnitude of the corresponding eigenvalue will be maximal. Having obtained eigenfunctions, the process will be repeated on the orthogonal complement, thus producing a sequence of orthogonal eigenfunctions and corresponding eigenvalues of decreasing magnitude. As this plan unfolds, it will also be possible to provide good answers to the apparently deeper questions (i)–(iii) above.

9.2   Selfadjoint integral operators

It will be convenient to develop results which apply to integral operators more general than Green's operator. The results of this section and the first result, Theorem 9.6, of the next section, are valid for *integral operators*

$$Gf(t) = \int_a^b g(t, s) f(s) \, ds$$

satisfying two conditions:
    (i) the kernel $g(t, s)$ is a continuous function on $I \times I$,
    (ii) $(Gf, h) = (f, Gh)$ for all $f, h \in C(I, \mathcal{C})$.

Integral operators satisfying condition (ii) will be called *selfadjoint*. The notation used for Green's operator will be retained in this more general setting.

Let B_1 denote the unit sphere in $C(I, \mathcal{C})$,

$$B_1 = \{f \in C(I, \mathcal{C}) \mid \|f\|_2 = 1\},$$

and define the *norm* $\|G\|$ of the integral operator G by

$$\|G\| = \sup\{\|Gf\|_2 \mid f \in B_1\}.$$

From this definition it follows that

$$\|Gf\|_2 \leq \|G\| \, \|f\|_2, \quad f \in C(I, \mathcal{C}). \tag{9.3}$$

This is clearly true if $f = 0$. If $\|f\|_2 > 0$, then $h = f/\|f\|_2 \in B_1$ and

$$\frac{\|Gf\|_2}{\|f\|_2} = \|Gh\|_2 \leq \|G\|,$$

so that (9.3) is true for all $f \in C(I, \mathcal{C})$.

Theorem 9.3: *The norm of the selfadjoint integral operator G satisfies*

(i) $$0 \leq \|G\| < \infty$$

and

(ii) $$\|G\| = \sup\{|\langle Gf, f \rangle| \mid f \in B_1\}.$$

Proof: Since $g(t, s)$ is continuous on $I \times I$ it follows that g is bounded on the square $I \times I$; that is, there is a constant $M > 0$ such that

$$|g(t, s)| \leq M, \quad t, s \in I.$$

Hence for each $f \in C(I, \mathcal{C})$,

$$|Gf(t)| = \left| \int_a^b g(t, s) f(s) \, ds \right| \leq \int_a^b |g(t, s)| |f(s)| \, ds \leq M \int_a^b |f(s)| \, ds.$$

Since the last integral is the inner product $\langle 1, |f| \rangle$ of the constant function 1 with $|f|$, the Schwarz inequality yields

$$\int_a^b |f(s)| \, ds = \langle 1, |f| \rangle \leq \|1\| \, \|f\|_2 = |I|^{1/2} \|f\|_2,$$

where $|I| = b - a$ is the length of I. Thus

$$|Gf(t)| \leq M|I|^{1/2} \|f\|_2, \quad t \in I, \tag{9.4}$$

and it follows that

$$\|Gf\|_2^2 = \int_a^b |Gf(t)|^2 \, dt \le M^2 |I|^2 \|f\|_2^2$$

or

$$\|Gf\|_2 \le M|I|\|f\|_2.$$

Since this inequality is valid for any $f \in C(I, \mathcal{C})$,

$$0 \le \|G\| \le M|I| < \infty,$$

proving (i).

Turning to a proof of (ii), note first that $\langle Gf, f \rangle \in \mathcal{R}$ for all $f \in C(I, \mathcal{C})$, since

$$2i\mathrm{Im}(\langle Gf, f \rangle) = \langle Gf, f \rangle - \overline{\langle Gf, f \rangle} = \langle Gf, f \rangle - \langle f, Gf \rangle = 0.$$

Let

$$\gamma = \sup\{|\langle Gf, f \rangle| \mid f \in B_1\}.$$

If $\|f\|_2 = 1$, then

$$|\langle Gf, f \rangle| \le \|Gf\|_2\|f\|_2 \le \|G\|,$$

so that $\gamma \le \|G\|$.

Observe that in general

$$|\langle Gf, f \rangle| \le \gamma\|f\|_2^2, \quad f \in C(I, \mathcal{C}).$$

If $f = 0$ this is clearly true, and, for $f \ne 0$, the function $h = f/\|f\|_2$ is in B_1 and

$$\frac{|\langle Gf, f \rangle|}{\|f\|_2^2} = |\langle Gh, h \rangle| \le \gamma.$$

The proof of the opposite inequality $\|G\| \le \gamma$ begins with the identity

$$\langle Gh, f \rangle = \langle h, Gf \rangle = \overline{\langle Gf, h \rangle}, \quad f, h \in C(I, \mathcal{C}),$$

which implies

$$4\mathrm{Re}\langle Gf, h \rangle = \langle G(f + h), f + h \rangle - \langle G(f - h), f - h \rangle.$$

This means that

$$4\mathrm{Re}\langle Gf, h \rangle \le \gamma(\|f + h\|_2^2 + \|f - h\|_2^2) = 2\gamma(\|f\|_2^2 + \|h\|_2^2).$$

If $\|Gf\|_2 = 0$, then clearly $\|Gf\|_2 \le \gamma$. For $\|f\|_2 = 1$ and $\|Gf\|_2 > 0$, we take $h = Gf/\|Gf\|_2$. Then $\|h\|_2 = 1$, and

$$4\|Gf\|_2 = \frac{4\langle Gf, Gf \rangle}{\|Gf\|_2} \le 4\gamma, \quad \|f\|_2 = 1.$$

Thus $\|G\| \le \gamma$, and (ii) is valid. $\square$

Two further properties of G will play an essential role in the proof of the existence of eigenvalues and eigenfunctions. A set $S \subset C(I, \mathcal{C})$ is said to be *uniformly bounded* if it is bounded in the uniform norm; that is, there is a constant $k > 0$, independent of $h \in S$, such that

$$\|h\|_\infty = \sup\{|h(t)| \mid t \in I\} \le k, \quad h \in S.$$

The set S is said to be *equicontinuous* if given any $\epsilon > 0$ there is a $\delta > 0$, depending only on ϵ, but not on $h \in S$, such that $|h(t_1) - h(t_2)| < \epsilon$ for all $h \in S$ whenever $|t_1 - t_2| < \delta$. Once we prove that the set

$$G(B_1) = \{Gf \mid f \in B_1\} \subset C(I, \mathcal{C})$$

is uniformly bounded and equicontinuous, the following result can be employed. (Recall that I is a compact interval.)

Ascoli–Arzela theorem: *Let $h_n \in C(I, \mathcal{C})$, $n = 1, 2, \ldots$, and suppose that the sequence $\{h_n\}$ is uniformly bounded and equicontinuous. Then there is a subsequence $\{h_{n(j)}\}$ and a continuous function h on I such that $\{h_{n(j)}\}$ converges uniformly to h,*

$$\lim_{j \to \infty} \|h_{n(j)} - h\|_\infty = 0.$$

A proof of this important result can be found in [17, p. 299] and [22, p. 144].

Theorem 9.4: *The set*

$$G(B_1) = \{Gf \mid f \in B_1\}$$

is uniformly bounded and equicontinuous.

Proof: Inequality (9.4) shows that for $f \in B_1$,

$$|Gf(t)| \le k, \quad k = M|I|^{1/2}, \quad t \in I$$

or

$$\|Gf\|_\infty \le k,$$

so that $G(B_1)$ is uniformly bounded.

Since the kernel $g(t, s)$ for G is continuous on the square $I \times I$, and since this is a closed bounded (compact) set, g is uniformly continuous on $I \times I$. This implies that given any $\epsilon > 0$ there is a $\delta > 0$, independent of $s \in I$, such that

$$|g(t_1, s) - g(t_2, s)| < \frac{\epsilon}{|I|^{1/2}} \quad \text{if} \quad |t_1 - t_2| < \delta.$$

If $f \in B_1$ and $|t_1 - t_2| < \delta$, then

$$|Gf(t_1) - Gf(t_2)| = \left| \int_a^b [g(t_1, s) - g(t_2, s)] f(s) \, ds \right|$$

$$< \frac{\epsilon}{|I|^{1/2}} \int_a^b |f(s)| \, ds \le \frac{\epsilon}{|I|^{1/2}} |I|^{1/2} \|f\|_2 = \epsilon,$$

where the Schwarz inequality has been used. This shows the equicontinuity of $G(B_1)$. $\square$

Applying the Ascoli–Arzela theorem to the set $G(B_1)$ yields the following result.

Corollary 9.5: *If $f_n \in B_1$, $n = 1, 2, \ldots$, then the sequence $\{Gf_n\}$ has a subsequence $Gf_{n(j)}$ which converges uniformly on I to an $h \in C(I, \mathcal{C})$,*

$$\lim_{j \to \infty} \|Gf_{n(j)} - h\|_\infty = 0.$$

We remark that if $h \in C(I, \mathcal{C})$, then

$$\|h\|_2^2 = \int_a^b |h(t)|^2 \, dt \le \|h\|_\infty^2 |I|$$

or

$$\|h\|_2 \le |I|^{1/2} \|h\|_\infty.$$

Thus h is bounded in the norm $\| \ \|_2$. Also, this implies that if $\|h_n - h\|_\infty \to 0$ as $n \to \infty$, then $\|h_n - h\|_2 \to 0$ as $n \to \infty$.

9.3   Eigenvalues for Green's operator

A key step toward Theorem 9.1 is showing the existence of at least one eigenvalue for G. A sequence of eigenvalues for L_β will then be found by an induction argument.

Theorem 9.6: *There exists an eigenfunction $x_1 \in B_1$ of G with eigenvalue μ_1, where $|\mu_1| = \|G\|$. No other eigenvalue for G has a larger magnitude.*

Proof: If $\|G\| = 0$ then $\|Gf\| = 0$, and so $Gf = 0$ for all $f \in C(I, \mathcal{C})$. In this case 0 is an eigenvalue. Assume now that $\|G\| > 0$. From Theorem 9.3 (ii), $\|G\| = \sup\{|\langle Gf, f \rangle| \mid f \in B_1\}$. There are two cases: either $\|G\| = \sup\{\langle Gf, f \rangle | f \in B_1\}$ or $\|G\| = \sup\{-\langle Gf, f \rangle | f \in B_1\}$. Assuming the first case, we will show $\mu_1 = \|G\|$ is an eigenvalue.

There is a sequence $\{f_n\} \subset B_1$ such that $\langle Gf_n, f_n \rangle \to \mu_1$. Corollary 9.5 implies that there is a subsequence $\{Gf_{n(j)}\}$ of $\{Gf_n\}$ which converges uniformly to a continuous function y_1. After a relabeling, this subsequence can be assumed to be $\{Gf_n\}$. Thus

$$\|Gf_n - y_1\|_\infty \to 0, \quad n \to \infty,$$

which implies

$$\|Gf_n - y_1\|_2 \to 0, \quad n \to \infty.$$

Since

$$\left| \ \|Gf_n\|_2 - \|y_1\|_2 \ \right| \le \|Gf_n - y_1\|_2 \to 0, \quad n \to \infty,$$

the inequality

$$\|Gf_n\|_2 \to \|y_1\|_2, \quad n \to \infty,$$

also follows. Moreover, by the Schwarz inequality,

$$0 < \mu_1 = \lim_n \langle Gf_n, f_n \rangle \le \lim_n \|Gf_n\|_2 = \|y_1\|_2,$$

implying that y_1 is not the zero function.

To show that y_1 is an eigenfunction of G with eigenvalue μ_1, Theorem 9.3 (ii) and the inequality $\|Gf_n\|_2 \le \|G\| \|f_n\|_2 = \mu_1$ are used in the following computation:

$$0 \le \|Gf_n - \mu_1 f_n\|_2^2 = \langle Gf_n - \mu_1 f_n, Gf_n - \mu_1 f_n \rangle = \|Gf_n\|_2^2 - 2\mu_1 \langle Gf_n, f_n \rangle + \mu_1^2$$

$$\le \mu_1^2 + \mu_1^2 - 2\mu_1 \langle Gf_n, f_n \rangle \to 2\mu_1^2 - 2\mu_1^2 = 0, \quad n \to \infty.$$

This means that

$$\|Gf_n - \mu_1 f_n\|_2 \to 0, \quad n \to \infty.$$

Applying the triangle inequality,

$$0 \le \|Gy_1 - \mu_1 y_1\|_2 \le \|Gy_1 - G(Gf_n)\|_2 + \|G(Gf_n) - \mu_1 Gf_n\|_2 + \|\mu_1 Gf_n - \mu_1 y_1\|_2$$

$$\le \|G\| \|y_1 - Gf_n\|_2 + \|G\| \|Gf_n - \mu_1 f_n\|_2 + |\mu_1| \|Gf_n - y_1\|_2 \to 0, \quad n \to \infty.$$

Thus $Gy_1 = \mu_1 y_1$, and y_1 is an eigenfunction of G with eigenvalue μ_1.

The case of $\|G\| = \sup\{-\langle Gf, f \rangle | f \in B_1\}$ is handled by constructing an eigenfunction y_1 for $-G$ instead of G. Then $-Gy_1 = \nu_1 y_1$ or $Gy_1 = -\nu_1 y_1$. Hence $\mu_1 = -\nu_1$ is an eigenvalue of G with eigenfunction y_1, and $|\mu_1| = \nu_1 = \|G\|$. In either case, letting $x_1 = y_1/\|y_1\|_2$, we see that $x_1 \in B_1$ is an eigenfunction of G with eigenvalue μ_1, where $|\mu_1| = \|G\|$.

Finally, suppose that μ is any eigenvalue of G with normalized eigenfunction $x \in B_1$. Then $|\mu| = \|Gx\| \le \|G\| = |\mu_1|$, so that μ_1 is an eigenvalue of G having the largest possible magnitude. $\square$

Theorem 9.6 can now be used to show that Green's operator has an infinite sequence of orthogonal eigenfunctions and eigenvalues. Together with Theorem 9.2 this will prove the existence of the infinite sequence of eigenvalues and orthonormal eigenfunctions for L_β, as asserted in Theorem 9.1, in the case where $N(L_\beta) = 0$.

Once again let G denote Green's operator. First note that Green's operator does not have norm 0. If $\|G\| = 0$, then (9.3) implies that $\|Gf\|_2 = 0$ for all $f \in C(I, \mathcal{C})$ or $Gf = 0$. But then $0 = L_\beta Gf = f$, a contradiction.

Theorem 9.7: *Green's operator G for L_β has an infinite sequence of eigenfunctions x_j, $j = 1, 2, \ldots$, which form an orthonormal set*

$$\langle x_i, x_j \rangle = \delta_{ij}, \quad i, j = 1, 2, \ldots,$$

with eigenvalues $\mu_1, \mu_2, \ldots$, satisfying

$$|\mu_1| \ge |\mu_2| \ge \cdots > 0.$$

Proof: Our first eigenfunction x_1, with eigenvalue μ_1 satisfying $|\mu_1| = \|G\|$, is obtained by applying Theorem 9.6 to Green's operator. An induction argument will be used to show that for any positive integer k the constructed orthonormal sequence of eigenfunctions $x_1, \ldots, x_k$ can be extended to an orthonormal sequence of eigenfunctions $x_1, \ldots, x_k, x_{k+1}$. As a consequence of the construction, the sequence of corresponding eigenvalues μ_k will have nonincreasing magnitudes.

Given orthonormal eigenfunctions $x_1, \ldots, x_k$ of G and corresponding eigenvalues $\mu_1, \ldots, \mu_k$ with nonincreasing magnitudes, define an integral operator

$$G_k f(t) = \int_a^b g_k(t, s) f(s) \, ds,$$

where

$$G_k f = Gf - \sum_{j=1}^{k} \langle Gf, x_j \rangle x_j = Gf - \sum_{j=1}^{k} \mu_j \langle f, x_j \rangle x_j.$$

The kernel of the operator G_k is

$$g_k(t, s) = g(t, s) - \sum_{j=1}^{k} \mu_j x_j(t) \overline{x_j(s)},$$

which is continuous on $I \times I$. Thus $\|G_k\| < \infty$. Moreover, the operator G_k is selfadjoint, for if $f, h \in C(I, \mathcal{C})$, then

$$\langle G_k f, h \rangle = \langle Gf, h \rangle - \sum_{j=1}^{k} \mu_j \langle f, x_j \rangle \langle x_j, h \rangle$$

$$= \langle f, Gh \rangle - \left\langle f, \sum_{j=1}^{k} \mu_j \langle h, x_j \rangle x_j \right\rangle = \left\langle f, Gh - \sum_{j=1}^{k} \mu_j \langle h, x_j \rangle x_j \right\rangle = \langle f, G_k h \rangle.$$

Note too that $\|G_k\| > 0$. If $\|G_k\| = 0$ then $G_k f = 0$ for all $f \in C(I, \mathcal{C})$. Since $Gf, x_j \in D_\beta$ and $\mu_j L_\beta x_j = x_j$ (using Theorem 9.2),

$$0 = L_\beta G_k f = L_\beta Gf - \sum_{j=1}^{k} \mu_j \langle f, x_j \rangle L x_j = f - \sum_{j=1}^{k} \langle f, x_j \rangle x_j.$$

This says that each $f \in C(I, \mathcal{C})$ can be written as a linear combination of $x_1, \ldots, x_k$,

$$f = \sum_{j=1}^{k} \langle f, x_j \rangle x_j,$$

which is not true by dimension count, proving $\|G_k\| > 0$.

Apply Theorem 9.6 to G_k, obtaining a normalized eigenfunction $x_{k+1} \in B_1$ and eigenvalue $\mu_{k+1} \in \mathcal{R}$ such that

$$G_k x_{k+1} = \mu_{k+1} x_{k+1}, \quad |\mu_{k+1}| = \|G_k\| > 0.$$

For any $f \in C(I, \mathcal{C})$ and $i = 1, \ldots, k$,

$$\langle G_k f, x_i \rangle = \langle Gf, x_i \rangle - \sum_{j=1}^{k} \langle Gf, x_i \rangle \langle x_j, x_i \rangle = \langle Gf, x_i \rangle - \langle Gf, x_i \rangle = 0.$$

In particular

$$\langle x_{k+1}, x_i \rangle = \langle Gx_{k+1}, x_i \rangle / \mu_{k+1} = 0, \quad i = 1, \ldots, k,$$

which shows that $x_1, \ldots, x_{k+1}$ is an orthonormal set.

Now x_{k+1} is an eigenfunction for G, since

$$\mu_{k+1} x_{k+1} = G_k x_{k+1} = Gx_{k+1} - \sum_{j=1}^{k} \mu_j \langle x_{k+1}, x_j \rangle x_j = Gx_{k+1}.$$

Moreover, x_{k+1} is an eigenfunction for G_{k-1}, for

$$\mu_{k+1} x_{k+1} = G_k x_{k+1} = G_{k-1} x_{k+1} - \langle G_{k-1} x_{k+1}, x_k \rangle$$

$$= G_{k-1} x_{k+1} - \langle x_{k+1}, G_{k-1} x_k \rangle = G_{k-1} x_{k+1} - \mu_k \langle x_{k+1}, x_k \rangle = G_{k-1} x_{k+1}.$$

This implies that

$$|\mu_{k+1}| = \|G_k x_{k+1}\|_2 = \|G_{k-1} x_{k+1}\|_2 \le \|G_{k-1}\| = |\mu_k|,$$

so that

$$|\mu_1| \ge \cdots \ge |\mu_{k+1}| > 0. \quad \square$$

9.4   Convergence of eigenfunction expansions

The following simple inequality is basic in understanding eigenfunction expansions.

Lemma 9.8: *If $\{x_j\}$ is an orthonormal sequence in $C(I, \mathcal{C})$ and $\langle f, x_j \rangle$ is the jth Fourier coefficient of an $f \in C(I, \mathcal{C})$ with respect to $\{x_j\}$, then*

$$\sum_{j=1}^{\infty} |\langle f, x_j \rangle|^2 \le \|f\|_2^2. \qquad \text{(Bessel's inequality)}$$

Proof: Making use of the identities $\langle x_j, f \rangle = \overline{\langle f, x_j \rangle}$ and $\langle x_i, x_j \rangle = \delta_{ij}$, we find that, for $k \ge 1$,

$$0 \le \|f - \sum_{j=1}^{k} \langle f, x_j \rangle x_j\|_2^2 = \left\langle f - \sum_{i=1}^{k} \langle f, x_i \rangle x_i, f - \sum_{j=1}^{k} \langle f, x_j \rangle x_j \right\rangle$$

$$= \langle f, f \rangle - \sum_{i=1}^{k} \langle f, x_i \rangle \langle x_i, f \rangle - \sum_{j=1}^{k} \langle f, x_j \rangle \langle x_j, f \rangle + \sum_{i=1}^{k} \sum_{j=1}^{k} \langle f, x_i \rangle \langle x_j, f \rangle \langle x_i, x_j \rangle$$

$$= \|f\|_2^2 - \sum_{j=1}^{k} |\langle f, x_j \rangle|^2.$$

Thus $\sum_{j=1}^{k} |\langle f, x_j \rangle|^2 \le \|f\|_2^2$, and this implies that the series $\sum_{j=1}^{\infty} |\langle f, x_j \rangle|^2$ is convergent and satisfies Bessel's inequality. $\square$

Theorem 9.9: *Let $\{x_j\}$ be the orthonormal sequence of eigenfunctions of G, with $G x_j = \mu_j x_j$, $j = 1, 2, \ldots$, as given in Theorem 9.7. Then $|\mu_j| \to 0$ as $j \to \infty$, and, if $x \in D_\beta$, its Fourier series with respect to this sequence $\{x_j\}$ is uniformly convergent to x,*

$$\|x - \sum_{j=1}^{k} \langle x, x_j \rangle x_j \|_\infty \to 0, \quad k \to \infty.$$

Proof: For $j = 1, 2, 3, \ldots$,

$$G x_j(t) = \int_a^b g(t, s) x_j(s) \, ds = \mu_j x_j(t), \quad t \in I,$$

and also

$$\int_a^b \overline{g(t, s) x_j(s)} \, ds = \mu_j \overline{x_j(t)}.$$

If $g_t(s) = \overline{g(t, s)}$ this is the same as

$$\langle g_t, x_j \rangle = \mu_j \overline{x_j(t)};$$

that is, $\mu_j \overline{x_j(t)}$ is the jth Fourier coefficient of g_t with respect to $\{x_j\}$. The Bessel inequality then implies that

$$\sum_{j=1}^{k} \mu_j^2 |x_j(t)|^2 \le \sum_{j=1}^{\infty} \mu_j^2 |x_j(t)|^2 \le \|g_t\|_2^2 = \int_a^b |g(t, s)|^2 \, ds. \qquad (9.5)$$

Since $|g(t, s)|$ is bounded, integrating both sides of (9.5) shows that there is a constant C such that

$$\sum_{j=1}^{k} \mu_j^2 \le \int_a^b \int_a^b |g(t, s)|^2 \, ds \, dt = C.$$

This inequality holds for all k, so

$$\sum_{j=1}^{\infty} \mu_j^2 \le C$$

and the series is convergent. In particular, $\mu_j^2 \to 0$ and $|\mu_j| \to 0$ as $j \to \infty$.

Moreover, the Fourier series of $x \in D_\beta$ is *absolutely uniformly convergent*; that is, the series

$$\sum_{j=1}^{\infty} |\langle x, x_j \rangle x_j(t)| \qquad (9.6)$$

is uniformly convergent on I. This will be demonstrated by verifying that the partial sums

$$s_k(t) = \sum_{j=1}^{k} |\langle x, x_j \rangle x_j(t)|$$

form a Cauchy sequence in the sup norm.

Each $x \in D_\beta$ can be written uniquely as $x = Gf$ for some $f \in C(I, \mathcal{C})$; in fact $f = L_\beta x$. Then

$$\langle x, x_j \rangle = \langle Gf, x_j \rangle = \langle f, Gx_j \rangle = \mu_j \langle f, x_j \rangle.$$

If $l > k$ the Schwarz inequality yields

$$|s_l(t) - s_k(t)|^2 = \left[\sum_{j=k+1}^{l} |\mu_j| |x_j(t)| |\langle f, x_j \rangle| \right]^2$$

$$\leq \left(\sum_{j=k+1}^{l} |\mu_j|^2 |x_j(t)|^2 \right) \left(\sum_{j=k+1}^{l} |\langle f, x_j \rangle|^2 \right). \tag{9.7}$$

Now (9.5) implies that

$$\sum_{j=k+1}^{l} |\mu_j|^2 |x_j(t)|^2 \leq \sum_{j=k+1}^{\infty} |\mu_j|^2 |x_j(t)|^2 \leq \int_a^b |g(t,s)|^2 \, ds \leq M^2 |I|,$$

where $M = \sup\{|g(t,s)| \mid t, s \in I\}$.

From Bessel's inequality for f we know that the series

$$\sum_{j=1}^{\infty} |\langle f, x_j \rangle|^2$$

is convergent, implying that

$$\sum_{j=k+1}^{l} |\langle f, x_j \rangle|^2 \to 0, \quad k, l \to \infty.$$

Therefore (9.7) yields

$$|s_l - s_k| \leq M |I|^{1/2} \left(\sum_{j=k+1}^{l} |\langle f, x_j \rangle|^2 \right)^{1/2} \to 0, \quad k, l \to \infty,$$

and the series (9.6) is uniformly convergent on I.

The final step is the proof that the Fourier series of $x \in D_\beta$ converges uniformly to x. If

$$u_k = \sum_{j=1}^{k} \langle x, x_j \rangle x_j,$$

it is easy to see that

$$0 \le \|u_l - u_k\|_\infty \le \|s_l - s_k\|_\infty \to 0, \quad k, l \to \infty.$$

Thus for some $h \in C(I, \mathcal{C})$, the sequence $\{u_k\}$ converges uniformly to h,

$$\|u_k - h\|_\infty \to 0, \quad k \to \infty,$$

and therefore

$$\|u_k - h\|_2 \to 0, \quad k \to \infty.$$

Since

$$\|x - u_k\|_2 = \|Gf - \sum_{j=1}^{k} \langle Gf, x_j \rangle x_j\|_2 = \|G_k f\|_2$$

$$\le \|G_k\|\|f\|_2 = |\mu_{k+1}|\|f\|_2 \to 0, \quad k \to \infty,$$

we finally have

$$\|x - h\|_2 = \|Gf - h\|_2 \le \|Gf - u_k\|_2 + \|u_k - h\|_2 \to 0, \quad k \to \infty,$$

or $x = h$. That is,

$$\|x - u_k\|_\infty = \|x - \sum_{j=1}^{k} \langle x, x_j \rangle x_j\|_\infty \to 0, \quad k \to \infty. \quad \square$$

Notice that the proof of Theorem 9.1 has now been completed. In fact, as a direct consequence of Theorems 9.2 and 9.9, the following important result is obtained.

Theorem 9.10: *When* $\dim(N(L_\beta)) = 0$, *the selfadjoint differential operator* L_β *has an infinite sequence of orthonormal eigenfunctions* $\{x_j\}$, *with eigenvalues* $\lambda_j = 1/\mu_j$, *where*

$$0 < |\lambda_1| \le |\lambda_2| \le \cdots, \quad |\lambda_j| \to \infty.$$

Each $x \in D_\beta$ *has an eigenfunction expansion*

$$x = \sum_{j=1}^{\infty} \langle x, x_j \rangle x_j,$$

the series converging uniformly to x *on* I.

Theorem 9.10 implies that

$$\|x - \sum_{j=1}^{k} \langle x, x_j \rangle x_j\|_2 \to 0, \quad k \to \infty, \tag{9.8}$$

if $x \in D_\beta$. From this we have the following result.

Theorem 9.11: *If $x, y \in D_\beta$, then*

$$\langle x, y \rangle = \sum_{j=1}^{\infty} \overline{\langle y, x_j \rangle} \langle x, x_j \rangle \tag{9.9}$$

and

$$\|x\|_2^2 = \sum_{j=1}^{\infty} |\langle x, x_j \rangle|^2. \qquad (\textit{Parseval equality}) \tag{9.10}$$

Proof: Using the Schwarz inequality for a slight modification of the proof of Bessel's inequality shows that

$$\left| \left\langle x - \sum_{i=1}^{k} \langle x, x_i \rangle x_i, \, y - \sum_{j=1}^{k} \langle y, x_j \rangle x_j \right\rangle \right| = \left| \langle x, y \rangle - \sum_{j=1}^{k} \overline{\langle y, x_j \rangle} \langle x, x_j \rangle \right|$$

$$\leq \left\| x - \sum_{i=1}^{k} \langle x, x_i \rangle x_i \right\|_2 \left\| y - \sum_{j=1}^{k} \langle y, x_j \rangle x_j \right\|_2 \to 0, \quad k \to \infty.$$

This proves (9.9) and taking $y = x$ gives (9.10). $\square$

One consequence of the Parseval equality is that the sequence of orthonormal eigenfunctions $\{x_j\}$ of G and L_β obtained in Theorem 9.7 is *complete*; that is, there is no strictly larger orthonormal set of eigenfunctions of L_β which contains $\{x_j\}$. For if x is an eigenfunction of L_β, $\|x\|_2 = 1$, such that $\langle x, x_j \rangle = 0$, $j = 1, 2, \ldots$, then the Parseval equality implies that $\|x\|_2 = 0$, a contradiction. Therefore, the sequence $\{\mu_j\}$ of eigenvalues of G from Theorem 9.7 contains all the eigenvalues of G, and the sequence $\{\lambda_j\}$, $\lambda_j = 1/\mu_j$ of Theorem 9.10 contains all the eigenvalues of L_β.

9.5   Extensions of the expansion results

In developing the results of this chapter, our attention has been restricted to the class of continuous functions $f \in C(I, \mathcal{C})$. For a number of important applied and theoretical reasons it is convenient to extend the development to the class of all (Riemann) integrable complex-valued functions on I (see [17, 22]). The collection of Riemann integrable functions $R(I, \mathcal{C})$ is a vector space over $\mathcal{C}$, which includes the piecewise continuous functions. The inner product and norm on $C(I, \mathcal{C})$ can be extended to $R(I, \mathcal{C})$:

$$\langle f, g \rangle = \int_a^b \overline{g}(t) f(t) \, dt, \quad \|f\|_2^2 = (f, f) = \int_a^b |f(t)|^2 \, dt, \quad f, g \in R(I, \mathcal{C}).$$

The *Schwarz inequality* remains valid,

$$|\langle f, g \rangle| \leq \|f\|_2 \|g\|_2, \quad f, g \in R(I, \mathcal{C}).$$

The eigenfunction expansion results of section 9.4 can be extended to include $f \in R(I, \mathcal{C})$. This makes use of the following approximation result, which we will assume (see exercise 14).

Lemma 9.12: *Given any $f \in R(I, \mathcal{C})$ and $\epsilon > 0$, there exists an $x \in D_0 \subset D_\beta$ such that*

$$\|f - x\|_2 < \epsilon.$$

Recall that

$$D_0 = \{x \in C^n(I, \mathcal{C}) \mid x^{(j-1)}(a) = x^{(j-1)}(b) = 0, \ j = 1, 2\}.$$

Because the lemma says that any $f \in R(I, \mathcal{C})$ may be approximated arbitrarily well with an $x \in D_0$, D_0 is said to be *dense* in $R(I, \mathcal{C})$ with the norm $\| \ \|_2$.

Let $\{x_j\}$ be the orthonormal sequence of eigenfunctions of the selfadjoint operator L_β, with $\dim(N(L_\beta)) = 0$, and $L_\beta x_j = \lambda_j x_j$. The following theorem extends the eigenfunction expansion result (9.10) to $f \in R(I, \mathcal{C})$.

Theorem 9.13: *If $f \in R(I, \mathcal{C})$ then*

$$f = \sum_{j=1}^{\infty} \langle f, x_j \rangle x_j,$$

where the series converges to f in the norm $\| \ \|_2$,

$$\left\| f - \sum_{j=1}^{k} \langle f, x_j \rangle x_j \right\|_2 \to 0, \quad k \to \infty. \tag{9.11}$$

If $x \in D_\beta$, then $L_\beta x = f \in C(I, \mathcal{C})$, and

$$x = \sum_{j=1}^{\infty} \langle x, x_j \rangle x_j, \quad L_\beta x = \sum_{j=1}^{\infty} \lambda_j \langle x, x_j \rangle x_j,$$

where these series converge to $x, L_\beta x$ in the norm $\| \ \|_2$.

Proof: Given $f \in R(I, \mathcal{C})$ and $\epsilon > 0$, Lemma 9.12 above guarantees an $x \in D_0$ such that

$$\|f - x\|_2 < \frac{\epsilon}{3}.$$

We know that

$$\left\| x - \sum_{j=1}^{k} \langle x, x_j \rangle x_j \right\|_2 \to 0, \quad k \to \infty,$$

so that for the given $\epsilon > 0$ there is an $N > 0$ such that

$$\left\| x - \sum_{j=1}^{k} \langle x, x_j \rangle x_j \right\|_2 < \frac{\epsilon}{3}, \quad k > N.$$

Now

$$\left\| f - \sum_{j=1}^{k} \langle f, x_j \rangle x_j \right\|_2 \leq \| f - x \|_2 + \left\| x - \sum_{j=1}^{k} \langle x, x_j \rangle x_j \right\|_2 + \left\| \sum_{j=1}^{k} [\langle x, x_j \rangle - \langle f, x_j \rangle] x_j \right\|_2.$$

The last term satisfies

$$\left\| \sum_{j=1}^{k} [\langle x, x_j \rangle - \langle f, x_j \rangle] x_j \right\|_2 \leq \| x - f \|_2 < \frac{\epsilon}{3}.$$

Since the sequence $\{x_j\}$ is orthonormal,

$$\left\| \sum_{j=1}^{k} [\langle x, x_j \rangle - \langle f, x_j \rangle] x_j \right\|_2^2 = \left\| \sum_{j=1}^{k} \langle x - f, x_j \rangle x_j \right\|_2^2 = \sum_{j=1}^{k} |\langle x - f, x_j \rangle|^2 \leq \| x - f \|_2^2,$$

where the last inequality is due to Bessel's inequality.

Taken together these estimates show that for any $\epsilon > 0$ there is an $N > 0$ such that

$$\left\| f - \sum_{j=1}^{k} \langle f, x_j \rangle x_j \right\|_2 < \epsilon, \quad k > N,$$

which is just (9.11).

If $x \in D_\beta$, then Theorem 9.10 shows that

$$x = \sum_{j=1}^{\infty} \langle x, x_j \rangle x_j,$$

where convergence is in the norms $\| \ \|_\infty$ and $\| \ \|_2$. The jth Fourier coefficient of $L_\beta x \in C(I, \mathcal{C})$ is

$$\langle L_\beta x, x_j \rangle = \langle x, l_\beta x_j \rangle = \langle x, \lambda_j x_j \rangle = \lambda_j \langle x, x_j \rangle,$$

and hence

$$L_\beta x = \sum_{j=1}^{\infty} \lambda_j \langle x, x_j \rangle x_j,$$

where the convergence is in the norm $\| \ \|_2$. $\square$

The proof of Theorem 9.11 extends to $f, h \in R(I, \mathcal{C})$ to give the following result.

Theorem 9.14: *If $f, h \in R(I, \mathcal{C})$, then*

$$\langle f, h \rangle = \sum_{j=1}^{\infty} \overline{\langle h, x_j \rangle} \langle f, x_j \rangle, \qquad (9.12)$$

and

$$\| f \|_2^2 = \sum_{j=1}^{\infty} |\langle f, x_j \rangle|^2. \quad (\text{Parseval equality}) \quad (9.13)$$

By (9.12) the sequence $c_j = \langle f, x_j \rangle$, $j = 1, 2, \ldots$, of Fourier coefficients of $f \in R(I, \mathcal{C})$ satisfies

$$\sum_{j=1}^{\infty} |c_j|^2 < \infty. \qquad (9.14)$$

Let us consider the set l^2 of all sequences $c = \{c_j\}$, $c_j \in \mathcal{C}$, satisfying (9.14). If $d = \{d_j\}$ is another such sequence in l^2, we can define an *inner product*

$$\langle c, d \rangle = \sum_{j=1}^{\infty} \overline{d_j} c_j.$$

This definition makes sense because the series is convergent. Indeed, the Schwarz inequality for $\mathcal{C}^k$ implies that

$$\left| \sum_{j=1}^{k} \overline{d_j} c_j \right|^2 \leq \left(\sum_{j=1}^{k} |c_j|^2 \right) \left(\sum_{j=1}^{k} |d_j|^2 \right) \leq \left(\sum_{j=1}^{\infty} |c_j|^2 \right) \left(\sum_{j=1}^{\infty} |d_j|^2 \right) < \infty, \quad (9.15)$$

so that the series in (9.15) converges absolutely. The *norm* of $c \in l^2$ is defined by

$$\|c\|_2^2 = \langle c, c \rangle.$$

The Schwarz inequality for l^2 is obtained by letting $k \to \infty$ on the left of (9.15),

$$|\langle c, d \rangle| \leq \|c\|_2 \|d\|_2.$$

The set l^2 is a vector space over $\mathcal{C}$, with addition and multiplication by a scalar defined componentwise,

$$c + d = \{c_j + d_j\}, \quad \alpha c = \{\alpha c_j\}, \quad \alpha \in \mathcal{C}.$$

We can think of $c, d \in l^2$ as $\infty \times 1$ matrices

$$c = \begin{pmatrix} c_1 \\ c_2 \\ \vdots \end{pmatrix}, \quad d = \begin{pmatrix} d_1 \\ d_2 \\ \vdots \end{pmatrix}, \quad d^* = (\overline{d_1}, \overline{d_2}, \ldots),$$

and then

$$\langle c, d \rangle = d^* c.$$

The Parseval equality for $f \in R(I, \mathcal{C})$ just says that if

$$X = \begin{pmatrix} x_1 \\ x_2 \\ \vdots \end{pmatrix},$$

then

$$\langle f, X \rangle = \begin{pmatrix} \langle f, x_1 \rangle \\ \langle f, x_2 \rangle \\ \vdots \end{pmatrix} = \begin{pmatrix} c_1 \\ c_2 \\ \vdots \end{pmatrix} \in l^2$$

and

$$\|f\|_2^2 = \|\langle f, X \rangle\|_2^2. \tag{9.16}$$

The equality (9.12) becomes

$$\langle f, h \rangle = \langle h, X \rangle^* \langle f, X \rangle = \langle \langle f, X \rangle, \langle h, X \rangle \rangle. \tag{9.17}$$

Because of (9.16), (9.17) the map

$$f \in R(I, \mathcal{C}) \to \langle f, X \rangle \in l^2 \tag{9.18}$$

preserves the norm and the inner product. It is clearly linear, and (9.16) implies that it is one-to-one, in the sense that if

$$\langle f, X \rangle = \langle h, X \rangle \quad \text{or} \quad \langle f - h, X \rangle = 0,$$

then $\|f - h\|_2 = 0$. In the case of $f, h \in C(I, \mathcal{C})$ this would imply $f = h$.

We may ask if the map (9.18) is onto l^2; that is, given any $c \in l^2$, is there an $f \in R(I, \mathcal{C})$ such that $c = \langle f, X \rangle$. The answer is no. It is clear that the results in Theorems 9.13 and 9.14 will be valid for any set of functions $F(I, \mathcal{C}) \supset R(I, \mathcal{C})$ such that D_0 is dense in $F(I, \mathcal{C})$ in the norm $\| \|_2$. The largest such set, interpreted appropriately, is the set $L^2(I, \mathcal{C})$ of all complex-valued functions f on I such that

$$\|f\|_2^2 = \int_a^b |f(t)|^2 \, dt < \infty,$$

where the integral here is the *Lebesgue integral*, a generalization of the Riemann integral. It can be shown that the map

$$f \in L^2(I, \mathcal{C}) \to \langle f, X \rangle \in l^2$$

is linear, one-to-one, onto l^2 and that the Parseval equality

$$\|f\|_2 = \|\langle f, X \rangle\|_2, \quad f \in L^2(I, \mathcal{C}),$$

is valid. This is the important *Riesz–Fischer theorem* [22, Theorem 10.43].

9.6   Notes

Much more detail on eigenfunction expansions, with generalizations to self-adjoint problems on infinite intervals and to nonselfadjoint problems, can be found in [3]. This material can also be productively examined from a more abstract point of view. The reference [21] provides a good introduction to the functional analytic viewpoint.

9.7   Exercises

1. Compute the Fourier coefficients of the following functions with respect to the orthonormal basis $x_n(t) = \sqrt{2/\pi}\sin(nt)$ on the interval $[0, \pi]$:

(a)
$$f(t) = \cos(5t);$$

(b)
$$f(t) = \chi_{[1,2]}(t) = \begin{array}{ll} 1, & 1 \le t \le 2, \\ 0, & \text{otherwise}; \end{array}$$

(c)
$$f(t) = e^t.$$

2. Following the development in section 8.1 as a model, use the separation of variables $u(x, t) = v(x)w(t)$ and expansion in eigenfunctions to formally solve the *wave equation*

$$\frac{\partial^2 u}{\partial t^2} = \frac{\partial^2 u}{\partial x^2}, \quad u(0, t) = 0 = u(\pi, t), \quad u(x, 0) = f(x), \quad \frac{\partial u(0, t)}{\partial t} = g(x).$$

Note that for each eigenvalue λ_j of the problem

$$-v'' = \lambda v, \quad v(0) = 0 = v(\pi)$$

there is now a second-order equation in t. (In keeping with the usual choice of variables for partial differential equations, x is now the spatial variable, not the solution of the differential equation.)

3. Consider the heat equation for a nonhomogeneous rod,

$$\frac{\partial u}{\partial t} = \frac{\partial}{\partial x} k(x) \frac{\partial u}{\partial x}, \quad u(0, t) = 0 = u(\pi, t), \quad u(x, 0) = f(x).$$

Using the development in section 8.1 as a model, express solutions of this problem using the eigenvalues λ_j and eigenfunctions $v_j(x)$ of the boundary value problem

$$-(k(x)v'(x))' = \lambda v, \quad v(0) = 0 = v(\pi).$$

4. Reconcile the negative result in the introduction on the Fourier series for $f(t) = t$ with Theorem 9.10.

5. (a) Use (8.24) and Theorem 8.13 to show that the kernel $g(t, s, \lambda)$ is a real-valued function if $\lambda \in \mathcal{R}$.

(b) Use the Green operator identity

$$(Gf, h) = (f, Gh), \quad f, h \in C(I, \mathcal{C}),$$

the continuity of the kernel $g(t, s)$, and part (a) to show that

$$g(t, s) = g(s, t).$$

6. Remove the restriction that $N(L_\beta) = 0$ by showing that there is some real number λ such that $N(L_\beta - \lambda) = 0$, and using the Green operator $G(\lambda)$. What is the relationship between eigenvalues of L_β and eigenvalues of $G(\lambda)$?

7. Suppose that $f \in C([a,b], \mathcal{R})$ and that $x_1, \ldots, x_N$ is a finite set of real orthonormal functions,

$$\int_a^b x_m(t) x_n(t) \, dt = \delta_{mn}.$$

Consider the problem of finding the best approximation to f by a real linear combination of the $x_n(t)$, in the sense that

$$E(a_1, \ldots, a_N) = \int_a^b \left[f(t) - \sum_{n=1}^N a_n x_n(t) \right]^2 dt, \quad a_n \in \mathcal{R},$$

is minimized.

(a) Show that if the function f is fixed, then $E(a_1, \ldots, a_n)$ has a minimum. (Hint: Write E as a quadratic polynomial.)

(b) Show that the minimum is attained when

$$a_n = \int_a^b f(t) x_n(t) \, dt.$$

8. Consider the eigenvalue problem

$$-x'' + a(t)x' + b(t)x = \lambda x, \quad x(0) = 0 = x(1), \tag{9.19}$$

where $a(t), b(t) \in C([0,1], \mathcal{R})$.

(a) Using exercise 2(b) of Chapter 8, express the eigenfunctions $x_n(t)$ of (9.19) as $x_n(t) = y_n(t) w(t)$, where

$$w(t) = \exp\left(\frac{1}{2} \int_0^t a(s) \, ds \right)$$

and y_n is an eigenfunction for a selfadjoint problem. What does this say about the eigenvalues for (9.19)?

(b) Show that the eigenfunctions for the problem (9.19) may not be orthogonal. Nonetheless, show that a function $f \in C^2([0,1], \mathcal{C})$ satisfying $f(0) = 0 = f(1)$ has a uniformly convergent expansion in eigenfunctions $x_n(t)$ of (9.19).

(c) Extend results (a), (b) to the case of general separated boundary conditions.

9. Suppose that $f \in C([0, \pi], \mathcal{C})$ has K continuous derivatives and that

$$f^{(k)}(0) = 0 = f^{(k)}(\pi), \quad k = 0, \ldots, K.$$

(a) Use integration by parts to show that the Fourier coefficients

$$a_n = \int_0^\pi f(t) \sin(nt) \, dt, \quad n = 1, 2, 3, \ldots,$$

satisfy $a_n = O(n^{-K})$; that is, there is a constant C such that $|a_n| \leq C/n^K$.

(b) Suppose that $K > 2M + 2$, that L_β is a selfadjoint differential operator defined on $I = [a, b]$, and that 0 is not an eigenvalue for L_β.

Show that $f, L_\beta f, \ldots, L_\beta^M f \in D_\beta$. If f has Fourier coefficients

$$a_n = \langle f, x_n \rangle$$

with respect to the eigenfunctions x_n of L_β, show that

$$a_n = O(\lambda_n^{-M}).$$

10. Recall that for each $n = 0, 1, 2, \ldots$, the Legendre equation

$$(1 - t^2)x'' - 2tx' + n(n+1)x = 0$$

has a polynomial solution $P_n(t)$ of degree n satisfying $P_n(1) = 1$, the nth Legendre polynomial (see section 5.6.2).

(a) Show that the Legendre equation can be rewritten in the form

$$-([1 - t^2]x')' = n(n+1)x.$$

Let $L = -([1 - t^2]x')'$. On the interval $I = [-1, 1]$ this operator has the selfadjoint form, except that the leading coefficient vanishes at the endpoints. Show that, for all $f, g \in C^2(I, \mathcal{C})$,

$$\langle Lf, g \rangle = \langle f, Lg \rangle. \tag{9.20}$$

(b) Use (9.20) to show that the Legendre polynomials form an orthogonal set. Compare with exercise 32 of Chapter 5.

(c) Show that the Legendre polynomials are a complete orthogonal set in $C(I, \mathcal{C})$ as follows. Suppose $f \in C(I, \mathcal{C})$ and $\langle f, P_n \rangle = 0$ for all $n = 0, 1, 2, \ldots$. Show that this implies that $\langle f, t^n \rangle = 0$ for all n. Conclude that $\langle f, \sin(at) \rangle = 0 = \langle f, \cos(at) \rangle$ for any real constant a. Finally, use the completeness of eigenfunctions for a suitable selfadjoint differential operator to conclude that $f = 0$.

11. Suppose that $f(t) \in C^2([0, 1], \mathcal{C})$ and $f(0) = 0 = f(1)$.

(a) Integration by parts gives

$$\int_0^1 [f'(t)]^2 \, dt = -\int_0^1 f(t) f''(t) \, dt.$$

Show that

$$\int_0^1 [f'(t)]^2 \, dt \leq (\|f\|_2^2 + \|f''\|_2^2)/2.$$

(b) Use the fundamental theorem of calculus and part (a) to show that

$$|f(t)| = \left| \int_0^t f'(s) \, ds \right| \leq \|f\|_2 + \|f''\|_2.$$

(c) Similarly, show that

$$f(t) = tf'(t) - \int_0^t (s-a)f''(s)\ ds$$

implies that for $t > 0$

$$|f'(t)| \leq \frac{1}{t}\left[|f(t)| + \left|\int_0^1 sf''(s)\ ds\right|\right]$$

$$\leq \frac{1}{t}\left[(\|f\|_2 + \|f''\|_2) + \left(\frac{1}{3}\right)^{1/2}\|f''(s)\|_2\right].$$

Obtain a similar inequality for $t < 1$ by writing

$$f(t) = -\int_t^1 f'(s)\ ds = (1-t)f'(t) - \int_t^1 (1-s)f''(s)\ ds.$$

Use these two inequalities to show that

$$|f'(t)| \leq 2\left[(\|f\|_2 + \|f''\|_2) + \left(\frac{1}{3}\right)^{1/2}\|f''(s)\|_2\right].$$

12. Suppose that $f_n \in C^2([a,b],\mathcal{C})$ satisfies $f_n(a) = 0 = f_n(b)$. In addition assume that $\{f_n\}$ and $\{f_n''\}$ are Cauchy sequences in the norm $\|\ \|_2$. Use exercise 11 to show that the sequences $\{f_n\}$ and $\{f_n'\}$ are Cauchy sequences in the norm $\|\ \|_\infty$.

13. Let $L_\beta = -x''$ with the domain

$$D_\beta = \{f \in C^2([a,b],\mathcal{C}) \mid f(a) = 0 = f(b)\}.$$

Use Theorem 9.10 and exercise 11 to show that if $x \in D_\beta$ then

$$\left\|\left(x - \sum_{j=1}^k (x,x_j)x_j\right)'\right\|_\infty \to 0.$$

14. The point of this exercise is to provide an indication of the proof for Lemma 9.12, which asserted that *given any $f \in R(I,\mathcal{C})$ and $\epsilon > 0$, there exists an $x \in D_0 \subset D_\beta$ such that*

$$\|f - x\|_2 < \epsilon.$$

A result often found in more advanced texts (see [22, Theorems 10.33 and 10.38] or [21, p. 39, example 2, and p. 48, Theorem II.9]) is that the above assertion is true if x is allowed to be a continuous function, and f is allowed to come from the larger class of functions which are square integrable in the Lebesgue sense. Assuming that the italicized statement is true for continuous functions x, prove that it then follows for $x \in D_0 \subset D_\beta$.

Chapter 10

Control of Linear Systems

10.1 Introduction

The modern technical culture expends enormous effort in attempts to manage complex engineering, ecological, and economic systems. Our last topic, control theory, provides a mathematical approach to some of these system management problems. Unlike most of the material in this book, much of control theory is of recent vintage. Although large parts of the subject were technically accessible for many years, the actual systematic development occurred in the middle of the twentieth century, with important work in the basic linear theory appearing as late as 1960 (see [13], [14]).

Two examples will initiate the exposition.

Example 1. Consider the following system of differential equations as a (simple) model describing the interrelationship of a population W of a predator, and the population R of its prey. Suppose that there are populations W_0, R_0 for the two species at which the system is in equilibrium. Letting $N_1 = W - W_0$ and $N_2 = R - R_0$, the deviation of the populations from the equilibrium value is modeled by the system of equations

$$\frac{dN_1}{dt} = a_{11}N_1 + a_{12}N_2, \qquad (10.1)$$

$$\frac{dN_2}{dt} = a_{21}N_1 + a_{22}N_2 + u(t).$$

Notice that the second equation includes a control $u(t)$ with which the population of the prey can be adjusted. This control is assumed to be largely at our disposal.

We are interested in the following types of questions. Suppose that the populations start near the equilibrium value. Is it possible, simply by inserting or removing prey, to bring the populations of both species to predetermined levels and keep them there? How will our ability to bring the populations to a target level be affected by constraints on the control $u(t)$? Is there a control strategy which will drive the populations to desired levels in minimal time, and how can such a strategy be characterized?

A superficial analysis shows that there must be some conditions on the matrix

$$\mathbf{A} = \begin{pmatrix} a_{11} & a_{12} \\ a_{21} & a_{22} \end{pmatrix}, \quad a_{ij} \in \mathcal{R},$$

before there is any hope of achieving such control. For instance, if $a_{12} = 0$ then the predator population will grow or decay without regard to what we do with the prey.

Example 2. The second example is a mechanical system (see Figure 10.1) consisting of N masses $m_1, \ldots, m_N$ at positions $x_1 < \cdots < x_N$ connected in a line by springs. In addition the first and last masses are connected to walls at $x = 0$ and $x = L$. Suppose first that the masses are equal and the spring constants are the same, so that there are no net forces acting on the masses if the springs are equally stretched, or m_n is located at $nL/(N+1)$. We introduce coordinates measuring the offset of each mass from its equilibrium position, $z_n = x_n - nL/(N+1)$. In these coordinates the equations of motion are

$$m\frac{d^2 z_1}{dt^2} = -kz_1 + k(z_2 - z_1), \tag{10.2}$$

$$m\frac{d^2 z_n}{dt^2} = -k(z_n - z_{n-1}) + k(z_{n+1} - z_n), \quad n = 2, \ldots, N-1,$$

$$m\frac{d^2 z_N}{dt^2} = -k(z_N - z_{N-1}) - kz_N,$$

$$z_n(0) = p_n, \quad z_n'(0) = v_n.$$

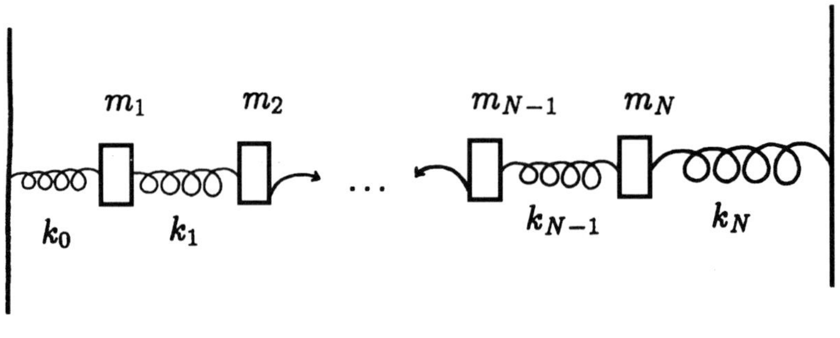

Figure 10.1: A system of springs and masses.

Our goal is to bring certain of the masses to rest, and hold them there, by moving other masses. We may ask whether all the masses can be brought to rest by applying forces to only a single mass or to the masses with odd index. Another question is whether our ability to control the positions of masses is affected by variation of the mass in time, such as might occur if some of the mass starts leaking out of the system like water from a hole in the bottom of a pail. Dropping the assumption of equal masses and spring constants, adjusting the coordinates to the initial equilibrium positions, and putting controlling forces on each mass (some of which might be zero), we obtain the linear system of equations

$$m_1\frac{d^2z_1}{dt^2} + m_1'(t)\frac{dz_1}{dt} = -k_0z_1 + k_1(z_2 - z_1) + u_1(t), \qquad (10.3)$$

$$m_n(t)\frac{d^2z_n}{dt^2} + m_n'(t)\frac{dz_n}{dt} = -k_{n-1}(z_n - z_{n-1}) + k_n(z_{n+1} - z_n) + u_n(t),$$

$$n = 2, \ldots, N - 1,$$

$$m_N\frac{d^2z_N}{dt^2} + m_N'(t)\frac{dz_n}{dt} = -k_{N-1}(z_N - z_{N-1}) - k_Nz_N + u_N(t).$$

These examples motivate consideration of the general system of equations with continuous matrix-valued coefficients,

$$X' = \mathbf{A}(t)X + \mathbf{B}(t)U(t) + F(t), \quad X(t_0) = X_0, \qquad (GLC)$$

where

$$\mathbf{A} \in C(I, M_N(\mathcal{R})), \quad \mathbf{B} \in C(I, M_{NM}(\mathcal{R})), \quad F \in C(I, \mathcal{R}^N).$$

The function $U(t)$ is assumed to be piecewise continuous with values in $\mathcal{R}^M$, $U \in PC(I, \mathcal{R}^M)$. For notational simplicity it will be convenient to let $t_0 = 0$. Usually the case $I = [0, \infty)$ will be assumed.

The function $U(t)$ is called a *control*. The values of the control may be restricted to a subset $\Omega \subset R^M$, called the *control region*. The controls $U(t)$ may be further constrained to belong to a class of functions called the *admissible control class* $\mathcal{U}$. Unless otherwise specified, the admissible control class will consist of all piecewise continuous functions with values in Ω. Equations with piecewise continuous coefficients are discussed in section 7.4. In general the goal will be to steer the solution $X(t)$ into a set $G(t)$ called the *target* by suitably choosing the control $U(t)$. To simplify the presentation the target set will simply be the vector $0 \in \mathcal{R}^N$.

The method of variation of parameters (Theorem 2.8) can be applied to the problem of control for the nonhomogeneous linear equation (GLC). The solution of the initial value problem has the form

$$X(t) = \mathbf{X}(t)X_0 + \mathbf{X}(t)\int_0^t \mathbf{X}^{-1}(s)[\mathbf{B}(s)U(s) + F(s)]\,ds, \qquad (10.4)$$

where $\mathbf{X}(t)$ is the basis satisfying the initial value problem

$$\mathbf{X}' = \mathbf{A}(t)\mathbf{X}, \quad \mathbf{X}(0) = I_N.$$

The approach to these problems will largely involve careful examination of the form of the solution (10.4). The notation $X = X(t; X_0, U)$ will be used to emphasize the dependence of the solution X on the initial data or the control.

In this chapter it will be convenient to use $\langle X, Y \rangle$ to denote the usual dot product for real vectors $X, Y \in \mathcal{R}^N$. The corresponding norm is $\|X\|$, and for a matrix $\mathbf{A}$,

$$\|\mathbf{A}\| = \sup_{\|X\|=1} \|\mathbf{A}X\|.$$

10.2   Convex sets

In the subsequent discussion, some ideas from the study of convex sets will be helpful. A subset Ω of a real vector space is *convex* if $U, V \in \Omega$ implies $wU + (1-w)V \in \Omega$ for all $w \in [0,1]$. Our convex sets are assumed to be nonempty. If $w_j \geq 0$ and $\sum_{j=1}^{J} w_j = 1$, say that $V = \sum_{j=1}^{J} w_j U_j \in \Omega$ is a *convex combination* of the points U_j.

Lemma 10.1: *Suppose that $\Omega \subset \mathcal{R}^M$ is convex and $U_1, \ldots, U_J \in \Omega$. If $w_j \geq 0$ and $\sum_{j=1}^{J} w_j = 1$, then $\sum_{j=1}^{J} w_j U_j \in \Omega$.*

Proof: Without loss of generality, assume that each $w_j > 0$. The proof is by induction on J, the case $J = 1$ being trivial. For $J > 1$ write

$$\sum_{j=1}^{J} w_j U_j = w_1 U_1 + \sum_{j=2}^{J} w_j U_j = w_1 U_1 + \left[\sum_{j=2}^{J} w_j\right] \sum_{j=2}^{J} w_j U_j / \left(\sum_{i=2}^{J} w_i\right).$$

Since $\sum_{j=2}^{J} w_j / (\sum_{i=2}^{J} w_i) = 1$, the term $\sum_{j=2}^{J} w_j U_j / (\sum_{i=2}^{J} w_i)$ is a convex combination of $J-1$ elements of Ω, hence is in Ω by the induction hypothesis. Since $\sum_{j=2}^{J} w_j = 1 - w_1$, the definition of convexity implies $\sum_{j=1}^{J} w_j U_j \in \Omega$. $\square$

If $\mathcal{S}$ is a vector subspace of $\mathcal{R}^M$ and $X \in \mathcal{R}^M$, then any set of points $\mathcal{T} = X + \mathcal{S} = \{X + S | S \in \mathcal{S}\}$ is called an *affine subspace* of $\mathcal{R}^M$. The dimension of $\mathcal{T}$ is defined to be the dimension of $\mathcal{S}$. If $\dim(\mathcal{S}) = M - 1$, then $\mathcal{T}$ is called a *hyperplane*.

Suppose $\mathcal{S}$ is an $(M-1)$-dimensional subspace and that $\mathcal{T} = X + \mathcal{S}$ is a hyperplane. Let $\beta \in \mathcal{R}^M$ be a nonzero vector which is orthogonal to $\mathcal{S}$, so that $\mathcal{S} = \{Y \in \mathcal{R}^M | \langle \beta, Y \rangle = 0\}$. Then if $X + S \in \mathcal{T}$ we have $\langle X + S, \beta \rangle = \langle X, \beta \rangle$. Conversely, suppose that $\beta \neq 0$. Define $\mathcal{S} = \{Y \in \mathcal{R}^M | \langle \beta, Y \rangle = 0\}$, and consider the set of $Y \in \mathcal{R}^M$ such that $\langle Y, \beta \rangle = \alpha$. Choose a particular Y_1 with $\langle Y_1, \beta \rangle = \alpha$. Then $\langle Y - Y_1, \beta \rangle = 0$ so that $Y - Y_1 \in \mathcal{S}$ and $Y \in Y_1 + \mathcal{S}$. Consequently the hyperplanes are exactly the level sets of the linear functions obtained by taking the inner product with some nonzero $\beta \in \mathcal{R}^M$.

Suppose that $X_1, X_2 \in \mathcal{R}^M$, $X_1 \neq X_2$. If $\beta = X_2 - X_1$, then $\langle X_2 - X_1, \beta \rangle \neq 0$, so that $\langle X_2, \beta \rangle \neq \langle X_1, \beta \rangle$. Thus there is a linear function given by inner product with β separating X_1 and X_2, or, geometrically, there is a hyperplane containing X_1 but not X_2.

Lemma 10.2: *If $\Omega \subset \mathcal{R}^M$ is convex, then either Ω lies in a hyperplane or Ω contains an open subset of $\mathcal{R}^M$. If $\Omega \subset \mathcal{R}^M$ is convex, is closed under* *multiplication by* -1, *and does not lie in a hyperplane, then* Ω *contains an open neighborhood of* 0.

Proof: For the first part, it will be convenient to have $0 \in \Omega$; this can always be achieved by subtracting a single vector from every element of Ω, and this translation process will not affect the result. If Ω does not lie in a proper subspace of $\mathcal{R}^M$ then there must be $U_1, \ldots, U_M \in \Omega$ which are linearly independent. There is an invertible linear map $\mathbf{L}$ taking U_j to the standard basis vector E_j. The function $\mathbf{L}$ is continuous, so the inverse image of any open set is open. Take the inverse image of the set of points K in $\mathcal{R}^M$ (called the standard open simplex) of the form

$$ K = \sum_{m=1}^{M} w_m E_m, \quad 0 < w_m < 1, \quad \sum_{m=1}^{M} w_m < 1. $$

The set $\mathbf{L}^{-1}(K)$ will be a open subset of the convex combinations (see exercise 4) of the points $0, U_1, \ldots, U_M$, which gives an open set contained in Ω.

For the second part, the linear map $\mathbf{L}$ will take the convex combinations of $\pm U_1, \ldots, \pm U_M \in \Omega$ to the cube $[-1,1]^M$, and $\mathbf{L}^{-1}((-1,1)^M)$ will be the desired neighborhood of 0. $\square$

Lemma 10.3: *If* $\Omega \subset \mathcal{R}^M$ *is a convex set which is not equal to* $\mathcal{R}^M$, *then there is a vector* $\beta \in \mathcal{R}^M$ *with* $\|\beta\| = 1$ *and a real number* α *such that every element* $X \in \Omega$ *satisfies* $\langle \beta, X \rangle \leq \alpha$. *For each* $Y \in \mathcal{R}^M$ *not in the closure of* Ω, β *and* α *may be chosen so that, in addition,* $\langle \beta, Y \rangle > \alpha$.

Proof: First, observe that there is an open set V disjoint from Ω. If a translate of Ω lies in a proper subspace of $\mathcal{R}^M$ this is easy. If Ω does not lie in a hyperplane, then by the previous lemma Ω contains an open set W. Let $Y_0 \in \mathcal{R}^M$ with $Y_0 \notin \Omega$.

Consider all the line segments

$$ (1-t)Y_0 + tP, \quad -1 \leq t \leq 1, $$

with midpoint Y_0 having one endpoint P in W. Since $Y_0 \notin \Omega$ the second endpoint $2Y_0 - P$ of each segment is not an element of Ω. This collection of endpoints forms an open set V disjoint from Ω. As a consequence of this first observation, there is a point Y which is not in the closure of Ω, which is also convex.

Let Z be a point in the closure of Ω which minimizes the distance to Y (see Figure 10.2). Let $X \in \Omega$ be different from Z, and consider the function

$$ \phi(t) = \langle tX + (1-t)Z - Y, tX + (1-t)Z - Y \rangle, \quad 0 \leq t \leq 1. $$

Since this function has a minimum at $t = 0$, it follows that $\phi'(0) \geq 0$. But $\phi'(0) = 2\langle X - Z, Z - Y \rangle$ so that $\langle X, Z - Y \rangle \geq \langle Z, Z - Y \rangle$.

Since Y is not in the closure of Ω, the vector $Y - Z$ satisfies $\langle Y - Z, Y - Z \rangle > 0$ and

$$\langle Y, Y - Z \rangle > \langle Z, Y - Z \rangle \geq \langle X, Y - Z \rangle.$$

Choose

$$\beta = \frac{Y - Z}{\|Y - Z\|}, \quad \alpha = \langle Z, \beta \rangle. \quad \square$$

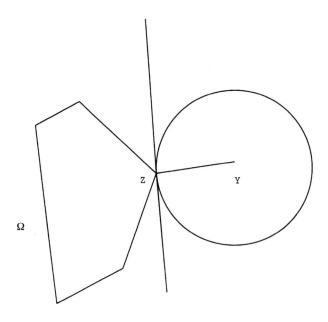

Figure 10.2: Constructing a separating hyperplane.

Recall that Z is a boundary point of a set S in $\mathcal{R}^M$ if every open ball centered at Z contains points of S and the complement of S.

Lemma 10.4: *Suppose that Z is a boundary point of the closed convex set $\Omega \subset \mathcal{R}^M$. Then there is a vector $\beta \in \mathcal{R}^M$ with $\|\beta\| = 1$ such that for all $X \in \Omega$*

$$\langle \beta, X \rangle \leq \langle \beta, Z \rangle.$$

Proof: Since Z is a boundary point of Ω there is a sequence of points $\{Y_n\}$ in the complement of Ω with $\|Y_n - Z\| \leq 1/n$ for $n = 1, 2, 3, \ldots$. By Lemma 10.3 there are vectors β_n with $\|\beta_n\| = 1$ and numbers α_n such that

$$\langle \beta_n, X \rangle \leq \alpha_n < \langle \beta_n, Y_n \rangle, \quad X \in \Omega.$$

Since $\langle \beta_n, Z \rangle \leq \alpha_n$, the Cauchy–Schwarz inequality implies that the sequence $\{\alpha_n\}$ is bounded below by $-\|Z\|$. Similarly, the inequalities

$$\alpha_n < \langle \beta_n, Y_n \rangle, \quad \|Y_n\| \leq \|Z\| + 1/n,$$

imply that the sequence $\{\alpha_n\}$ is bounded above by $\|Z\| + 1$. In addition, the set of vectors in $\mathcal{R}^M$ with norm 1 is compact. Thus the sequence $\{\beta_n\}$ has a subsequence $\{\beta_k\}$ which converges to $\beta \in \mathcal{R}^M$ with $\|\beta\| = 1$ and such that $\lim_{k \to \infty} \alpha_k = \alpha$.

For every $X \in \Omega$,

$$\langle \beta, X \rangle = \lim_{k \to \infty} \langle \beta_k, X \rangle \leq \lim_{k} \alpha_k = \alpha.$$

On the other hand,

$$\langle \beta, Z \rangle = \lim_{k \to \infty} \langle \beta_k, Y_k \rangle \geq \lim_{k} \alpha_k = \alpha.$$

Since $Z \in \Omega$,

$$\langle \beta, Z \rangle = \alpha. \quad \square$$

For any (nonempty) set $\Lambda \subset \mathcal{R}^M$, define the *convex hull* of Λ to be the intersection of all convex sets containing Λ, so that the convex hull of Λ is the smallest convex set containing Λ. Suppose that $\mathcal{O}$ is the set of all convex combinations of points from Λ. It is easily seen that $\mathcal{O}$ is convex and that every convex set containing Λ must contain $\mathcal{O}$. Thus $\mathcal{O}$ is the convex hull of Λ.

If Ω is convex, then we say that U is an *extreme point* of Ω if

$$U = wV + (1 - w)Z, \quad V, Z \in \Omega, \quad w \in (0, 1),$$

implies $U = V = Z$.

Lemma 10.5: *If $\Omega \subset \mathcal{R}^M$ is a compact convex set which is contained in an affine subspace of dimension N, then every $X \in \Omega$ may be written as the convex combination of at most $N + 1$ extreme points of Ω.*

Proof: If Ω has only one point or if X is an extreme point, there is nothing to show.

The proof proceeds by induction on the dimension N. If $N = 1$ then a compact convex set is a point or a compact interval, so the result is immediate. Suppose the result is true for $K < N$ and let Ω be a compact convex subset contained in an affine subspace $\mathcal{T}$ of dimension N.

It will be convenient to work in $\mathcal{R}^N$ instead of $\mathcal{T}$. First (see exercises 3 and 4), there is an $X_p \in \Omega$ and an $M \times N$ matrix $\mathbf{L}$ such that the function $Y \to \mathbf{L}Y + X_p$ maps $\mathcal{R}^N$ one to one and onto $\mathcal{T}$. The set $\Lambda = \mathbf{L}^{-1}(\Omega - X_p) \subset \mathcal{R}^N$ is convex, and W is an extreme point of Ω if and only if $\mathbf{L}^{-1}(W - X_p)$ is an extreme point of Λ.

We first show that Λ has an extreme point. If Λ has at least two points, Z_1 and Z_2, then there is a vector $\beta \in \mathcal{R}^N$ such that $\langle Z_1, \beta \rangle < \langle Z_2, \beta \rangle$. Since this function is continuous and Λ is compact, there is a subset of Λ where this function takes its maximum value. The set of all $Y \in \mathcal{R}^N$ where the function takes on this maximum is an affine subspace C of dimension $N - 1$. By the induction hypothesis, $\Lambda \cap C$ has at least one extreme point Y_1, and because $\langle Y_1, \beta \rangle$ is maximal, Y_1 must also be an extreme point of Λ (exercise 5).

For any $Y \in \Lambda$ the segment $(1 - t)Y_1 + tY, 0 \le t < 1$, lies in Λ. Since Λ is compact it has a boundary point $Z = (1-t_1)Y_1 + t_1 Y$ for some $t_1 \ge 1$. Since Λ is a closed set, Lemma 10.4 guarantees that there is a $\beta \in \mathcal{R}^N$ with $\|B\| = 1$, and the function

$$Y \to \langle Y, \beta \rangle, \quad Y \in \Lambda,$$

is maximal at Z. Again using the induction hypothesis, Z may be written as a convex combination of N extreme points $Y_2, \ldots, Y_{N+1}$, and then Y may be written as a convex combination of the extreme points $Y_1, \ldots, Y_{N+1}$.

Finally, $X = \mathbf{L}Y + X_p$ is a convex combination

$$X = X_p + \sum_{j=1}^{N+1} w_j \mathbf{L}Y_j. \quad \square$$

10.3   Control of general linear systems

Returning to the system (GLC), define the *controllable set at time* $t \ge t_0$,

$$C(t) = \{X_0 \in \mathcal{R}^N; X(t; X_0, U) = 0 \text{ for some } U \in \mathcal{U}\},$$

and define the *controllable set* to be $C = \bigcup_{t \ge 0} C(t)$. Similarly, define the *attainable set at time* $t \ge t_0$ as

$$K(t; X_0) = \bigcup_{U \in \mathcal{U}} X(t; X_0, U),$$

and let the *attainable set* be $K(X_0) = \bigcup_{t \ge 0} K(t; X_0)$. Say that (GLC) is *unconstrained* if the admissible control class $\mathcal{U} = PC([0, \infty), \mathcal{R}^M)$. We begin with a simple observation regarding the problem (GLC).

Lemma 10.6: *Suppose that the admissible control class*

$$\mathcal{U} \subset PC([0, \infty), \mathcal{R}^N)$$

is convex. Then for equation (GLC) the sets $C(t)$ and $K(t; X_0)$ are convex subsets of $\mathcal{R}^N$.

If $\mathcal{U}$ is a vector space, then for the problem

$$X' = \mathbf{A}(t)X + \mathbf{B}(t)U(t), \quad X(0) = X_0,$$

C(t) is a vector space.

Proof: Suppose that $X_0, Y_0 \in C(t_1)$. Then there are piecewise continuous functions $U_0(t), U_1(t) \in \mathcal{U}$ and solutions $X(t), Y(t)$ of (GLC) such that

$$X' = \mathbf{A}(t)X + \mathbf{B}(t)U_0(t) + F(t), \quad X(0) = X_0, \quad X(t_1) = 0,$$

$$Y' = \mathbf{A}(t)Y + \mathbf{B}(t)U_1(t) + F(t), \quad Y(0) = Y_0, \quad Y(t_1) = 0.$$

Multiplying the first equation by w, the second by $(1-w)$, and adding yields

$$(wX + (1-w)Y)' = \mathbf{A}(t)(wX + (1-w)Y) + \mathbf{B}(t)(wU_0(t) + (1-w)U_1(t)) + F(t).$$

Observe that $Z(t) = wX + (1-w)Y$, with initial data $wX_0 + (1-w)Y_0$, is driven by the control $wU_0(t) + (1-w)U_1(t)$ to the target, i.e., $Z(t_1) = wX(t_1) + (1-w)Y(t_1) = 0$.

The arguments for the convexity of $K(t; X_0)$ and the second part about $C(t)$ being a vector space are analogous. $\square$

The main subject for this section is the system of equations

$$X' = \mathbf{A}(t)X + \mathbf{B}(t)U(t), \quad X(0) = X_0. \tag{LC}$$

The first problem is to find a piecewise continuous control $U(t)$ which will steer an initial value X_0 to the value 0 for some $t > 0$.

Let $\mathbf{X}(t)$ be the matrix-valued function satisfying

$$\mathbf{X}' = \mathbf{A}(t)\mathbf{X}, \quad \mathbf{X}(0) = I_N.$$

Using the variation of parameters formula (Theorem 2.8), the solution of (LC) can be written as

$$X(t) = \mathbf{X}(t)X_0 + \mathbf{X}(t) \int_0^t \mathbf{X}^{-1}(s)\mathbf{B}(s)U(s) \, ds. \tag{10.5}$$

If $X(t_1) = 0$, then

$$0 = X_0 + \int_0^{t_1} \mathbf{X}^{-1}(s)\mathbf{B}(s)U(s) \, ds.$$

Take the inner product in $\mathcal{R}^N$ with any vector Z to get

$$0 = \langle Z, 0 \rangle = \left\langle Z, X_0 \right\rangle + \left\langle Z, \int_0^{t_1} \mathbf{X}^{-1}(s)\mathbf{B}(s)U(s) \, ds \right\rangle \tag{10.6}$$

$$= \langle Z, X_0 \rangle + \int_0^{t_1} \langle Z, \mathbf{X}^{-1}(s)\mathbf{B}(s)U(s) \rangle \, ds.$$

In view of (10.6) it is not surprising to find a relationship between the sets $C(t_1)$ and the ranges of the linear mappings $\mathbf{X}^{-1}(s)\mathbf{B}(s)$. For notational convenience, define $\mathbf{Y}(s) = \mathbf{X}^{-1}(s)\mathbf{B}(s)$. Let $V(t_1)$ be the vector space spanned by the range of $\mathbf{Y}(s)$ (that is, the column spaces) for $0 \le s < t_1$. That is,

$$V(t_1) = span(Ran(\mathbf{Y}(s))), \quad 0 \le s < t_1, \quad V(t_1)^\perp = \bigcap_{0 \le s < t_1} Ran^\perp(\mathbf{Y}(s)).$$

Theorem 10.7: *For the unconstrained control problem (LC),*

$$C(t_1) = V(t_1).$$

Proof: Suppose that $Z \in V(t_1)^\perp$; that is,

$$Z \in \operatorname{Ran}^\perp \mathbf{Y}(s), \quad 0 \le s < t_1.$$

Then the integrand for

$$\int_0^{t_1} \langle Z, \mathbf{X}^{-1}(s)\mathbf{B}(s)U(s) \rangle \, ds$$

is zero. Consequently, if $X_0 \in C(t_1)$, then by (10.6) it follows that $X_0 \in V(t_1)^{\perp\perp} = V(t_1)$.

Conversely, suppose that $X_0 \in V(t_1)$. We want to show that there is a control $U(s)$ such that

$$\int_0^{t_1} \mathbf{Y}(s)U(s) \, ds = -X_0.$$

In fact the control $U(s)$ can be found among the piecewise constant functions. Let $V_1, \ldots, V_K$ be a basis for $V(t_1)$. By hypothesis there are vectors $U_{j(k)}$ such that

$$V_k = \sum_{j(k)} \mathbf{Y}(t_{j(k)})U_{j(k)}, \quad k = 1, \ldots, K.$$

For $\epsilon > 0$ define vectors

$$W_k(\epsilon) = \sum_{j(k)} \frac{1}{\epsilon} \int_{t_{j(k)}}^{t_{j(k)}+\epsilon} \mathbf{Y}(s)U_{j(k)} \, ds, \quad k = 1, \ldots, K. \tag{10.7}$$

Notice that by the continuity of $\mathbf{Y}(s) = \mathbf{X}^{-1}(s)\mathbf{B}(s)$ we have

$$\lim_{\epsilon \to 0^+} W_k(\epsilon) = V_k.$$

We claim that for ϵ sufficiently small the vectors $W_k(\epsilon)$ are a basis for $V(t_1)$. Writing the integrals defining $W_k(\epsilon)$ as limits of Riemann sums, note first that the Riemann sums are elements of $V(t_1)$, and since $V(t_1)$ is a closed subset of $\mathcal{R}^N$, so too are $W_k(\epsilon)$.

To see that the vectors $W_k(\epsilon)$ are linearly independent for ϵ sufficiently small, write each $W_k(\epsilon)$ as a linear combination of the vectors V_j,

$$W_k(\epsilon) = \sum_{j=1}^K c_{j,k}(\epsilon)V_j.$$

The coefficients $c_{j,k}(\epsilon)$ are continuous functions of ϵ, and letting $\delta_{j,k}$ be the Kronecker δ,

$$\lim_{\epsilon \to 0^+} c_{j,k}(\epsilon) = \delta_{j,k}.$$

This means that $\det(c_{j,k}(\epsilon)) \to 1$; by continuity $\det(c_{j,k}(\epsilon)) > 0$ for ϵ sufficiently small. This implies that for ϵ small the dimension of the span of $\{W_k(\epsilon); k = 1, \ldots, K\}$ is K, so the vectors are linearly independent.

Picking an $\epsilon > 0$ with $W_k(\epsilon)$ a basis for $V(t_1)$, we can then write

$$-X_0 = \sum_{j,k} b_{j,k} W_k(\epsilon),$$

so there is a piecewise constant function $U(s)$ such that

$$\int_0^{t_1} \mathbf{X}^{-1}(s)\mathbf{B}(s)U(s)\,ds = -X_0.$$

It now follows from (10.5) that $X(t_1) = 0$. $\square$

As an example, consider the forcing of a simple harmonic oscillator,

$$x'' = -k^2 x + u, \quad k > 0,$$

which is equivalent to the first-order system

$$\begin{pmatrix} x_1 \\ x_2 \end{pmatrix}' = \begin{pmatrix} 0 & 1 \\ -k^2 & 0 \end{pmatrix} \begin{pmatrix} x_1 \\ x_2 \end{pmatrix} + \begin{pmatrix} 0 \\ 1 \end{pmatrix} u.$$

In this case

$$M = 1, \quad \mathbf{B}(s) = \begin{pmatrix} 0 \\ 1 \end{pmatrix}, \quad \mathbf{X}(\mathbf{t}) = \exp\left(\begin{pmatrix} 0 & 1 \\ -k^2 & 0 \end{pmatrix} t \right).$$

It is easy to verify that

$$\mathbf{X}(t) = \begin{pmatrix} \cos(kt) & \sin(kt)/k \\ -k\,\sin(kt) & \cos(kt) \end{pmatrix}, \quad \mathbf{X}^{-1}(t) = \begin{pmatrix} \cos(kt) & -\sin(kt)/k \\ k\,\sin(kt) & \cos(kt) \end{pmatrix},$$

and

$$\mathbf{Y}(t) = \begin{pmatrix} -\sin(kt)/k \\ \cos(kt) \end{pmatrix}.$$

For any $t > 0$ the vector space $V(t) = \mathcal{R}^2$, which means that the harmonic oscillator can be forced to the position 0 and velocity 0 state as quickly as you like, provided of course that there are no constraints on the control u. This example will also serve to illustrate the next set of ideas.

Say that the system *(GLC)* is *proper* if

$$Z^t \mathbf{Y}(s) = (0, \ldots, 0) \in \mathcal{R}^M, \quad Z \in \mathcal{R}^N,$$

for all s in some open interval implies that $Z = 0$. This condition is equivalent to saying that the N rows of $\mathbf{Y}(s) = \mathbf{X}^{-1}(s)\mathbf{B}(s)$ are linearly independent $\mathcal{R}^M$-valued functions on every open interval.

Corollary 10.8: *If the unconstrained system (LC) is proper, then $C(t) = \mathcal{R}^N$ for all $t > 0$.*

Proof: If the system (LC) is proper, there cannot be any nonzero vector Z which is orthogonal to the range of $\mathbf{Y}(s)$ for all s in an open interval. Thus for all $t > 0$ we have

$$V(t)^{\perp} = \bigcap_{0 \le s < t} Ran^{\perp}\mathbf{Y}(s) = 0,$$

so $V(t) = C(t) = \mathcal{R}^N$. □

If the coefficients $\mathbf{A}(t), \mathbf{B}(t)$ of *(LC)* are analytic, then so too will be any of the component functions of $Z\mathbf{Y}(s)$. Since an analytic function can vanish on an interval only if it is identically zero, a system LC with analytic coefficients will be proper if the rows of $\mathbf{Y}(s)$ are linearly independent on some open interval. Such a system can be shown to be proper by examining the behavior of derivatives of $\mathbf{A}(t)$ and $\mathbf{B}(t)$ at $t = 0$ (see exercise 10).

There is another approach by which the system *(LC)* may be analyzed. Define $\mathbf{W}(t) : \mathcal{R}^N \to \mathcal{R}^N$ by

$$\mathbf{W}(t) = \int_0^t \mathbf{Y}(s)\mathbf{Y}^t(s)\, ds. \tag{10.8}$$

$\mathbf{W}(t)$ is a real symmetric $N \times N$ matrix, and

$$\langle Z, \mathbf{W}(t)Z \rangle = \int_0^t Z^t \mathbf{Y}(s)\mathbf{Y}^t(s)Z\, ds \ge 0,$$

so that all eigenvalues of $\mathbf{W}(t)$ are nonnegative. Notice that $\mathbf{Y}^t(s) : \mathcal{R}^N \to \mathcal{R}^M$ and that if $\mathbf{B}(s)$ is (piecewise) continuous, so is $\mathbf{Y}(s)$. Thus we can construct controls $U(s)$ by fixing $Z \in \mathcal{R}^N$ and letting

$$U(s) = \mathbf{Y}^t(s)Z.$$

It follows easily that the range of $\mathbf{W}(t)$ is a linear subspace of $C(t)$.

Suppose that $Z \in C(t)$, which is orthogonal to the range of $\mathbf{W}(t)$. Then

$$0 = Z^t \int_0^t \mathbf{Y(s)}\mathbf{Y^t(s)}\, ds\, Z = \int_0^t Z^t\mathbf{Y}(s)\mathbf{Y}^t(s)Z\, ds = \int_0^t \|Z^t\mathbf{Y}(s)\|^2\, ds,$$

so that

$$Z^t\mathbf{Y}(s) = 0, \quad 0 \le s \le t.$$

As a consequence,

$$Z^t \int_0^t \mathbf{Y}(s)U(s)\, ds = \int_0^t Z^t\mathbf{Y}(s)U(s)\, ds = 0$$

for all piecewise continuous $U(s)$. But since $Z \in C(t)$ there is some control U such that

$$0 = Z + \int_0^t \mathbf{Y}(s)U(s)\, ds.$$

Taking the inner product with Z yields

$$0 = Z^t Z + Z^t \int_0^t \mathbf{Y}(s)U(s) \ ds,$$

and thus $Z = 0$. This establishes the following result.

Theorem 10.9: *For the unconstrained problem (LC), the range of*

$$\mathbf{W}(t) = \int_0^t \mathbf{Y}(s)\mathbf{Y}^t(s) \ ds$$

is equal to $C(t)$.

As an example, the function $\mathbf{W}(t)$ may be computed for the harmonic oscillator system,

$$\begin{pmatrix} x_1 \\ x_2 \end{pmatrix}' = \begin{pmatrix} 0 & 1 \\ -k^2 & 0 \end{pmatrix} \begin{pmatrix} x_1 \\ x_2 \end{pmatrix} + \begin{pmatrix} 0 \\ 1 \end{pmatrix} u.$$

In this case

$$\mathbf{Y}(t) = \begin{pmatrix} -\sin(kt)/k \\ \cos(kt) \end{pmatrix},$$

so that

$$\mathbf{W}(t) = \frac{1}{k^2} \int_0^t \begin{pmatrix} \sin^2(ks) & -k\sin(ks)\cos(ks) \\ -k\sin(ks)\cos(ks) & k^2\cos^2(ks) \end{pmatrix} \ ds$$

$$= \frac{1}{k^2} \begin{pmatrix} t/2 - \sin(2kt)/(4k) & [\cos(2kt)-1]/4 \\ [\cos(2kt)-1]/4 & k^2 t/2 + k\sin(2kt)/4 \end{pmatrix}.$$

It is not difficult to check that $\det(\mathbf{W(t)}) > 0$ for $t > 0$. Thus the range of $\mathbf{W(t)} = \mathcal{R}^2$ for all $t > 0$.

So far the controls $U(s)$ have been unconstrained. Having gained some understanding of when it is possible to drive every initial vector X_0 to zero, we ask when it is possible to do this with piecewise continuous controls lying in a bounded set $\Omega \subset \mathcal{R}^M$ which contains a closed ball of radius $R > 0$ centered at the origin. One approach, which can be useful in many cases, divides the problem of steering initial data to 0 into two parts. The first part is to understand when a small ball of initial values at time t_0 can be steered to 0 at time $t_0 + t$. The second part is to let the equation with control $U = 0$ drive the solution close to zero.

Let

$$w(t) = \min \langle Z, \mathbf{W}(t)Z \rangle, \quad Z \in \mathcal{R}^N, \quad \|Z\| = 1;$$

$w(t)$ is of course a nonnegative nondecreasing function of t. Let

$$y(t) = \max \|\mathbf{Y^t}(s)Z\|, \quad \|Z\| = 1, \quad 0 \le s \le t.$$

Theorem 10.10: *For the problem (LC), if Ω contains a closed ball of radius $R > 0$ centered at the origin and $w(t) > 0$, then $C(t)$ contains the closed ball of radius $w(t)R/y(t)$.*

Proof: Let $Z \in \mathcal{R}^N$ be in the closed ball of radius $R/y(t)$. Use as controls the functions $U(s) = \mathbf{Y}^t(s)Z$, which by the definition of $y(t)$ are admissible. Using such controls gives

$$\int_0^t \mathbf{Y}(s)U(s) \, ds = \mathbf{W}(t)Z.$$

Now $\mathbf{W}(t)$ is a linear transformation on $\mathcal{R}^N$ with an orthonormal basis of eigenvectors and eigenvalues at least as large as $w(t)$. Thus the set of vectors $\mathbf{W}(t)Z$ includes a closed ball of radius $w(t)R/y(t)$. Consequently, any $X_0 \in \mathcal{R}^N$ in this ball can be steered to 0 by an admissible control in time t. $\square$

In connection with this last result, two observations are noteworthy. First, suppose that $\mathbf{A}(t)$ and $\mathbf{B}(t)$ are constant, and that $C(t)$ does contain a ball of positive radius. Then the size of the ball of initial values at time t_0 which can be steered to 0 at time $t_0 + t$ is independent of t_0. Similarly, if $\mathbf{A}(t)$ and $\mathbf{B}(t)$ are periodic and $C(t)$ contains a ball of positive radius, then for arbitrarily large t_0 the same ball of initial values at time t_0 can be steered to 0 at time $t_0 + t$.

Within this circle of ideas, the second ingredient is a set of conditions on $\mathbf{A}(t)$ implying that all solutions of

$$X' = \mathbf{A}(t)X, \quad X(t_0) = X_0, \tag{10.9}$$

converge to the zero vector as $t \to \infty$. Transposing equation (10.9) leads to

$$[X^t]' = X^t \mathbf{A}^t(t). \tag{10.10}$$

Multiplying equation (10.9) on the left by $X^t(t)$ and equation (10.10) on the right by $X(t)$ yields, respectively,

$$\langle X(t), X'(t) \rangle = \sum_{n=1}^N x_n(t)x_n'(t) = \langle X(t), \mathbf{A}(t)X(t) \rangle,$$

$$\langle X'(t), X(t) \rangle = \sum_{n=1}^N x_n'(t)x_n(t) = \langle X(t), \mathbf{A}^t(t)X(t) \rangle.$$

Adding these equations we find that

$$2\sum_{n=1}^N x_n'(t)x_n(t) = \frac{d}{dt}\|X(t)\|^2 = \langle X(t), [\mathbf{A}(t) + \mathbf{A}^t(t)]X(t) \rangle. \tag{10.11}$$

Appropriate conditions on $\mathbf{A}(t)$ will guarantee that every solution has limit zero as $t \to \infty$.

Theorem 10.11: *Suppose that there is an $\epsilon > 0$ such that the real symmetric matrix $\mathbf{A}(t) + \mathbf{A}^t(t)$ satisfies*

$$\langle Z, [\mathbf{A}(t) + \mathbf{A}^t(t)]Z \rangle \le -\epsilon\|Z\|^2, \quad t \ge 0, \quad Z \in \mathcal{R}^N.$$

Then every solution $X(t)$ of

$$X' = \mathbf{A}(t)X, \quad X(t_0) = X_0,$$

satisfies

$$\lim_{t \to \infty} X(t) = 0.$$

Proof: From equation (10.11),

$$\frac{d}{dt}\|X(t)\|^2 = \langle X(t), [\mathbf{A}(t) + \mathbf{A}^t(t)]X(t) \rangle$$

$$\leq -\epsilon\|X(t)\|^2, \quad t \geq 0.$$

There is no loss of generality in assuming that $\|X(t)\| \neq 0$. Dividing both sides of the inequality by $\|X(t)\|^2$ yields

$$\left[\frac{d}{dt}\|X(t)\|^2 \right] / \|X(t)\|^2 = \frac{d}{dt}\log(\|X(t)\|^2) \leq -\epsilon.$$

Thus

$$\log(\|X(t)\|^2) - \log(\|X(t_0)\|^2) \leq -\epsilon(t - t_0), \quad t > t_0,$$

and exponentiating we find that

$$\|X(t)\|^2 \leq \|X(t_0)\|^2 e^{-\epsilon(t-t_0)}. \quad \square$$

10.4 Constant coefficient equations

As we demonstrated in Chapter 3, linear systems with constant coefficients are explicitly solvable. It is not surprising then that detailed information about control can be extracted for the linear autonomous control problem,

$$X' = \mathbf{A}X + \mathbf{B}U(t), \quad X(0) = X_0, \qquad (LAC)$$

where the matrices $\mathbf{A}, \mathbf{B}$ are constant. Notice that example (10.1) is of this type, as is (10.2) after it is converted to a first-order system. For the problem *(LAC)* the variation of parameters formula has the more explicit form

$$X(t) = \exp(\mathbf{A}t)X_0 + \exp(\mathbf{A}t) \int_0^t \exp(-\mathbf{A}s)\mathbf{B}U(s) \, ds. \qquad (10.12)$$

Thanks to this explicit form it is possible to develop a more explicit description of the cases where $C = \mathcal{R}^N$. Define the *controllability matrix*

$$\mathcal{M} = (\mathbf{B} : \mathbf{AB} : \cdots : \mathbf{A^{N-1}B}).$$

Theorem 10.12: *The controllable set of the unconstrained problem (LAC) is $C = Ran(\mathcal{M})$, and in particular $C = \mathcal{R}^N$ if and only if $\mathcal{M}$ has rank N.*

Proof: By Lemma 10.6 the controllable set C is a vector space, so it suffices to understand $C^{\perp}$. Suppose that $V \perp Ran(\mathcal{M})$, that is $V \perp Ran(\mathbf{A}^k\mathbf{B})$ for $k = 0, \ldots, N - 1$. By the Cayley–Hamilton theorem, the matrix $\mathbf{A}$ satisfies its characteristic polynomial, which implies that $\mathbf{A}^N$ is a linear combination of $\mathbf{I}, \mathbf{A}, \ldots, \mathbf{A}^{N-1}$. This in turn implies that $V \perp Ran(\mathbf{A}^k\mathbf{B})$ for $k = 0, 1, 2, \ldots$. Writing

$$e^{\mathbf{A}t}\mathbf{B} = \sum_{k=0}^{\infty} \frac{\mathbf{A}^k t^k}{k!} \mathbf{B},$$

and taking the inner product with V, we find that for all $Z \in \mathcal{R}^M$

$$\langle V, e^{\mathbf{A}t}\mathbf{B}Z \rangle = \sum_{k=0}^{\infty} \left\langle V, \frac{\mathbf{A}^k t^k}{k!} \mathbf{B}Z \right\rangle = 0$$

for all t, and so $V \in C^{\perp}$ by Theorem 10.7.

Suppose conversely that $V \in C^{\perp}$, which by Theorem 10.7 means $V \in Ran(e^{-\mathbf{A}t}\mathbf{B})^{\perp}$ for all $t > 0$. Then

$$\langle V, e^{-\mathbf{A}t}\mathbf{B}Z \rangle = \sum_{k=0}^{\infty} \left\langle V, \frac{\mathbf{A}^k (-t)^k}{k!} \mathbf{B}Z \right\rangle$$

$$= \sum_{k=0}^{\infty} \langle V, \mathbf{A}^k \mathbf{B}Z \rangle \frac{(-t)^k}{k!} = 0, \quad t > 0, \quad Z \in \mathcal{R}^M.$$

But if a convergent power series vanishes on an open interval, then each coefficient is zero, so $V \perp Ran(\mathbf{A}^k\mathbf{B})$ for $k = 0, \ldots, N - 1$. $\square$

Trying to drive initial data to 0 is a more interesting problem when the control U must sit inside a bounded set. To get a feel for the problem it will help to consider the simple case when $\mathbf{A}$ is a diagonal matrix. Then (10.12) takes the form

$$X(t) = \begin{pmatrix} \exp(\lambda_1 t) & \ldots & 0 \\ \vdots & \ldots & \vdots \\ 0 & \ldots & \exp(\lambda_N t) \end{pmatrix} X_0$$

$$+ \int_0^t \begin{pmatrix} \exp(\lambda_1[t-s]) & \ldots & 0 \\ \vdots & \ldots & \vdots \\ 0 & \ldots & \exp(\lambda_N[t-s]) \end{pmatrix} \mathbf{B}U(s) \, ds.$$

Suppose that $\operatorname{Re}(\lambda_1) = \mu_1 > 0$ is maximal. Since $U(s)$ is bounded, there is a constant c_1 such that

$$\left| \int_0^t \exp(\mathbf{A}[t-s])\mathbf{B}U(s) \, ds \right| \le c_1 \int_0^t \exp(\mu_1[t-s]) \, ds = \frac{c_1}{\mu_1}[\exp(\mu_1 t) - 1].$$

There is, however, no constraint on the vector X_0. By taking

$$X_0 = (c_2, 0, \ldots, 0)^t$$

for c_2 sufficiently large, we can ensure that $X(t)$ never vanishes. At least in this case, a bounded control cannot drive arbitrary initial vectors to 0 if the eigenvalues of $\mathbf{A}$ have positive real part.

Within the context of constant coefficient equations, the next theorem will describe exactly when it is possible to drive arbitrary initial data to 0 with bounded controls U. It will be convenient to first prove a technical lemma. Suppose that $p_j(t)$ denotes a polynomial.

Lemma 10.13 : *If a linear combination*

$$g(t) = \sum_{j=1}^{K} p_j(t) \exp(-\lambda_j t)$$

is not identically zero and $\text{Re}(\lambda_j) \leq 0$, *then there are* $\delta, \epsilon > 0$ *and an infinite sequence of pairwise disjoint intervals* $\{I_k\} \subset (0, \infty)$, *with* $length(I_k) > \delta$, *such that* $|g(t)| \geq \epsilon$ *for all* $t \in \cup I_k$.

Proof: Write $\lambda_j = \mu_j + i\nu_j$, with μ_j, ν_j real. First extract from g the function g_1 consisting of those terms for which $\text{Re}(-\lambda_j)$ is maximal. This function may be assumed to be nonzero somewhere, and so we may extract from g_1 the function g_2 which is not identically zero and which contains all powers t^r of maximal degree. This function has the form

$$g_2(t) = t^r \exp(-\mu t) \left[\sum_{j=1}^{J} c_j \exp(i\nu_j t) \right].$$

Since $-\mu \geq 0$ and

$$\lim_{t \to \infty} \frac{g(t) - g_2(t)}{t^r \exp(-\mu t)} = 0,$$

it clearly suffices to prove the lemma for

$$f(t) = \sum_{j=1}^{J} c_j \exp(i\nu_j t),$$

which is not identically zero.

Notice that the derivative

$$f'(t) = \sum_{j=1}^{J} i c_j \nu_j \exp(i\nu_j t)$$

is uniformly bounded for $t \in \mathcal{R}$. By the fundamental theorem of calculus,

$$|f(t) - f(s)| = \left| \int_s^t f'(u) \, du \right| \leq \int_s^t |f'(u)| \, du.$$

Thus for any $\epsilon > 0$ there is a $\delta > 0$ such that $|t - s| < \delta$ implies $|f(t) - f(s)| < \epsilon$, with δ independent of t. To show that $|f(t)|$ is large on a sequence of pairwise disjoint intervals, it thus suffices to show that $|f(t_n)| > 3\epsilon/2$ for a sequence of points $t_n \to \infty$.

Let t_1 be such that

$$2\epsilon = \left| \sum_{j=1}^{J} c_j \exp(i\nu_j t_1) \right| > 0.$$

If $\nu_j = 0$, then $\exp(i\nu_j t)$ is constant, and such terms will not influence the selection of the sequence $\{t_n\}$. For $\nu_j \neq 0$, each function $\exp(i\nu_j t)$ is periodic with period $2\pi/\nu_j$. Because of this periodicity, the value of the function $f(t)$ is determined by the vector

$$V(t) = (\nu_1 t \bmod 2\pi/\nu_1, \ldots, \nu_J t \bmod 2\pi/\nu_J), \quad t \in (0, \infty),$$

which has values in the Cartesian product of circles $S_1 \times \cdots \times S_J$, where circle S_j has radius $1/\nu_j$

Since this product of circles is compact, the sequence of vectors $V(m)$, $m = 0, 1, 2, \ldots$, has a convergent subsequence. Thus for any $\delta > 0$ there is a positive integer m_0, and an increasing sequence of positive integers m_n such that the distance from $V(m_n)$ to $V(m_0)$ is less than δ. In other words $V(m_n - m_0)$ is no more than δ from $(0, \ldots, 0)$. Since this implies that each of the exponentials $\exp(i\nu_j[t_1 + m_n - m_0])$ can be made as close as we like to $\exp(i\nu_j t_1)$, it follows that $f(t_1 + m_n - m_0)$ can be made as close to $f(t_1)$ as we like. $\square$

Theorem 10.14: *Suppose that the control region Ω is a bounded set. If $C = \mathcal{R}^N$, every eigenvalue λ of $\mathbf{A}$ satisfies $\mathrm{Re}(\lambda) \leq 0$. Conversely, if Ω contains an open ball centered at 0, every eigenvalue λ of $\mathbf{A}$ satisfies $\mathrm{Re}(\lambda) \leq 0$, and the controllability matrix $\mathcal{M}$ has rank N, then $C = \mathcal{R}^N$.*

Proof: The proof is based on a careful analysis of (10.12). The first simplification is to show that $\mathbf{A}$ can be assumed to be in Jordan canonical form. Writing $\mathbf{A} = \mathbf{Q}^{-1}\mathbf{J}\mathbf{Q}$, equation (10.12) becomes

$$X(t) = \mathbf{Q}^{-1}\exp(\mathbf{J}t)\mathbf{Q}X_0 + \mathbf{Q}^{-1}\exp(\mathbf{J}t)\mathbf{Q}\int_0^t \mathbf{Q}^{-1}\exp(-\mathbf{J}s)\mathbf{Q}BU(s)\,ds$$

or

$$\mathbf{Q}X(t) = \exp(\mathbf{J}t)\mathbf{Q}X_0 + \exp(\mathbf{J}t)\int_0^t \exp(-\mathbf{J}s)\mathbf{Q}BU(s)\,ds. \qquad (10.13)$$

This is just the solution for the equation for $Z = \mathbf{Q}X$,

$$Z'(t) = \mathbf{J}Z + \mathbf{B}_1 U, \quad \mathbf{B}_1 = \mathbf{Q}B, \quad Z(0) = Z_0 = \mathbf{Q}X_0.$$

The form of $\exp(\mathbf{J}t)$ was established in Theorem 3.10:

$$\exp(\mathbf{J}t) = \mathrm{diag}(\exp(\mathbf{J}_1 t), \ldots, \exp(\mathbf{J}_q t)),$$

where

$$\exp(\mathbf{J}_i t) = \exp(\lambda_i t) \begin{pmatrix} 1 & t & t^2/2! & \dots & t^{r_i-1}/(r_i-1)! \\ 0 & 1 & t & \dots & t^{r_i-2}/(r_i-2)! \\ 0 & 0 & \vdots & \dots & \vdots \\ 0 & \dots & 0 & 1 & t \\ 0 & \dots & 0 & 0 & 1 \end{pmatrix}.$$

Suppose that the eigenvalue λ_1 of $\mathbf{A}$ has positive real part μ_1 and that no other eigenvalue has larger real part. First we get a growth estimate on the integral term in (10.13). Since Ω is bounded there is a constant K such that $|U(s)| \leq K$. Thus there is a constant c_1 such that

$$\left| \int_0^t \exp(\mathbf{J}[t-s])\mathbf{Q}\mathbf{B}U(s)\, ds \right| \leq \int_0^t |\exp(\mathbf{J}[t-s])|\, |\mathbf{Q}\mathbf{B}|K\, ds$$

$$\leq c_1 \int_0^t \exp(\mu_1[t-s])[(t-s)^{r_i-1} + 1]\, ds.$$

Repeated integration by parts shows that the last term is bounded by

$$c_2 \exp(\mu_1 t) t^{r_i-1}, \quad t > 1.$$

By increasing c_1 the inequality

$$\left| \int_0^t \exp(\mathbf{J}[t-s])\mathbf{Q}\mathbf{B}U(s)\, ds \right| \leq c_1 \exp(\mu_1 t)[1 + t^{r_i-1}]$$

will hold for all $t \geq 0$.

By virtue of the form of $\exp(\mathbf{J}t)$, there is vector

$$Z_0 = \mathbf{Q}X_0 = c_2\, (0, \dots, 0, 1, 0, \dots, 0)$$

such that

$$\exp(\mathbf{J}t)\mathbf{Q}X_0 = c_2 \exp(\lambda_1 t)(0, \dots, 0, t^{r_i-1}/(r_i-1)! + \cdots + t + 1, 0, \dots, 0).$$

If c_2 is large enough then for all $t \geq 0$

$$|\exp(\mathbf{J}t)\mathbf{Q}X_0| > \left| \int_0^t \exp(\mathbf{J}[t-s])\mathbf{Q}\mathbf{B}U(s)\, ds \right|$$

so that $Z(t) \neq 0$ for all $t \geq 0$. Since $X = \mathbf{Q}^{-1}Z$, the same conclusion holds for X. This establishes the first part of the theorem.

Suppose, conversely, that Ω contains an open ball Ω_1 centered at 0, every eigenvalue λ of $\mathbf{A}$ satisfies $\mathrm{Re}(\lambda) \leq 0$, and the controllability matrix $\mathcal{M}$ has rank N. We will show that for any initial value Z_0 there is a control $U(s) \in \Omega_1$ and a $t > 0$ such that $Z(t) = 0$.

Replace the control region Ω with the smaller convex control region Ω_1. By Lemma 10.6 the sets $C(t)$ are convex for $t > 0$. Also note that if $0 < s < t$, then $C(s) \subset C(t)$, since once a solution $X(t)$ hits 0 we can keep it there by extending the control U to be 0 thereafter. Since the sets $C(t)$ are an increasing collection of convex sets, C is a convex subset of $\mathcal{R}^N$. If $C \neq \mathcal{R}^N$ then there must be a vector $\beta \in \mathcal{R}^N$ such that $Z \in C$ implies $\langle \beta, Z \rangle < \alpha$ for some $\alpha > 0$. We will show that the hypotheses preclude the existence of such a vector.

From (10.12) the vectors $Z \in C$ are just the vectors having the form

$$Z = - \int_0^t \exp(-\mathbf{A}s)\mathbf{B}U(s)\ ds$$

for some $t > 0$ and U an admissible control. We want to show that the condition

$$\alpha > \langle \beta, Z \rangle = - \int_0^t \langle \beta, \exp(-\mathbf{A}s)\mathbf{B}U(s) \rangle\ ds$$

$$= - \int_0^t \beta^T \exp(-\mathbf{A}s)\mathbf{B}U(s)\ ds$$

cannot be satisfied for all admissible controls U. The hypothesis that the controllability matrix $\mathcal{M}$ has rank N is equivalent to saying that

$$span_{s \geq 0}(Ran(\exp(-\mathbf{A}s)\mathbf{B})) = \mathcal{R}^N.$$

This means that $V(s) = \beta^T \exp(-\mathbf{A}s)\mathbf{B} \in \mathcal{R}^M$ is not identically zero. Moreover the form of $\exp(-\mathbf{A}s)$ implies that each component of V is a real function which is a linear combination $\sum_j p_j(s) \exp(-\lambda_j s)$, where the $p_j(s)$ are polynomials and the λ_j are the eigenvalues of $\mathbf{A}$.

An admissible control $U(s)$ is constructed as follows. Suppose the mth component of $V(s)$ is not identically zero. Take all components of $U(s)$ except the mth identically zero. Appealing to Lemma 10.13, the remaining component of U is chosen to be 0 off the intervals I_k and is a small nonzero constant on each I_k. By suitable choice of these constants the function $-\beta^T \exp(-\mathbf{A}s)\mathbf{B}U(s)$ can be made positive and bounded below uniformly on the intervals I_k; thus

$$- \int_0^t \beta^T \exp(-\mathbf{A}s)\mathbf{B}U(s)\ ds$$

can be made arbitrarily large. $\square$

10.5   Time-optimal control

Given that there is a control steering an initial vector X_0 to 0, one is naturally interested in understanding the control which is best in some suitable sense. It is also important to characterize the optimal control, since from an engineering viewpoint the mere theoretical existence of an optimal control has limited utility. This section introduces the subject of time-optimal control, in which the goal is to steer X_0 to 0 in minimal time.

In most treatments of control theory these problems are addressed with mathematically sophisticated tools, including Lebesgue's theory of integration (see [6, 14, 15]). Such a treatment allows controls $U(t)$ which are more general than the piecewise continuous functions we have considered. Since this section is intended to be introductory, the results are described in a more elementary context, with the intent to give an indication of the types of results and the techniques which may be employed.

Throughout this section the control region Ω is a compact set.

To begin, consider controls $U(s)$ which are continuous functions, $0 \leq s \leq t$. Recall that a collection of functions is (sequentially) *compact* if every sequence has a convergent subsequence. The Ascoli–Arzela theorem, stated in section 9.2 (see [17, p. 273]), describes the compact sets of continuous functions on the closed interval $[0, t]$.

Compactness of a family of continuous functions is often encountered when there are uniform bounds on the derivatives of the family. This idea can be relevant for control. Think of a problem such as controlling the motion of a car by turning the steering wheel. Here a bound on the derivative is a bound on how fast we can change the position of the wheel.

For technical reasons it will be convenient to introduce an idea a bit more general than differentiability. Assume that the admissible class of functions $\mathcal{U}$ consists of the continuous functions $U : [0, \infty) \to \mathcal{R}^M$ such that on each interval $[0, t]$ there are constants K_1 and K_2, possibly depending on t but independent of U, such that our controls satisfy

$$|U(s)| \leq K_1, \quad |U(s_2) - U(s_1)| \leq K_2|s_2 - s_1|. \tag{10.14}$$

The first condition says that Ω is the closed ball of radius K_1 in $\mathcal{R}^M$. The second condition is called a *Lipschitz condition* on the function U. Notice that if U is differentiable at a point s, then the Lipschitz condition implies that the derivative has norm bounded by K_2. It is not difficult to verify that on each interval $[0, t]$ a collection of functions satisfying these conditions with fixed constants K_1, K_2 is bounded, closed under uniform limits, and equicontinuous, so by the Ascoli–Arzela theorem the collection of functions is compact. It is also a simple exercise to show that $\mathcal{U}$ is convex.

For this class of controls it is not difficult to show that if there is a control steering X_0 to 0, then there is a control in the class which does this in minimal time.

Theorem 10.15: *Suppose that there is some $t_1 > 0$ such that the problem (GLC) can be steered to 0 by a Lipschitz control $U \in \mathcal{U}$. Then there is a minimal t^* and a control $U^*(t) \in \mathcal{U}$ such that the control U^* steers X_0 to zero at time t^*.*

Proof: Let t^* be the infimum of the set of t such that there is a control $U \in \mathcal{U}$ steering X_0 to 0. Thus there is a sequence of times t_n decreasing to t^* and controls $U_n \in \mathcal{U}$ such that U_n steers X_0 to 0 at time t_n. If we restrict the set of controls to the interval $[0, t_1]$, then this set of controls is compact by hypothesis, so the sequence U_n has a subsequence $U_{n(k)}$ which converges uniformly to a function $U^* = \lim_{k \to \infty} U_{n(k)} \in \mathcal{U}$. Since the value $X(t)$ depends continuously on U (Theorem 7.6), it follows that $X(t^*; X_0, U^*) = 0$. □

It is interesting to try to understand the structure of such an optimal control. In this case the optimal control $U(t)$ either lies on the boundary of Ω or else U is moving as quickly as possible between points on the boundary of Ω. To establish this result, a technical lemma is first proved. Recall that the attainable set at time t is

$$K(t; X_0) = \bigcup_{U \in \mathcal{U}} X(t; X_0, U).$$

Lemma 10.16: *If the attainable set $K(t_1; X_0)$ at time t_1 contains an open neighborhood of 0, then so does $K(t; X_0)$ for $t < t_1$ and $t_1 - t$ sufficiently small.*

Proof: The neighborhood of zero in $K(t_1)$ contains a set S (an N-simplex), which is the convex hull of points $X_n(t_1)$ for $n = 1, \ldots, N+1$, and with the origin in its interior. Let $U_n(t)$ be controls steering X_0 to $X_n(t)$. With these controls fixed, the $X_n(t)$ are continuous functions of t, so the origin remains in the interior of the simplex with vertices $\{X_n(t)\}$ for t near t_1. By Lemma 10.6, the simplex is therefore in $K(t, X_0)$. □

Theorem 10.17: *Suppose the problem (GLC) is proper and has a time-optimal Lipschitz control $U \in \mathcal{U}$ with optimal time t^*. Then, for each $0 < t < t^*$, either $|U(t)| = K_1$ or, for any open interval $I \subset (0, t^*)$ containing t,*

$$\sup_{s \in I, s \neq t} \frac{|U(t) - U(s)|}{|t - s|} = K_2.$$

Proof: Suppose to the contrary that U is a control steering X_0 to 0 at time $t^* > 0$ and that for some $0 < t < t^*$ the function $U(t_0)$ satisfies both $|U(t)| < K_1$, and there is a $K_3 < K_2$ such that, for all s sufficiently close to t, $|U(t) - U(s)| \leq K_3|t - s|$.

The control U will be modified by the addition of terms $\tilde{U}$. Notice that any sum $U + \tilde{U}$ will be admissible if we pick a sufficiently small neighborhood I_1 of t such that $\tilde{U}$ vanishes outside of I_1 and if there are sufficiently small positive constants C_1, C_2 such that $|\tilde{U}| < C_1$ and $|\tilde{U}'| < C_2$. Let $\tilde{\mathcal{U}}$ be such a set of controls $\tilde{U}$, and note that $\tilde{\mathcal{U}}$ can be chosen to be convex and closed under multiplication by -1. We want to show that the addition of these controls in $\tilde{\mathcal{U}}$ means that $K(t^*)$ includes an open neighborhood of 0.

If $X(t)$ is the solution of (GLC) with the optimal control U, then since $X(t^*) = 0$, the attainable set at time t^* contains all vectors of the form

$$W(t^*) = \mathbf{X}(t^*) \left[X_0 + \int_0^{t^*} \mathbf{X}^{-1}(s)[\mathbf{B}(s)[U(s) + \tilde{U}(s)] + F(s)] \right] ds$$

$$= \mathbf{X}(t^*) \int_0^{t^*} \mathbf{X}^{-1}(s)\mathbf{B}(s)\tilde{U}(s) \, ds.$$

Since $\mathbf{X}(t^*)$ is an invertible linear transformation, it suffices to show that the convex set of vectors

$$\int_0^{t^*} \mathbf{Y}(s)\tilde{U}(s) \, ds, \quad \tilde{U} \in \tilde{\mathcal{U}},$$

includes an open neighborhood of the origin.

If $\eta \in \mathcal{R}^N$, then

$$\eta^t \int_0^{t^*} \mathbf{Y}(s)\tilde{U}(s) \, ds = \int_0^{t^*} \eta^t \mathbf{Y}(s)\tilde{U}(s) \, ds.$$

Since the system is proper, if $\eta \neq 0$ the continuous function $\eta^t Y(s)$ is not identically zero on the small neighborhood of t, and thus there is an admissible $\tilde{U}(s)$ so the integral is not zero. Since $\tilde{\mathcal{U}}$ is closed under multiplication by -1, it follows from Lemma 10.2 that the set $K(t^*)$ includes a neighborhood of the origin.

Finally, by Lemma 10.16, for $t < t^*$ and sufficiently close, we still hit an open set containing 0, so that our control is not optimal, a contradiction. $\square$

If all constraints on the rate of change of controls were removed, then Theorem 10.17 would suggest that an optimal control $U(t)$ should always lie on the boundary of the control region. Our final result is along this line. Assume now that the control region Ω is a compact convex set. A control U is called a *bang-bang control* if $U(t)$ is always an extreme point of Ω. Say that $U \in PCBB$ if U is a piecewise constant bang-bang control.

Theorem 10.18: *Suppose that Ω is a compact convex set, and the problem GLC has a piecewise continuous control steering X_0 to 0 at time T. Then for every $\epsilon > 0$ there is a piecewise constant bang-bang control such that $|X(T)| < \epsilon$.*

Proof: The proof begins with several simplifying observations. Since X_0 can be steered to 0 in time T with a piecewise continuous control U, the variation of parameters formula (10.4) for (GLC) leads to

$$0 = X_0 + \int_0^T \mathbf{X}^{-1}(s)[\mathbf{B}(s)U(s) + F(s)] \, ds.$$

Since $\mathbf{X}(t)$ is continuous if U is piecewise continuous, (10.4) also shows that the theorem will be established if for every $\epsilon > 0$ there is a piecewise constant bang-bang control $\tilde{U}$ such that

$$\left| X_0 + \int_0^T \mathbf{X}^{-1}(s)[\mathbf{B}(s)\tilde{U}(s) + F(s)] \, ds \right| < \epsilon.$$

Since $U(t)$ is piecewise continuous, there are points T_j, $j = 0, \ldots, J$ such that $0 = T_0 < T_1 < \cdots < T_J = T$, with $U(t)$ continuous on $[T_j, T_{j+1}]$. Since the collection of intervals $[T_j, T_{j+1}]$ is finite, the proof can be further reduced to showing that for every $\epsilon > 0$ there is a $\tilde{U} \in PCBB$ such that

$$\left| \int_{T_j}^{T_{j+1}} \mathbf{X}^{-1}(s)\mathbf{B}(s)[U(s) - \tilde{U}(s)] \, ds \right| < \epsilon.$$

Recall [17, p. 215] that on a compact interval $[T_j, T_{j+1}]$ the continuous function $U(t)$ is *uniformly continuous*; that is, for every $\epsilon > 0$ there is a $\delta > 0$ such that for all $s, t \in [T_j, T_{j+1}]$, $|t - s| < \delta$ implies $|U(t) - U(s)| < \epsilon$. This implies that for any $\mu > 0$ there is a set of points t_k, $k = 0, \ldots, K$ with $T_j = t_0 < t_1 < \cdots < t_K = T_{j+1}$ such that if $U_1(s) = U(t_k)$, $t_k \le s \le t_{k+1}$, then

$$\left| \int_{T_j}^{T_{J+1}} \mathbf{X}^{-1}(s)\mathbf{B}(s)[U(s) - U_1(s)] \, ds \right|$$

$$= \left| \sum_{k=0}^{K-1} \int_{t_k}^{t_{k+1}} \mathbf{X}^{-1}(s)\mathbf{B}(s)[U(s) - U(t_k)] \, ds \right| < \mu.$$

By Lemma 10.5, $U(t_k)$ may be written as a convex combination of extreme points $E_m \in \Omega$,

$$U(t_k) = \sum_{m=0}^{M-1} w_m E_m.$$

Thus

$$\int_{t_k}^{t_{k+1}} \mathbf{X}^{-1}(s)\mathbf{B}(s)U(t_k) \, ds$$

$$= \sum_{m=0}^{M-1} w_m \int_{t_k}^{t_{k+1}} \mathbf{X}^{-1}(s)\mathbf{B}(s)E_m \, ds.$$

Now the interval $[t_k, t_{k+1}]$ is partitioned into subintervals $[\tau_m, \tau_{m+1}]$ such that

$$\frac{\tau_{m+1} - \tau_m}{t_{k+1} - t_k} = w_m.$$

On the interval $[t_k, t_{k+1}]$ replace the constant control $U(t_k)$ with the control which stays in the state E_m for a time determined by the weight w_m. That is,

$$\tilde{U}(t) = \sum_{m=0}^{M-1} E_m \chi_{[\tau_m, \tau_{m+1}]}(t), \quad t_k \le t < t_{k+1}.$$

Observe that

$$\int_{t_k}^{t_{k+1}} \mathbf{X}^{-1}(t_k)\mathbf{B}(t_k)\tilde{U}(s) \, ds \tag{10.15}$$

$$= \sum_{m=0}^{M-1} \int_{\tau_m}^{\tau_{m+1}} \mathbf{X}^{-1}(t_k)\mathbf{B}(t_k)E_m \, ds = \sum_{m=0}^{M-1} [\tau_{m+1} - \tau_m]\mathbf{X}^{-1}(t_k)\mathbf{B}(t_k)E_m$$

$$= \sum_{m=0}^{M-1} [t_{k+1} - t_k]w_m\mathbf{X}^{-1}(t_k)\mathbf{B}(t_k)E_m = \int_{t_k}^{t_{k+1}} \mathbf{X}^{-1}(t_k)\mathbf{B}(t_k)U(t_k) \, ds.$$

The term which we want to be small is

$$\int_{T_j}^{T_{j+1}} \mathbf{X}^{-1}(s)\mathbf{B}(s)[U(s) - \tilde{U}(s)] \, ds$$

$$= \sum_{k=0}^{K-1} \int_{t_k}^{t_{k+1}} \mathbf{X}^{-1}(s)\mathbf{B}(s)U(s) - \mathbf{X}^{-1}(t_k)\mathbf{B}(t_k)U(t_k) \ ds$$

$$+ \sum_{k=0}^{K-1} \int_{t_k}^{t_{k+1}} [\mathbf{X}^{-1}(t_k)\mathbf{B}(t_k) - \mathbf{X}^{-1}(s)\mathbf{B}(s)]\tilde{U}(s) \ ds,$$

where the identity (10.15) has been used. By refining the partition $T_j = t_0 < t_1 < \cdots < t_K = T_{j+1}$ if necessary, this can be made as small as desired. $\square$

10.6   Notes

The reader looking for more information on control theory can consult [15] or [28]. A more comprehensive treatment is [14]. The short book [6] is also interesting.

10.7   Exercises

1. Suppose the target set is a curve, $G(t) = g(t)$ for some continuously differentiable function $g(t)$. Reduce this problem to one of the form (GLC), where the target is $G(t) = \{0\}$.

2. Show that if $\Omega \subset \mathcal{R}^M$ is convex, then so is the closure of Ω.

3. Suppose that V is a K-dimensional linear subspace of R^N. Show that there is an $N \times K$ matrix $\mathbf{L}$ which maps R^K one to one and onto V. (Hint: Let the columns of $\mathbf{L}$ be a basis for V.)

4. Suppose that $\mathbf{L}$ is an $N \times K$ matrix with linearly independent columns and $X_p \in R^N$.

(a) Show that $\Lambda \subset R^K$ is convex if and only if

$$\Omega = \{\mathbf{L}Y + X_p \,|\, Y \in \Lambda\}$$

is convex in R^N.

(b) Show that $Y \in \Lambda$ is a convex combination

$$Y = \sum_{j=1}^{m} w_j Z_j$$

if and only if $\mathbf{L}Y$ is a convex combination of the points $\mathbf{L}Z_j$.

(c) Show that Y is an extreme point of Λ if and only if $\mathbf{L}Y + X_p$ is an extreme point of Ω.

5. Suppose that $\beta \in R^N$, and $\Omega \subset R^N$ is convex. Assume that

$$\alpha = \max_{X \in \Omega} \langle \beta, X \rangle.$$

Show that if Z is an extreme point of

$$\Omega \cap \{X \in \Omega \,|\, \langle \beta, X \rangle = \alpha\},$$

then Z is an extreme point of Ω.

6. Suppose that

$$\Delta = \{(x, y) \in \mathcal{R}^2 \mid x^2 + y^2 \leq 1\}.$$

(a) Show that Δ is convex and the extreme points of Δ are the points (x, y) such that $x^2 + y^2 = 1$.

(b) Show that every point of Δ can be written as a convex combination of two extreme points.

(c) Show that Δ cannot be written as the convex hull of a finite set of its extreme points.

7. Consider the scalar equation

$$x' = a(t)x + b(t)u, \quad a, b \in C([0, \infty), \mathcal{R}),$$

with admissible controls $\mathcal{U} = \{u \in C([0, \infty), \mathcal{R})\}$.

(a) Give necessary and sufficient conditions for $C(t) = \mathcal{R}$ for all $t > 0$.

(b) When is the equation proper?

(c) Suppose that $\mathcal{U} = \{u \in C([0, \infty), \mathcal{R}) \mid |u(t)| \leq 1\}$. Give conditions implying $C = \mathcal{R}$.

8. Consider the scalar equation

$$x'' = -kx + u, \quad k > 0.$$

(a) Write the equivalent first-order system.

(b) Find the controllability matrix and its rank. Apply Theorem 10.12.

(c) Apply Theorem 10.14 to analyze this problem if

$$\mathcal{U} = \{u \in C([0, \infty), \mathcal{R}) \mid |u(t)| \leq 1\}.$$

9. Discuss the control of Example 1.

(a) When can the population be driven back to equilibrium with unconstrained controls?

(b) What do solutions of (10.1) look like when u is absent and the eigenvalues of $\mathbf{A}$ have nonpositive real parts?

(c) Suppose that the control satisfies $|u(t)| \leq K$. When do we then expect to be able to drive any perturbation of the population from equilibrium back to the equilibrium?

10. Consider the system LC with analytic coefficients. Showing that the system is proper requires showing that

$$Z^t \mathbf{Y}(\mathbf{t}) = Z^t \mathbf{X}^{-1}(t)\mathbf{B}(t) = 0, \quad Z \in \mathcal{R}^N,$$

for all t in some open interval implies that $Z = 0$.

(a) Show that if the rank of the matrix

$$(\mathbf{Y}(0) : \mathbf{Y}'(0) : \cdots : \mathbf{Y}^{(N-1)}(0))$$

is N, then the system *(LC)* is proper. (Hint: Use Taylor's theorem.)

(b) Show that the matrix

$$(\mathbf{Y}(0) : \mathbf{Y}'(0) : \cdots : \mathbf{Y}^{(N-1)}(0))$$

depends only on the values of $\mathbf{A}^{(n)}(t)$ and $\mathbf{B}^{(n)}(t)$ for $0 \le n \le N - 1$.

11. Suppose that the coefficient $\mathbf{A}(t)$ for the system (LC) is a diagonal matrix function and that $\mathbf{B}(t) = \mathbf{B}$ is a constant $N \times 1$ matrix. Compute the value of $\mathbf{W}(t)$ in (10.8).

12. If the matrix $\mathbf{A} \in M_N(\mathcal{R})$ is symmetric, show that the hypothesis of Theorem 10.11 is satisfied if and only if all the eigenvalues of $\mathbf{A}$ are negative.

13. Interpret the Lipschitz condition (10.13) graphically when $U : \mathcal{R} \to \mathcal{R}$.

14. Let $\mathcal{U}$ be the set of functions U on $[0, t]$ such that there are constants K_1 and K_2, independent of U, and

$$|U(s)| \le K_1, \quad |U(s_2) - U(s_1)| \le K_2|s_2 - s_1|, \quad s_1, s_2 \in [0, t].$$

Show that $\mathcal{U}$ is convex.

15. Suppose that the admissible class of controls consists of functions $p(t)$ which on each interval $[j, j + 1)$, $j = 0, 1, 2, \ldots$, are polynomials of degree at most K,

$$p(t) = \sum_{k=0}^{K} c_{j,k} t^k, \quad j \le t < j + 1,$$

and such that $|c_{j,k}| \le 1$. If $T > 0$, show that there is a control U^* in $\mathcal{U}$ and a solution $X^*(t)$ such that

$$|X^*(T, U^*)| = \min |X(T; X_0, U)|, \quad U \in \mathcal{U}, \quad |X_0| \le 1.$$

Bibliography

[1] H. Anton and C. Rorres. *Elementary Linear Algebra: Applications Version.* John Wiley, New York, 1994.

[2] P. Bennett. *Advanced Circuit Analysis.* Saunders, Fort Worth, TX, 1992.

[3] E.A. Coddington and N. Levinson. *Theory of Ordinary Differential Equations.* McGraw-Hill, New York, 1955.

[4] P. Doucet and P. Sloep. *Mathematical Modeling in the Life Sciences.* Ellis Horwood, New York, 1992.

[5] R. Feynman, R. Leighton, and M. Sands. *The Feynman Lectures on Physics.* Addison–Wesley, Reading, MA, 1989.

[6] H. Hermes and J. LaSalle. *Functional Analysis and Time Optimal Control.* Academic Press, New York, 1969.

[7] I.N. Herstein and D.J. Winter. *Matrix Theory and Linear Algebra.* Macmillan, New York, 1961.

[8] E. Hille. *Ordinary Differential Equations in the Complex Domain.* John Wiley, New York, 1976.

[9] K. Hoffman and R. Kunze. *Linear Algebra.* Prentice-Hall, Englewood Cliffs, NJ, 1961.

[10] R.A. Horn and C.R. Johnson. *Matrix Analysis.* Cambridge University Press, Cambridge, U.K., 1987.

[11] J. Jacques. *Compartmental Analysis in Biology and Medicine.* Elsevier, Amsterdam, 1972.

[12] M. Kline. *Mathematical Thought from Ancient to Modern Times.* Oxford University Press, New York, 1972.

[13] J.P. LaSalle. *The time optimal control problem*, in Contributions to the Theory of Nonlinear Oscillations, Solomon Lefschetz, ed., Periodicals Service, Germantown, NY, 1960, pp. 1–24.

[14] E.B. Lee and L. Markus. *Foundations of Optimal Control Theory.* John Wiley, New York, 1967.

[15] J. Macki and A. Strauss. *Introduction to Optimal Control Theory.* Springer-Verlag, New York, 1982.

[16] W. Magnus and S. Winkler. *Hill's Equation.* Dover, New York, 1979.

[17] J. Marsden and M. Hoffman. *Elementary Classical Analysis.* W.H. Freeman, New York, 1993.

[18] M. Meerschaert. *Mathematical Modeling.* Academic Press, CA, 1993.

[19] M. Pinsky. *Partial Differential Equations and Boundary Value Problems with Applications.* McGraw-Hill, New York, 1991.

[20] R. Rosen, ed. *Foundations of Mathematical Biology.* Academic Press, New York, 1972.

[21] M. Reed and B. Simon. *Methods of Modern Mathematical Physics.* Academic Press, New York, 1972.

[22] W. Rudin. *Principles of Mathematical Analysis.* McGraw-Hill, New York, 1964.

[23] W. Siebert. *Circuits, Signals and Systems.* MIT Press, Cambridge, MA, 1986.

[24] W. Strauss. *Partial Differential Equations: An Introduction.* John Wiley, New York, 1992.

[25] F. Wan. *Mathematical Models and Their Analysis.* Harper and Row, New York, 1989.

[26] W. Wasow. *Asymptotic Expansions for Ordinary Differential Equations.* Dover, New York, 1976.

[27] Whittaker and Watson. *A Course of Modern Analysis.* Cambridge University Press, Cambridge, U.K., 1927.

[28] J. Zabczyk. *Mathematical Control Theory: An Introduction.* Birkhäuser–Boston, Cambridge, MA, 1992.

Index

335